Fisheries Science Series

This series presents cutting-edge studies and accumulated wisdom from Japan and Asia, with titles addressing many of the aqua-bioscience and aquatic environment fields that form the basis of fishery science. Most of our food originates on land, but we cannot overlook food from the sea. Fishery science provides us with substantial knowledge – knowledge of living organisms in water, their habitats and environment; the knowledge to utilize these organisms; the political and administrative knowledge to organize social activities and systems to distribute fishery products; and a technical and engineering knowledge of ships, fishing equipment, sea ports and harbors. The discipline covers a huge variety of subjects, and each subject includes basic and applicative aspects that are both related and essential to one another. For fishery science to prosper, none of them can be ignored. As a country with a long history of fish eating, Japan has created unique, world-class cultures and technologies in fisheries, aquaculture, aquatic environments, seafood science, and other fishery-related sciences. Through carefully selected state-of-the-art works on topics in the field of fishery science, this series aims to contribute to the development of fishery and the welfare of people around the globe. This is an official book series of the Japanese Society of Fisheries Science.

More information about this series at http://www.springer.com/series/13529

Toshio Takeuchi
Editor

Application of Recirculating Aquaculture Systems in Japan

Springer

JSFS

Editor
Toshio Takeuchi
Tokyo University of Marine Science and Technology
Minato-ku, Tokyo, Japan

ISSN 2522-0470 ISSN 2522-0489 (electronic)
Fisheries Science Series
ISBN 978-4-431-56583-3 ISBN 978-4-431-56585-7 (eBook)
https://doi.org/10.1007/978-4-431-56585-7

Library of Congress Control Number: 2017958819

Printed on acid-free paper

This Springer imprint is published by Springer Nature
The registered company is Springer Japan KK
The registered company address is: Chiyoda First Bldg. East, 8-1 Nishi-Kanda Chiyoda-ku, 101-0065 Tokyo, Japan

Foreword to the Series

We all have to survive, and most of our food originates from that grown on land, but we can't overlook food from the sea. We catch creatures living in the water ecosystem by fishing techniques and eat them raw or cooked. That whole process and related activities are collectively called "fishery," and fishery is supported by fishery science that relates to a vast range of fields.

Fishery science brings us much knowledge—biological knowledge of the life in water; knowledge about their habitats and environment; knowledge to utilize these lives; political and administrative knowledge to organize social activities and system to distribute fishery products; technical and engineering knowledge of ships, fishing equipment, seaports, and harbors; and so on. It covers a great variety of subjects, and each subject contains both basic and applicative aspects relating to and essential to one another. To have fishery science prosper in human society, none of them can be ignored.

This series includes many of the aqua-bioscience fields and aquatic environment fields as the base of fishery science.

In this Fisheries Science Series, we provide you with carefully selected up-todate topics of excellent works in the fields of fishery science. We hope our series can contribute to the development of fishery and the welfare of people worldwide.

Tokyo, Japan
July 2017

Katsumi Aida
Series Editor-in-Chief

Preface

In 2006, a shocking report was published in *Science* that stated that the world's fish stocks would be depleted by 2048. Other reports followed on the declining populations of bluefin tuna, eels, and other fish species one after the other. The prospect of disappearing fish stocks took on a realistic tone.

On the other hand, the global human population is expected to surpass 9 billion in 2050, leading concerned experts to warn us of an imminent food supply crisis. In such a situation, demand for fisheries products has been growing in Europe, North America, and East Asia, with the annual per capita consumption of seafood reaching 20 kg in 2014. As an increasing amount of protein is taken from the sea, fish farming has been developing so much as to almost surpass wild fisheries as the main sector within the fishing industry, accounting for 44% of the industry's total output in 2014.

However, aquaculture poses many problems that must be solved. For example, the massive importation of cultured shrimps and other marine products from developing countries into Japan causes coastal environmental deterioration due to the felling of mangrove necessitated by aquaculture. In Japanese coastal areas, self-contamination has been spreading due to net cage fish culture. The document titled "The Future We Want," adopted at the United Nations Conference on Sustainable Development (UNCSD or "Rio+20") held in June 2012, states in its section on agriculture, forestry, and fisheries that sustainable agricultural production and productivity must be boosted in consideration of diverse agricultural practices in the respective countries. Along this line of thought, we believe that initiatives for resource-recycling and a more sustainable growth-oriented culture of aquatic resources will become far more important in the future.

Unlike in the West, aquaculture in Japan is mainly conducted on the sea surface. In other words, it is mariculture that is particularly active in Japan. The mariculture of yellowtail *Seriola quinqueradiata*, red sea bream *Pagrus major*, and coho salmon *Oncorhynchus kisutch*, which accounts for almost 90% of the country's total mariculture output, is entirely net cage-based, while a land-based running water system is used for Japanese flounder *Paralichthys olivaceus* and kuruma prawn *Marsupenaeus japonicus*. Since these farming methods all involve feeding, they tend to cause red tide and a deterioration of water quality in fish farms in inner bays and surrounding coastal areas. It is therefore necessary to prevent eutrophication and the accumulation of sediment in these coastal areas. The preventive measures that have been taken thus far include water quality control through the adjustment of population density within net cages and the improvement of feed forms and feeding methods so as to reduce feed remnants and undigested feed. In more recent years, active research has been carried out on integrated fish farming, in which nitrogen and phosphorus released by fish are used for the culture of algae, and closed recirculating land-based aquaculture systems, in which fish excrement is not released into the sea.

This book introduces the application of recirculating aquaculture systems in Japan from the viewpoint of bioscience, in the hope of contributing to the further development of aquaculture, whose output has been growing on a global scale. The book also outlines problems to be solved for future generalization, economic and business aspects of land-based aquaculture, as well as business opportunities. In concrete terms, closed recirculating aquaculture systems are outlined, and examples of their application to various marine fish types are presented. Also discussed are ecologically integrated fish farming, aquaponics, and closed ecological recirculating aquaculture systems, in which fish excrement is put to effective use, and systems for the eventual industrialization of closed recirculating land-based fish farming.

In July 2013, the Fisheries Agency of Japan organized a meeting on the future of fish farming, where proposals were made regarding land-based aquaculture. The meeting concluded that fish farmers were expected to actively engage in land-based aquaculture because, as stated in the Basic Plan for Fisheries, it is useful in diversifying fish farms, promoting new local industries in fishing villages, and creating jobs for which specialist knowledge and expertise may be utilized. Accordingly, it is expected that land-based aquaculture will be promoted as a national policy.

It is my hope that this book will serve not only as a textbook for undergraduate and postgraduate students but also as a useful reference for corporate researchers considering the fish farming business. I would like to express my deep gratitude to the members of the editing committee, who carefully reviewed my manuscript, and the staffs at the Japanese Society of Fisheries Science and Springer Japan for their kind cooperation regarding publication. I dedicate this book to my beloved wife, "Le Vien (Eri)," who has supported me for 40 years.

Tokyo, Japan
August 2017

Toshio Takeuchi

Contents

Part III Tackle of Recirculating Aquaculture in Fishes

Part I
What Is Recirculating Aquaculture Systems?

Chapter 1
Overview of Land-Based Recirculating Aquaculture

Toshio Takeuchi

Abstract Land-based aquaculture systems can be divided into two types, those that use running water and those that recirculate water in a closed system. In Japan, recirculating aquaculture systems (RASs) that have a water exchange rate of below 5% in total volume are further specified as closed recirculating aquaculture systems (CRASs). This section gives an overview of the recent developments in research on land-based aquaculture, and on CRAS in particular, as a preparation for the case studies from Chap. 2 onward. In short, this section will describe the current state of CRAS, its features, advantages, and development history, as well as the challenges, economics, feasibility, and business opportunities for CRAS in the future.

Keywords Land-based aquaculture • Closed recirculating aquaculture system • History • Economics

1.1 Introduction

Conventional aquaculture has its own disadvantages, which led to the development of CRAS. Currently, most aquaculture fisheries, such as red sea bream *Pagrus major*, yellowtail *Seriola quinqueradiata*, and Pacific bluefin tuna *Thunnus orientalis*, are conducted as mariculture using net cages. Figure 1.1 illustrates a disadvantage of this net cage culture (Hall et al. 1992), using rainbow trout *Oncorhynchus mykiss* as an example species. Although rainbow trout is a freshwater fish, it is an appropriate example, because it is a well-studied fish culture that also can be kept in saltwater, where it is known as steelhead. In the past, fish in net cages were fed either raw fish or moist pellets, which combined raw fish and mash feed. Currently, fish are more often given only dry pellets. As soon as fish have reached a marketable size, they are harvested, and 27–28% of the nitrogen given as feed is redeemed as fish. Death accounts for 2–5% of nitrogen; however, the main disadvantage is that the rest is lost to the environment, as shown at the bottom of the

T. Takeuchi (✉)
Tokyo University of Marine Science and Technology, 4-5-7 Konan, Minato, Tokyo 108-8477, Japan
e-mail: take@kaiyodai.ac.jp

T. Takeuchi (ed.), *Application of Recirculating Aquaculture Systems in Japan*, Fisheries Science Series, https://doi.org/10.1007/978-4-431-56585-7_1

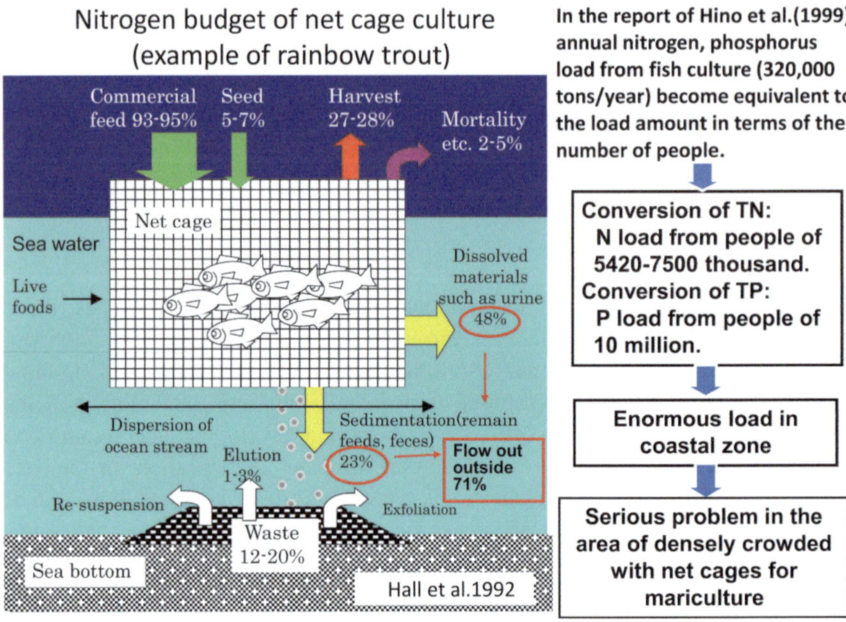

Fig. 1.1 Problems of net cage mariculture (Modification of Hall et al. 1992 in left side and Hino et al. 1999 in right side)

figure. Effluents such as urine and soluble nitrogen account for 48% of the nitrogen, and fecal and feed scrap sediments account for 23%. These add up to 71% of nitrogen being wasted, which clearly contributes to marine pollution. According to calculations regarding aquaculture in Japan published in 1999 (Hino et al. 1999), the quantity of total nitrogen wastes derived from both inland water aquaculture and mariculture is equivalent to that produced by 5.42 to 7.5 million people, and total phosphorus waste is equivalent to that of 10 million people; these figures represent significant amounts of pollution. This would not pose as large of a concern if the effluent were spread around the whole coastline; however, aquaculture is conducted in a limited number of locations concentrated in coastal zones such as the Seto Inland Sea and Ise Bay. Essentially, mariculture will inevitably lead to massive self-pollution.

This section will introduce land-based aquaculture, especially CRAS, in order to further stimulate the global growth of aquaculture. It also provides an overview on the challenges regarding the popularization, economics, and feasibility of land-based aquaculture as well as the business opportunities that it can provide.

1.2 Current Status of CRAS

Due to the limited number of marine resources, the global production capacity of marine fisheries will inevitably be reached as the global demand for aquatic products continues to swell. It is therefore necessary to turn to aquaculture to meet this increasing demand. However, mariculture can be deployed at only a limited number of suitable locations. These limited conditions led to the fast development and industrialization of land-based aquaculture, such as CRAS, which does not require a location with specific conditions and has a higher productivity potential than offshore sea and inland water aquaculture.

Two factors have contributed to this fast development. The first factor is the effect of legal restrictions concerning pollution load, in the form of total effluent regulations and the Water Pollution Prevention Act. Until recently, aquaculturists regarded themselves as the victims of water pollution, as pollution in lakes, rivers, and enclosed bays has a negative impact on fish. With the discovery that the effluent from net cage mariculture causes self-pollution, as mentioned earlier, this opinion has changed. Now, aquaculturists are held responsible for the damages caused by eutrophication, and solutions are demanded of them, especially for reducing freshwater and saltwater levels of nitrogen and phosphorus (Takeuchi 2002; Takeuchi et al. 2002).

The second factor is the shifting trend of consumer awareness toward safety in aquaculture products. Consumers now demand transparent, completely regulated production systems that provide safe, responsible, and stable products as efficiently as possible. One result of this shifting awareness is the application of the hazard analysis critical control point (HACCP) system to the production processes of aquatic products. HACCP has already been implemented in other areas of food production, and between 1998 and 2000, Japan's national fishery agency tasked the Japan Fisheries Association with implementing HACCP in aquaculture. This first resulted in the HACCP manual for yellowtail, followed by the manual for seaweed *Porphyra tenera* and scallop *Patinopecten yessoensis* (Japan Fisheries Association 2000, 2001). Furthermore, the March 2007 meeting of the Japanese cabinet resulted in the introduction and standardization of the good aquaculture practice (GAP) methods in its basic fishery plan, and by March 2010, the handbook for the GAP method was created. The March 2011 great east Japan earthquake further increased consumers' concerns for safety. With the dispersal of radioactive substances, especially cesium-137, following the accident at the Fukushima Daiichi Nuclear Power Plant, assuring the safety of aquatic products from the Fukushima area has become even more difficult. These events led to a further realization that the best way to pursue the implementation of the GAP method is through CRASs for both freshwater and saltwater cultures.

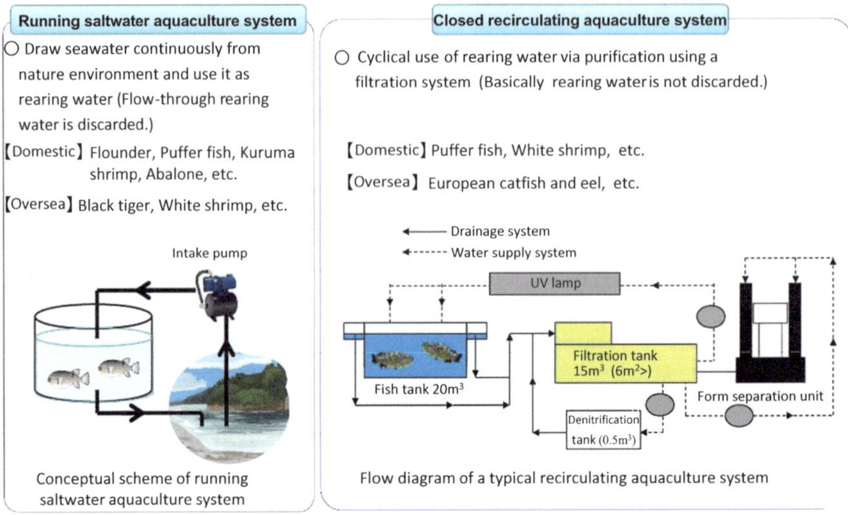

Fig. 1.2 What is land-based aquaculture (Modification of Fisheries Agency HP, http:www.jfa.maff.go.jp/j/saibai/yousyoku/arikata/pdf/4-3-1docu.pdf)

1.2.1 What Is Land-Based Aquaculture?

Figure 1.2 shows a diagram of land-based aquaculture, adapted from the homepage of the fishery agency (http://www.jfa.maff.go.jp/). Land-based aquaculture is the culture of aquatic products in an artificial environment constructed on land. There are two main approaches to land-based aquaculture, the running water system, which continuously pumps in rearing water from an external source and pumps out used water, and the closed recirculating system, which filters and recirculates the rearing water within a closed system.

In general, the running water aquaculture system involves setting up the culture tank or pond near the seashore or a river. The rearing water is drawn from the environment, and the used water is pumped back, thus polluting the environment with feed residue, feces, and urine. In Japan, this method is often employed for the culture of Japanese flounder *Paralichthys olivaceus* and kuruma shrimp *Marsupenaeus japonicus*, and in Southeast Asia, it is especially popular for the culturing of tiger shrimp *Penaeus monodon* and white shrimp *Litopenaeus vannamei*. In Thailand, a running water aquaculture system to culture white shrimp adapted for use in rice fields has caused a large harmful shifts in soil salinity.

In comparison, a RAS does not significantly expel effluents into the environment. In Europe, a system in which 10–50% of the rearing water is refreshed per day is designated a RAS. In Japan, a system that requires less than 5% of the rearing water to be refreshed and that has total water loss from normal evaporation and waste removal of less than 1% is defined as a CRAS.

1.2.2 The Current State of Land-Based Aquaculture

The total production volume of marine products (including fish, shellfish, crustaceans, and mollusks) from land-based aquaculture in Japan is estimated to be 6300 metric tons, equating to 66 million yen in profit. The majority of land-based aquaculture employs the running water system, and their production is limited to commercially profitable products such as Japanese flounder, tiger puffer *Takifugu rubripes*, kuruma shrimp, and abalone *Haliotis* spp. There are currently no statistics on the production from RAS aquaculture; however, it is estimated to be less than 100 metric tons. The reasons for the unpopularity of RAS are the high initial costs of facility construction and the high running costs due to factors such as electricity usage. These costs cause the price of the products to exceed the market price, making RAS an unfeasible system (this point will be discussed in more detail later). In addition, fish farmers' lack of experience with this system may result in massive deaths caused by inappropriate disinfection and oxygen supply, mismanagement of fish seed, and poor estimation of nitrifying capacity, all of which would force the facility to shut down. Other problems include the lack of backups (e.g., of materials) in case of emergencies.

1.2.3 A History of Land-Based Aquaculture

RAS was invented surprisingly early; the concept of a recirculating filter-equipped aquarium for freshwater and saltwater fish was developed in the late 1950s by Dr. Saeki, Faculty of Agriculture, University of Tokyo. The criteria for the setup were defined not from experience, but theoretically. In the 1960s, there were great advances in the study of the microbiology and sanitization of filtration tanks, and the RAS seemed near realization. These technologies were applied to aquariums, leading to dramatic improvements in fish culture. However, their implementation in aquaculture was very limited, and the technologies were only partly incorporated in the RAS for eel culture, which was developed later.

Meanwhile in Europe and the United States, the research on RASs progressed from the 1970s, and its application in recirculating systems for fish culture was actively pursued through methods such as involving manufacturing companies and other industries. Japan lagged behind greatly in this practical phase. Research in Japan on CRAS for marine fish progressed in the late 1980s (Hirayama et al. 1988), influenced by the trends abroad. Especially significant contribution was made by the Central Research Institute of Electric Power Industry toward the creation of a system that was compatible for industrialization, which was designed for Japanese flounder (Hino et al. 1999). Following practical implementation trials with eels (Maruyama 2002), pejerrey *Odontesthes bonariensis* (Yoshino et al. 1999), and tiger puffer (Marino-Forum 21 1999–2003), full-scale facilities have been set up for species such as white shrimp (Nohara 2012) and production has begun. See Pet III and Chap. 10 for details on the current filtration system used in tanks and on CRAS for various aquatic products.

1.2.4 Advantages of CRAS

The advantages of CRAS include (1) artificially controllable culture environment (less impact from meteorological phenomena such as global warming and typhoons); (2) promoting the brand of the products, such as by raising their quality and avoiding the use of chemicals; (3) the location of facilities can be more freely decided, as there are no restrictions on placement posed by the fishery act such as the demarcated private fishery areas; (4) possibilities for the employments of a greater range of workers, such as the elderly, because the work can be performed on land; and (5) contribution to a better environment by not releasing effluents. Taken together, these advantages allow CRAS to avoid the impacts of natural hazards such as red tides and typhoons, and high-density production and uniform product size can be achieved all year round, as water temperature can be controlled. This type of system will also enable the production of aquatic products in an environmentally friendly, safe, stable, and efficient manner. Absolute product traceability would be possible, as would the introduction and branding of new breeds, avoidance of aquaculture chemicals that prevent disease outbreaks, avoidance of contamination by heavy metals and dioxins in the environment, and reduction of the environmental impact of nitrogen and phosphorus pollution. Additionally, we have reported possibilities for even further-enhanced productivity through the adjustment of culture conditions. For instance, changing the photoperiod and changing the salinity of culture water in a previous study led to enhanced growth of fish and regulated reproduction (Takeuchi and Endo 2004).

1.2.5 Features of Representative Systems

Table 1.1 shows features of representative RASs and CRASs in Europe, the United States, and Japan (Hino et al. 1999). It is worth noting that the table has taken into consideration various factors, such as the kind of system implemented at the facility, type of fish (freshwater or saltwater), method of solid waste removal and (de)nitrification, water exchange rate, and final culture density. This means that a single, uniform system is not sufficient, as different systems are intermingled. The reality is that trial and error in optimizing the system is inevitable.

Table 1.1 Characteristics of certain land-based aquaculture systems (Hino et al. 1999)

Item	Certain European system	Central Research Institute of Electric Power system	Israel system	Fully equipped system	Miyazaki university system
Purpose	Full scale	Experimental	Experimental	Experimental	Experimental
Target fish	Seabass etc. (saltwater)	Japanese flounder (saltwater)	Carp (freshwater)	Pejerrey (0.7% brackish water)	Eel (freshwater)
Rearing period	1–2 years	About 1 year	About 80 days	About 1 year	About 8 months
Scale of the system	Variously	10 kL	50 kL	2.1 kL	1.1 kL
Filtration	Mesh filter	Settling tank and drum filter	Settling tank	Drum filter with foam separation unit	Foam separation unit
Nitrification	Biofilter	Immersed filtration	Trickling filtration	Rotary disk fluid bed	Upward flow-type immersed filtration
Ammonia-N	<2 mg-N/L	<1.05 ± 1.22 mg-N/L	<2 mg-N/L	<0.2 mg-N/L	<Av. 1 mg-N/L
Nitrite-N	<2 mg-N/L	<1.53 ± 1.24 mg-N/L	0–0.4 mg-N/L	<0.1 mg-N/L	<Av. 0.1 mg-N/L
Denitrification	Non	Non	Anaerobiotic fluid bed (sand, sludge product)	Immersed filtration (fiber type filter, methanol)	Upward flow type (cylinder type filter, methanol)
Nitrate-N	–	–	Max 40 mg-N/L → 0–15 mg-N/L	900 mg-N/L → 150 mg-N/L	150 mg-N/L → 40–50 mg-N/L
Oxygen supply	Oxygen-generating and oxygen-dissolving device	Oxygen-generating and oxygen-dissolving device/aeration	Trickling and aeration	Oxygen-generating and oxygen-dissolving device	Foam separation apparatus
Saturation degree/DO density	>90%	>10–130%	6–7 mg/L	Av. 103%	Av. 80%
Water temp.	16–24 °C	20–25 °C	22–27 °C	20 °C	28 °C
pH	6.5–8.3	7–7.5	7–7.8	7–8.2	7.5–8

(continued)

Table 1.1 (continued)

Item	Certain European system	Central Research Institute of Electric Power system	Israel system	Fully equipped system	Miyazaki university system
Sterilization of rearing water	UV irradiation	UV irradiation	None	UV irradiation	None
Exchange water rate	100%/day	150%/year	6%/day	8.5 L/day (0.4%/day)	None
Survival rate	–	95%	69%	92%	91%
Final culture density	>100 kg/m^3	39 kg/m^3	15 kg/m^3	18 kg/m^3	33 kg/m^3

1.3 Challenges in the Popularization of Land-Based Aquaculture

1.3.1 Comparisons of Mariculture and Land-Based Aquaculture

The advantages of CRAS are especially obvious in the category of issues involving the environment. Recent changes in climate leading to increased saltwater temperature and the subsequent alteration in the quantity and type of catch, the prevalence of red tides, the aging of the fisherman population, population decline in fishery regions, and regional radioactive contamination are all problems that mariculture must face. In addition, contact with external water creates a high risk of disease in mariculture and in running water systems. Running water systems for kuruma shrimp culture have suffered considerable damages from infections of penaeid acute viremia, *Vibrio*, and *Fusarium*, among others. As in a CRAS, the concentration of probiotics can be stabilized; it is possible to suppress the proliferation of *Vibrio* bacteria in culture water as well as in shrimp guts. This also has the extra advantages of strengthening the shrimps' immune system and enhancing their growth (Mochizuki and Takeuchi 2008). This is just one example of the ability of CRASs to improve productivity, viability, safety, and marketability of aquatic products. This is the so-called "sixth sector ($1 \times 2 \times 3$) industrialization" policy. For the reasons listed above, the superiority of CRAS in realizing these policies is immeasurable.

The difference between CRAS and mariculture is also apparent in the quality of their products. In the case of tiger puffer, those grown in CRASs have less damage in regions such as the caudal fin and score higher in taste tests (more translucent and firmer meat, thicker skin, etc.), compared to their counterparts grown in net cage cultures (Japan Aqua Tec Co. Ltd: Marino-Forum 21). A minor, unforeseen benefit is that tiger puffers grown in CRAS from fish seed and given a formulated diet are free of tetrodotoxin, as under natural conditions this neurotoxin is the product of bioaccumulation in the fish's organs. This means that CRAS-reared tiger puffer can be consumed without the risk of poisoning.

1.3.2 Development of New Fish Culture for RASs

In 2010, the US Food and Drug Administration declared the genetically modified Atlantic salmon *Salmo salar* containing a king salmon growth hormone gene and eelpout gene safe to eat. In 2012, they reported that the same genetically modified organism would not endanger the fish species in nature, as the female fish are triploid and therefore sterile. Currently, approval for the genetically modified Atlantic salmon is pending, having reached the final stage of public comment. This strain looks similar to ordinary Atlantic salmon, but grows twice as fast as its counterpart and can be grown more efficiently with less feed. Such a fish is ideal for culturing in CRAS, as it would have reduced production time and burden on components such as the filtration and denitrification tanks, resulting in a higher efficiency of the overall system.

Similarly, we have successfully inserted a growth hormone gene of medaka *Oryzias latipes* in tilapia *Oreochromis niloticus*, a freshwater fish. This quadrupled the growth rate, increased the feed efficiency 1.6-fold, halved the nitrogen excretion, and led to a 40% reduction in phosphorus excretion (Fig. 1.3) (Lu et al. 2009). Marine fish such as Japanese flounder, tiger puffer, and red sea bream cannot grow when given feed containing only plant-derived lipids and require polyunsaturated fatty acids such as fish-derived eicosapentaenoic acid (EPA) and docosahexaenoic acid (DHA), which increases the cost of fish feed. Genetically modified fish have been developed that can circumvent this dietary limitation by internally converting linolenic acid, the plant-derived precursor, to EPA and DHA (Alimuddin et al. 2008; Yamamoto et al. 2010). The use of such fish will decrease the feed cost and is therefore favorable to CRAS; however, consumers generally express a strong resistance to genetically modified organisms. This resistance is causing a prolonged approval time in the case of Atlantic salmon, and thus the trend of opposition to genetically modified organisms must be followed closely. In November 2013, Environment Canada approved the use of genetically modified salmon egg, with the condition that they only be produced within CRAS.

Quite recently, the targeting-induced local lesion in genome (TILLING) method was applied to fish. This method has been widely used for the selective breeding of plants. Using the TILLING method, Dr. Yasutoshi Yoshiura developed the "double

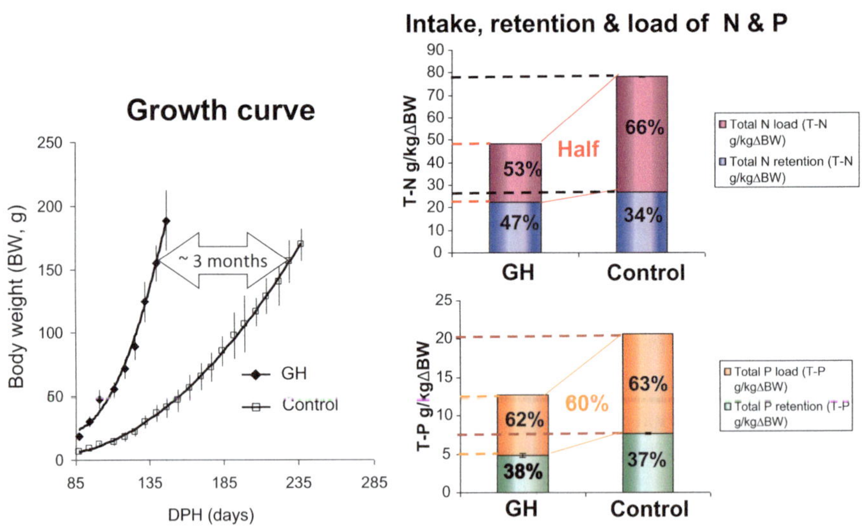

Fig. 1.3 Efficient productivity and lowered nitrogen and phosphorus discharge load from growth hormone (GH)-transgenic tilapia under visual satiation feeding (Drawing figures from Lu et al. 2009)

muscle medaka," which is 1.3-fold heavier and has twice the muscle volume of an ordinary medaka (from a grant from the National Agriculture and Food Research Organization, Enterprise for the Stimulation of Innovative Creation in Fundamental Research). The technique involved in the strain's creation was the identification of myostatin, a muscle-regulating protein, from the Belgian Blue breed of cattle, which arose in the nineteenth century by spontaneous mutation and is heavily muscled (referred to as "double muscling"). Using the TILLING method, a medaka containing the same myostatin mutation as a Belgian Blue was developed.

The TILLING method differs from genetic modification in that this technique artificially induces a mutation that might otherwise have occurred naturally. The technique has already been applied for the creation of novel variants of rice, wheat, soy, corn, tomato, potato, melon, and other species. The application in medaka is the first example in fish, and efforts are being made for its further application in tiger puffer for aquaculture purposes. This is the first step toward the creation of "domesticated fish," and its acceptance by consumers will contribute to the further advancement of CRAS industrialization.

1.4 Economics and Feasibility of Land-Based Aquaculture

1.4.1 Economics-Based Approach Toward Planning, Facility Design, and Profitability

To successfully establish land-based aquaculture as an enterprise, it is necessary to develop a competitive culture technique and propose a business model. In doing so, the following aspects become of importance: reduction of initial costs and variable costs such as feed and energy, establishing a stable and safe production technology, and choice of location for maximum geographic advantage.

The facility of saltwater fish in Japan must be able to hold an aquarium of at least 1000 metric tons for it to be profitable, even after the six-tenth factor rule and 20% reduction in the initial costs have been applied by employing common units, simple facilities, and uniform design. The final culture density in the aquarium, which must be at least 5%, is also important for profitability. It is preferable to obtain specific pathogen-free fish seed. To maximize profitability and productivity, the time period between the introduction of fish seed and the shipment of the product should be as short as possible, and the facility should be prepared for a daily shipping of uniformly sized fish, which will aid marketability, including in online vending. Transparency in the implementing body and the establishment of a consortium would also contribute to the stability of the enterprise.

As for the quality aspects of the fish, factors such as the development of novel feeds that improves the flavor of the fish, the choice of brandable fish, and the promotion of the fish's traceability so as to create an image of safe (e.g., chemical and heavy metal-free, radioactive contamination-free) product become of importance.

The location must be carefully considered. An ideal location should be close to consumers and have easy access to an energy source such as thermal energy. For example, we have established a CRAS facility for the experimental validation of tiger puffer culture in a farm in the Obihiro region of Hokkaido, which is far from the usual production area of tiger puffer. This location has the advantages of having an uncompetitive market for the selected product and access to high-purity bio-methane from cow manure, which contributes to a reduction in operating costs (see Chap. 14 for details). Such reductions in operating costs from optimal use of the facilities surrounding the location are essential and can be achieved by the use of thermal energy and exhaust heat from other facilities for the heating of culture water, by utilization of the cold water generated from the heat pump for processing, by harnessing renewable energy sources such as offshore wind power, and other such methods. The organic linking of mechanical elements, such as heat retention, filtration, and waste disposal, to the surrounding facilities can enable the construction of a low-cost system with optimal technology and equipment, which is crucial to successful implementation.

1.4.2 Costs Involved in Land-Based Aquaculture

Figure 1.4 shows the trial balance of tiger puffer production from CRAS. It is clear that the bulk of the costs are incurred by the facility and the costs of the electricity required for temperature regulation and power supply. The production cost of tiger puffer in the current trial balance is 3300 yen/kg, which is far from the market value of about 2500 yen/kg. This deficit underscores the infeasibility of the enterprise; therefore, methods of cost reduction are required. First, the installation of aquarium larger than 1000 metric tons becomes necessary. To reduce electrical power consumption, power-saver settings and the incorporation of renewable energy will be essential. Another trial calculation predicts a 10% increase in the internal rate of return by increasing the final culture density from 3% to 5%. A final necessary measure is the education of future technicians who would be able to keep the fish at high growth and survival rates at high culture density. However, if subsidies can be obtained from instructions such as the government for the initial costs of construction, feasibility would improve sufficiently even if the final culture density remained at 3%.

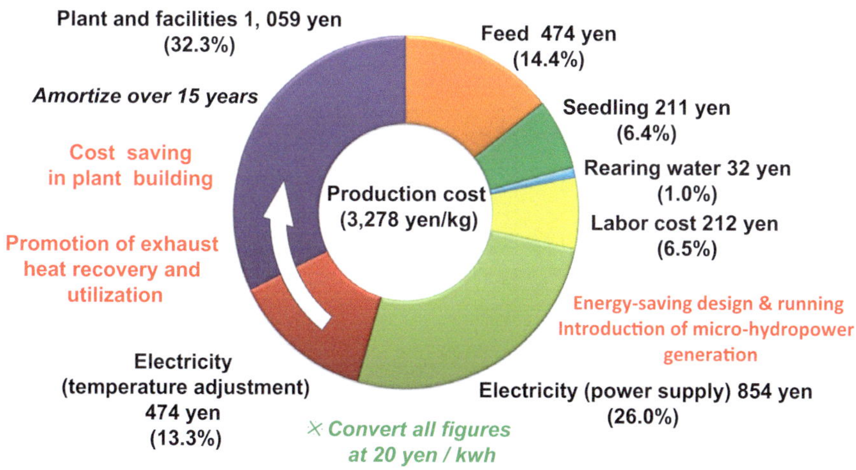

Fig. 1.4 Trial balance of tiger puffer production from closed recirculating aquaculture system

1.5 Possible Business Opportunities

Currently, the Japanese government is promoting the implementation of the "sixth sector industrialization" policy. This policy, which is aimed at revitalizing remote regions and stimulating employment, would also provide business opportunities for CRAS. Provided that traceability and quality control of the products are satisfactory, export of CRAS products would also become possible. The opportunity to export the CRAS technology itself, for instance, to desert regions, should also be considered. To this end, the creation of a consortium of industry, academia, and government would be effective. Figure 1.5 presents the industrialization scheme of CRAS and biomass energy (Takeuchi et al. 2013), as a reference for future directions.

Finally, the use of the nitrogen and phosphorus wastes from CRAS deserves mention, as environmentally friendly waste disposal is one of many future challenges for implementing CRASs. Aquaponics, the system of CRAS combined with hydroponics, would provide a way to effectively utilize these organic materials. To date, the United States has chiefly led the industrialization of aquaponics through advancements in freshwater systems. As in Japan, the main aquaculture product is marine fish; the development of a saltwater system is necessary for the application of aquaponics. Some candidate plants include ice plant, glasswort, and tetragon, to be grown with Japanese flounder, tiger puffer, and some kinds of grouper, respectively.

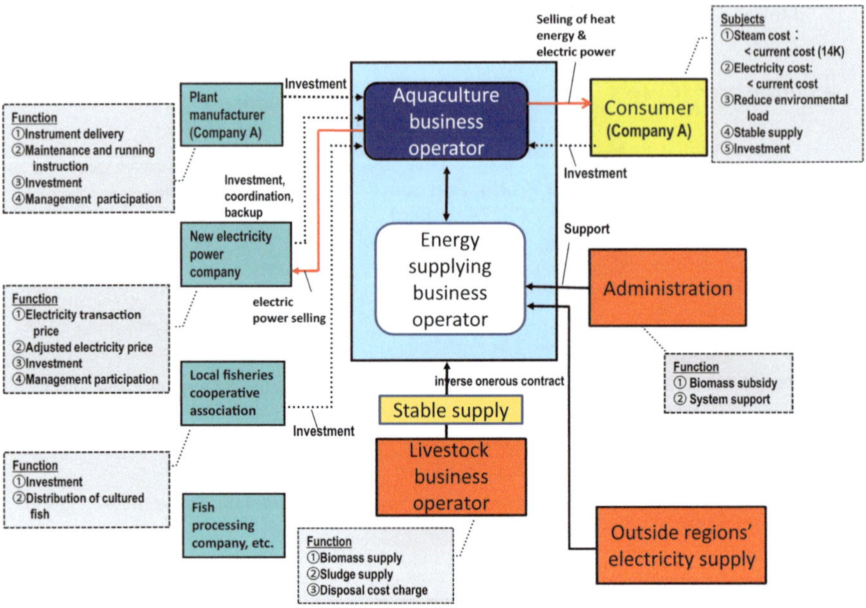

Fig. 1.5 Proposal of the industrialization scheme of the closed recirculating aquaculture and biomass energy

1.6 Conclusions

According to the Food and Agriculture Organization, the world's total fishery production was 150 million metric tons in 2004 and is predicted to reach 172 million metric tons in 2015. Given such data, the realization sets in that there is a limit to the productivity of our oceans. This issue will ignite serious discussion on how we can continue to effectively use our marine resources in the future.

Until now, there was little exchange of information between aquaculturists and industries, which operated independently of each other. As the development of the individual equipment and machinery for the setup of a complete aquaculture system is almost complete, the time has come for these entrepreneurs to confer and proceed together in joint efforts toward establishing a total aquaculture system that includes marketing.

Japan is currently in possession of advanced key technologies that can be applied to our existing RAS. Through further technological advance and development, especially in the field of clean energy, the industrialization of CRAS would become highly plausible. The additional development of "domesticated fish" adapted to CRAS conditions (and in fact, only suitable for use in CRAS, as their use in open systems would disturb existing gene pools) and a reduction of fish-derived components in feed will improve the feed conversion efficiency and growth rate, increasing the system's profitability. Similarly, the development of a novel feed that prevents the leaching of feces would also aid the successful implementation of CRAS. Lastly, collective support of further development by a consortium of academics, industry, and the government is crucial.

One of CRAS' recent slogans is "aiming for the creation of sustainable society through eco-engineering." This includes the developments of aquaponics (see Chap. 11) and a closed (controlled) ecological life support system (CELSS) to be used in space. CRAS may lead to the development of a closed ecological recirculating aquaculture system (CERAS; see Chap. 13), a type of CELSS. In this way, CRAS represents a system that can help to sustainably improve current fishery output and to serve as a foundation for future technological developments in aquaculture.

References

Alimuddin KV, Satoh S, Takeuchi T, Yoshizaki G (2008) Cloning and over-expression of masu salmon (*Oncorhynchus masou*) fatty acid elongase-like gene. Aquaculture 282:13–18

FAO (2012) The States of World Fisheries and Aquaculture, 209p

Fisheries Agency HP (2013) http://www.jfa.maff.go.jp/

Hall POJ, Holby O, Kollberg S, Samuelsson MO (1992) Chemical fluxes and mass balances in a marine fish cage farm. IV. Nitrogen. Mar Ecol Prog Ser 89:81–91

Hino A, Maruyama T, Kurokura H (eds) (1999) Reduction of environmental emission from aquaculture, Fisheries science series 123. Kouseisha-kouseikaku, Tokyo. (In Japanese)

Hirayama K, Mizuma H, Mizue Y (1988) The accumulation of dissolved organic substances in closed recirculation culture systems. Aquac Eng 7:73–87

Japan Fisheries Association (2000, 2001) Yellow tail, Seaweed, Scallop, Cultured management manual by the HACCP method. 30p, 37p in 2000 and 38p in 2001

Lu J, Li J, Furuya Y, Yoshizaki G, Sun H, Endo M, Haga Y, Satoh S, Takeuchi T (2009) Efficient productivity and lowered nitrogen and phosphorus load from GH-transgenic tilapia *Oreochromis niloticus* under visual satiation feeding. Aquaculture 293:241–247

Marino-Forum 21 (1999–2003) Report of development of closed recirculating aquaculture system. 153p; 248p; 97p; 85p; 109p

Maruyama T (2002) Developmental examination of the recirculating aquaculture system in eel. Report of follow-up commissioned projects for sustainable culture promotion, Fisheries Agency (National Federation Inlandwater Fish. Coop.), pp 115–185

Mochizuki H, Takeuchi T (2008) Effect of probiotics treatment on growth, digestibility and *Vibrio* control in Kuruma shrimp *Marsupenaeus japonicus* reared in a closed recirculating system. Aquacult Sci 56:281–294

Nohara S (2012) White shrimp. Aquacult Bus 49(4):54–58

Takeuchi T (2002) Reduction of nitrogen and phosphorus from freshwater fish farm. J Water Waste 44(7):611–619

Takeuchi T, Endo M (2004) Recent advances in closed recirculating aquaculture systems. Eco-Engineering 16(1):15–20

Takeuchi T, Akiyama T, Yamagata Y (2002) Trend occurring high reduction technology of organic compound loading in marine aquaculture areas. J Res Environ 38(8):801–810

Takeuchi T, Endo M, Mouri K (2013) Aquaculture factory and closed ecological recirculating aquaculture system. In: Land aquaculture-sales strategy, management skill and fish culture for commercialization and distribution. Johokikou, Tokyo, pp 269–280

Yamamoto Y, Kobeya N, Takeuchi Y, Alimuddin HY, Satoh S, Takeuchi T, Yoshizaki G (2010) Cloning and nutritional regulation of polyunsaturated fatty acid desaturase and longase of a marine teleost, the nibe croaker *Nibea mitsukurii*. Fish Sci 76:463–472

Yoshino H, Gruenberg ED, Watanabe I, Miyajima K, Satoh O (1999) Changes in water quality and performance of a closed recirculating saltwater aquaculture system for rearing Pejerrey, *Odonthestes bonariensis*. Aquacult Sci 47(2):289–297

Part II
Basic Information of Closed Recirculating Systems from the View Point of Bioscience

Chapter 2
Characteristics of Closed Recirculating Systems

Yoshihisa Yamamoto

Abstract The most important requirement for a closed recirculating aquaculture systems (CRAS) is the maintenance of the water quality to ensure normal development and growth of the cultured fish. Since cultured fish discharges about 70% of their total dietary nitrogen into the culturing water, the essential functions of CRAS are as follows: (1) removal of nitrogen, especially ammonia nitrogen which is toxic to the fish and (2) removal of organic matter discharged by the fish. The standard components of CRAS used for grow-out are as follows: a culturing tank, sedimentation tank and physical clarification units (drum filter unit and foam separation unit) used to remove the organic matter, biofiltration units used mainly to remove ammonia nitrogen, disinfection units (e.g., ultraviolet irradiation unit, electrolysis unit, copper treatment unit, and so on), denitrification unit, oxygen supply unit, wastewater treatment units, recirculation pump, CO_2 removal units, and temperature control units. In this chapter, the functions, principles, and practical examples of each CRAS component are discussed.

Keywords CRAS • Drum filter unit • Foam separation unit • Biofiltration unit • Disinfection unit • Oxygen supply unit • Nitrification • Intermittent filter type

2.1 Introduction

All aquatic animals, including fish, excrete ammonium nitrogen that is quite toxic to the aquatic animals through their metabolism and consume oxygen and emit carbon dioxide thorough their aspiration (Fig. 2.1). The most important requirement for a closed recirculating aquaculture system (CRAS) is the maintenance of water quality to ensure normal development and growth of the cultured fish. Therefore, efficient water purification units are required to maintain a stable and favorable water conditions. The water purification functions of CRAS include the removal of suspended and sedimentary organic matter by the physical clarification unit, the

Y. Yamamoto (✉)
Japan Fisheries Research and Education Agency, National Fisheries University, 2-7-1 Nagata-Honmachi, Shimonoseki, Yamaguchi 759-6595, Japan
e-mail: yama1215@fish-u.ac.jp

T. Takeuchi (ed.), *Application of Recirculating Aquaculture Systems in Japan*,
Fisheries Science Series, https://doi.org/10.1007/978-4-431-56585-7_2

removal of dissolved ammonium nitrogen by the biofilter, and the reduction of pathogens by the disinfection unit. This chapter introduces the functions, principles, and management of these CRAS components and presents practical examples of each (Fig. 2.2).

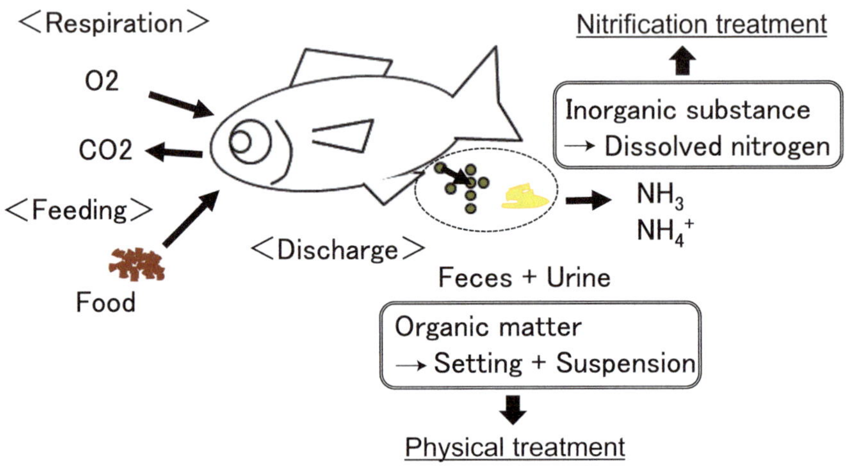

Fig. 2.1 Schematic diagram of nitrogen balance and metabolism and respiration of fish

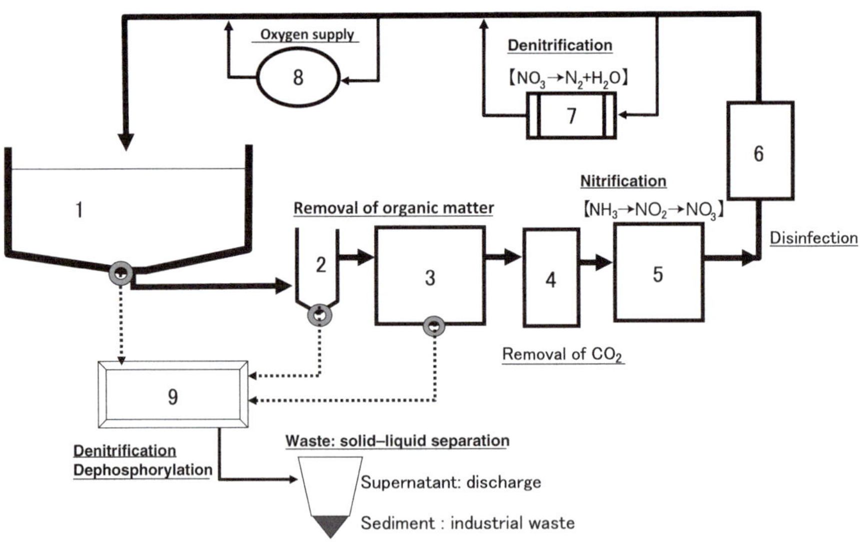

Fig. 2.2 Schematic diagram of equipment and its function in CRAS: *1*, rearing tank; *2*, sedimentation tank; *3*, physical filter unit; *4*, degasifier; *5*, biofilter unit; *6*, disinfection unit; *7*, denitrification unit; *8*, oxygen supply unit; *9*, waste treatment unit

2.2 Physical Clarification

The physical clarification unit in a CRAS removes solid organic matter such as feces and remaining feed. If the physical clarification unit malfunctions, the suspended organic matter flows directly into the downstream biofilter unit; blockage of the filter ensues because the clearance gap of the biofilter medium is quite small. This blockage decreases the nitrification ability of the biofilter and also causes the tank to overflow. Even if the blockage does not occur suddenly, this situation shortens the lifetime of the biofilter. Further, the recirculating water which is of high turbidity decreases the disinfection capability of the ultraviolet unit. Thus, the function of the physical clarification unit directly affects various CRAS processes.

Suspended matter with a particle size over 100 μm can be easily removed by the physical clarification system, while that less than 100 μm is not effectively removed (Chen et al. 1993). In fish culturing, about 70% of the discharged suspended matters (feces and remaining feed) are smaller than 30 μm (Chen et al. 1993). Therefore, an important feature of CRAS design is an efficient system that can remove such a small suspended matter.

2.2.1 *Methods of Physical Clarification*

The methods used for physical clarification in general water treatment systems are as follows (Fujita et al. 1994; Kikuchi 1999): (1) sedimentation, (2) screen separation, (3) coagulation-sedimentation, (4) granulated filtration, and (5) foam separation. The details of each system are introduced in Sects. 2.2.2, 2.2.3, 2.2.4, 2.2.5, and 2.2.6.

2.2.2 *Sedimentation Treatment*

Sedimentation treatment is the simplest method requiring only a sedimentation tank. Whenever this method is applied to the CRAS, the flow speed of the water should be slow enough to let the suspended matter sink efficiently, since the sedimentation velocity is usually quite slow. This limitation makes the collection and deposition of suspended matter smaller than 100 μm difficult in a small-scale tank (Kikuchi 1999). A larger sedimentation tank is necessary to slow the water flow; however, the use of a larger tank would be impractical. The sedimentation tank is usually set directly downstream of the culturing tank and is connected by drain pipes (Fig. 2.3). The sediments that gather in the center of the culturing tank enter the sediment tank through the drain vent. A well-designed commercial product named Eco-Trap (Aqua Optima, Norway) (Losordo et al. 2000) has a

Fig. 2.3 Sediment tank in CRAS at Okinawa Prefectural Sea Farming Center

dual-drain construction that is effective in removing larger-sized sediment in the culturing tank. The majority of the water flows out through the center drain, which is connected to a reservoir tank; the rest of the water, which contains the sediment, flows through a side drain that is connected to the sediment tank. This dual-drain system effectively removes suspended matter from the CRAS.

2.2.3 Screen Separation

The drum filter unit is widely used for screen separation in land-based aquaculture plants. The characteristics of this unit include the capacity to treat large amounts of water and easy maintenance performed by automatic backwash. The drum filter produced by Hydrotech, Inc. (Sweden), is well known for its excellent performance and is used worldwide (Fig. 2.4) (Twarowska et al. 1997; Yamamoto et al. 2011). However, it requires large amounts of water for backwashing (10% of total culturing water daily). Therefore, the use of this drum filter in a CRAS requires additional water to maintain the water level. The pore size of the filter screen ranges from 10 to 1000 μm. For aquaculture, a filter screen with a pore size of about 100 μm is commonly used in order to balance screen blockage and backwash frequency with the ability to remove small suspended matter. As a result, removing suspended matter smaller than 50 μm in diameter is difficult (Losordo et al. 2000). To prevent

Fig. 2.4 Drum filter (Hydrotech, Inc.) using microscreen for physical treatment in CRAS

blockage of the biofilter medium by suspended matter, an additional foam separation unit (see Sect. 2.2.6) is used along with the drum filter unit in some CRASs.

2.2.4 Coagulation-Sedimentation Treatment

This treatment applies the flocculation agents used in sewage disposal to aid in the sedimentation of suspended solids (Ebeling et al. 2003, 2005; Rishel and Ebeling 2006). This treatment is commonly used to treat organic wastewater discharged from CRAS (Ermukdakul et al. 2013). Flocculation agents such as polyaluminum chloride (PAC) and polymers can negatively affect the fish health and eventually caused respiratory impairment (Tabata and Ishibashi 1984). Fish have been reported to die of suffocation due to gill blockage when cultured in recycled water containing flocculation agents. Flocculation agents remain powerful tools for removing suspended solids from effluent of CRAS, including the one discharged by microscreen backwash.

2.2.5 Granulated Filtration

In Japan, sand filtration is the most common granulated filtration method used in flow-through culturing systems. Other materials used in CRAS include anthracite, sand, and gravel. Sand filtration is rapid and can remove suspended solid matter that is larger than 5 μm; however, the backwash treatment requires large amounts of water (Huguenin and Colt 1989). The CRAS at the Palavas Research Center of Ifremer in France has a daily discharge rate of 20% of the total culturing water (Yamamoto et al. 2011). This rate is about the average for CRAS using granulated filtration. Therefore, this filtration system can be effectively used for CRAS if a large amount of high-quality water is available at a low cost. A floating bead filter is another popular and effective granulated filter; details are presented in Sect. 2.3.2.6.

2.2.6 Foam Separation Treatment

The principle of foam separation treatment (also known as a protein skimmer) is the floatation separation of suspended matter after being adsorbed by fine bubbles through electrostatic interaction (Takahashi 2005) (Fig. 2.5). Foam separation involves the following steps: (1) creation of the microbubbles, (2) adsorption of the suspended matter onto the bubbles, (3) enlargement of the bubble size by merging and unification, (4) cohesion of suspended matter by the enlarged bubbles, (5) formation of foam by the previous process, and (6) discharge of the foam by air pressure. This process can discharge mucus secreted from fish and extremely small suspended matter from the circulating water (Suzuki and Maruyama 1999) and is important in CRAS for its effectiveness at removing suspended organic matter

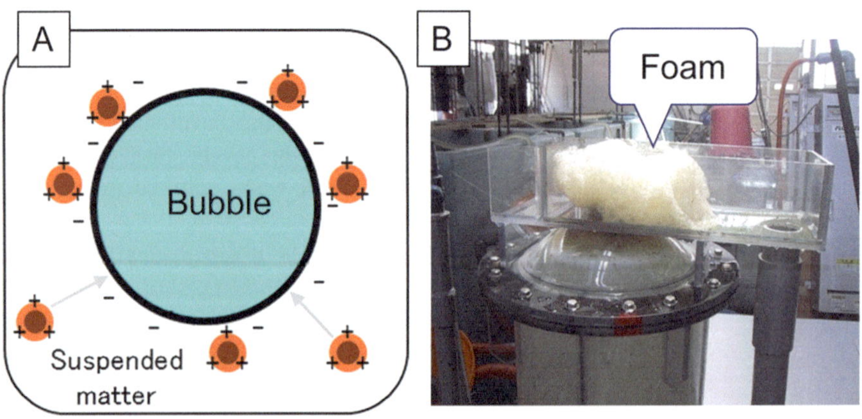

Fig. 2.5 Image of characteristic surface charge and absorption with microbubble and suspended matter (**a**) and foam separation (**b**)

smaller than 30 μm (Suzuki and Maruyama 2000; Suzuki et al. 2000; Maruyama et al. 1999). However, the volume of wastewater discharge from a form separation unit is difficult to control, especially in units with only one drain controller. During the gas-liquid mixing, the organic matter is crushed into tiny pieces by the pump with negative pressure and a rapid water current. The ammonium nitrogen in the water increases after this treatment because it is eluted from the tiny pieces of organic matter. The early foam separation units used bubbles produced by aeration via a porous air stone or cork, but this method did not remove the small suspended matter efficiently. Therefore, a lot of improvements have been made in foam separation units. Some of the new products are made by Japanese companies. In Japan, three types of foam separation units are mainly in use: (1) wing shear or KA type (Fig. 2.6) (Maruyama et al. 1999), which produces bubbles by mixing the gas with the water via the high-speed rotation of a rotary wing driven by a motor; (2) jet stream or YBM type, which produces bubbles by mixing via high-speed water current produced by a high pressure of water jet stream; and (3) combination of in-line mixer and venture tube type or FRA type (Fig. 2.7) (Yamamoto 2009),

Fig. 2.6 Setting of KA-type foam separation unit in CRAS at Yashima Station in FRA

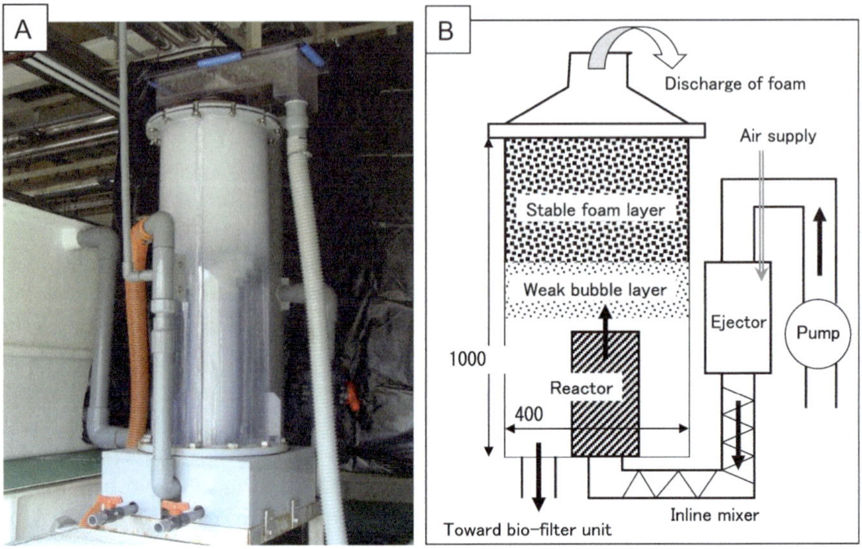

Fig. 2.7 Setting of FRA-type foam separation unit (**a**) in CRAS at Yashima Station in FRA and its inner structure (**b**)

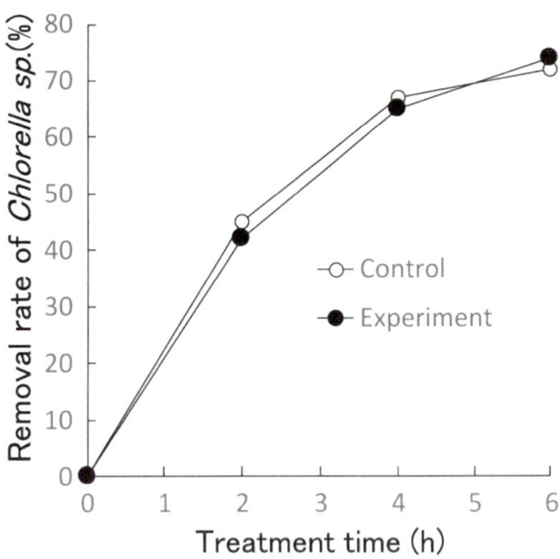

Fig. 2.8 Comparison of removal rate of suspended matter (indication, *Chlorella* sp.) between two types of foam separation units. *Control, KA type; Experiment, FRA type

which produces bubbles by mixing the gas and water at a high speed using both a venture tube and in-line mixer.

Studies comparing the effects of these foam separation units measured the density of *Chlollera* sp. (3–10 μm) before and after the foam separation treatment. The results showed that both types of units were able to remove 70–80% of the *Chlollera* sp. (Fig. 2.8). Both have stable performance in removing the organic matter, so users should choose the most suitable type based on the location of the aquaculture plant, target species, utility, maintenance, and costs.

2.2.7 *Effectiveness of the Foam Separation Unit in CRAS*

The main purpose of a foam separation unit is to remove organic matter from the water (Suzuki et al. 1996). In addition, dissolved oxygen also increases after this treatment because of the great extent of gas-liquid mixing during microbubble production. The accumulation of CO_2 decreases the water pH, hindering fish growth. Therefore, decreasing the concentration of dissolved CO_2 in the water is necessary. With the extreme gas-liquid mixing in the foam separation unit, the dissolved CO_2 gasifies quickly. Thus, the foam separation unit is an efficient piece of equipment, as it has three functions: removing organic matter, increasing dissolved O_2, and removing CO_2.

The effectiveness of this equipment can be seen most during seed production, when large amounts of suspended matter are discharged into the water. Indeed the 8-day post-hatching survival rate in red sea bream was 10% higher with the application of a foam separation unit to CRAS than without it. Additionally, the low level of accumulation of a suspended matter in the biofilter prevents blockage of the filter medium (Fig. 2.9) (Yamamoto 2013).The effectiveness of foam separation units is also reported by Maruyama et al. (1999) and Suzuki et al. (1999), (2000) in their studies of CRASs for cultivating the Japanese flounder, *Paralichthys olivaceus*, and the Japanese eel, *Anguilla japonica*.

2.3 Biofiltration Unit

In CRAS, the removal of ammonium nitrogen is the most important requirement for water purification. Previous research on nitrogen balance in farmed fish has shown that about 30% of the nitrogen in feed is incorporated into the fish body, while the rest is discharged as feces and urine via nitrogen metabolism or as remaining feed (Hall et al. 1992). In the case of teleosts, more than 50% of the nitrogen in feces and urine is in the form of ammonium nitrogen (Kikuchi 1995). Further, the remaining feed is gradually decomposed by bacteria, also releasing ammonium nitrogen. Therefore, over 35% of the nitrogen in feed ends up in the water as ammonium nitrogen (Fig. 2.10), which is highly toxic to fish. Hence, a CRAS requires a component that promptly treats and removes the ammonium nitrogen discharged by fish. Several methods in addition to biofiltration are available for removing ammonium nitrogen, including ammonia stripping method, chlorine resolution treatment method, and ozonization. All methods except biofiltration present problems in their application to CRAS. For example, the ammonia stripping method requires the pH level to be above 11, which is not realistic in CRAS. Harmful

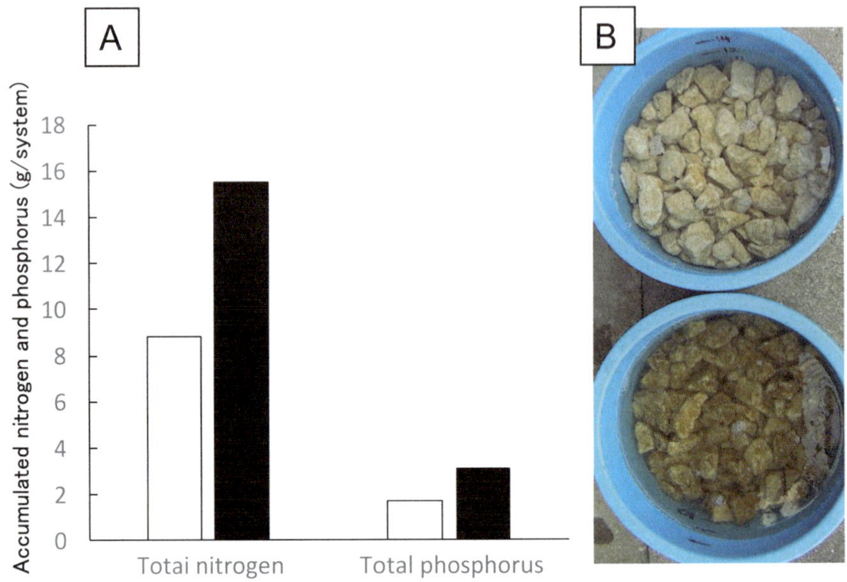

Fig. 2.9 Comparison of accumulated TN and TP in biofilter after seed production of red sea bream by systems with or without a foam separation unit (**a**) and photo of biofilter medium after experiment (**b**: top, system with foam separation unit; bottom, system without foam separation unit). □, System with foam separation unit; ■, system without foam separation unit

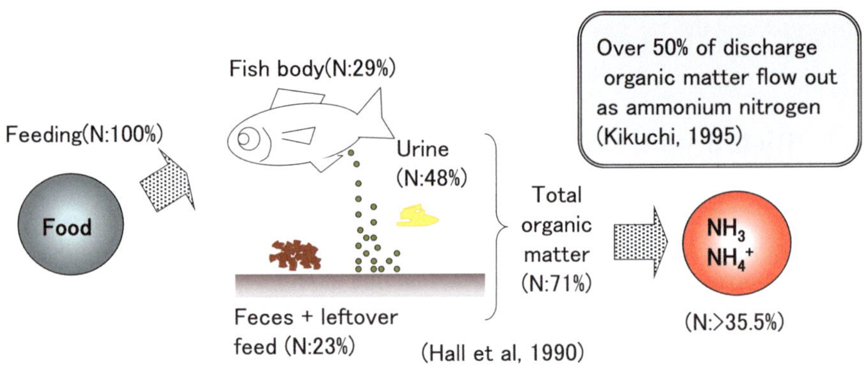

Fig. 2.10 Schematic diagram of nitrogen balance and discharge of ammonium nitrogen

byproducts can be produced during the chlorine resolution treatment and ozonization. Since removing the harmful byproducts by absorption is very expensive, these three methods are not suitable for use in fish and shellfish farming. Therefore, biofiltration is at present the only one practical and reliable method for removing ammonium nitrogen for fish farming because it utilizes the nitrification bacteria under natural water conditions and does not use any hazardous materials. The

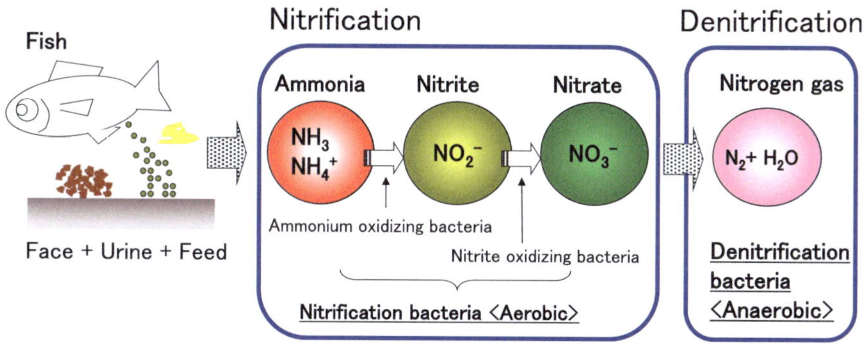

Fig. 2.11 Schematic diagram of flow of nitrification and denitrification by bacteria

biofilter converts the ammonium nitrogen (NH_3) discharged by fish into nitrite nitrogen and nitrate nitrogen. Through these processes, the toxicity of nitrogen is greatly decreased (Fig. 2.11).

2.3.1 Nitrification Function of Biofiltration

Biofiltration involves the process of nitrification, the conversion of highly toxic ammonium nitrogen (NH_4^+) into nitrite nitrogen (NO_2^-) and nitrate nitrogen (NO_3^-). This process can be shown as the following stoichiometric equations with the energy of oxidation in parenthesis (Fujita et al. 1994):

1. $NH_4^+ + 1.5O_2 \rightarrow 2H^+ + H_2O + NO_2^- (-\Delta G = 65\ \text{kcal})$
2. $NO_2^- + 0.5O_2 \rightarrow NO_3^- (-\Delta G = 18\ \text{kcal})$

These equations show that the process of changing 1 mol ammonium nitrogen into nitrate nitrogen consumes 2 mol oxygen and 2 mol alkali. In other words, oxygen is required for nitrification along with the decrease in pH. The oxidation of ammonium nitrogen into nitrite nitrogen requires more energy than that of nitrite nitrogen into nitrate nitrogen. In an actual reaction, the following stoichiometric equations is suggested with the consideration of the nitrogen utilization for composition of bacterial cells (Fujita et al. 1994):

$$NH_4^+ + 1.83O_2 + 1.98HCO_3^- \rightarrow 0.021C_5H_7NO_2 + 0.98NO_3^- + 1.041H_2O + 1.88H_2CO_3$$

These equations show that the process of nitrification of 1 mg ammonium nitrogen consumes 4.18 mg oxygen and 7.07 mg alkali. In addition, this process uses 0.17 mg of nitrification bacteria (Fujita et al. 1994).

Nitrification bacteria are autotrophic bacteria that use energy produced by the oxidation process of inorganic substances. The substrate of 1 mg ammonium nitrogen is incorporated into only 0.17 mg of bacterial cell. The efficiency of nitrogen incorporation into bacterial cells (bacterial cells/substrate, based on weight) is 0.17, although other reports suggest 0.02–0.13 g/gN, a lower value than that of yeast or heterotrophic bacteria, such as colon bacillus (0.72–0.84) (Fujita et al. 1994). Therefore, the growth rate of heterotrophic bacteria is much faster than that of autotrophic bacteria.

To maintain efficient nitrification in the biofilter, nitrification bacteria must be propagated at a high density in the filtration medium (Yamamoto 2013). This maturation of biofilter requires a 2–3-month incubation period. Thanks to the recent progress of molecular biological approaches, identification of nitrification bacteria has been progressed significantly (Schreier et al. 2010; Kumara et al. 2013; Kitamura et al. 2016; Lee et al. 2016). Recently, the significance of archaebacterial nitrification in the submerged biofilters has been shown (Sakami et al. 2012). Therefore, it is important to conduct more precise research on the activity of nitrification bacteria in biofilters under various environmental conditions.

2.3.2 *Types of Biofiltration Methods*

There are several types of biofiltration including submerged filter, trickling filter, moving bed filter, rotating disk filter, intermittent filter, and so on (Wheaton et al. 1991; Gutierrez-Wing and Malone 2006; Timmons and Ebeling 2013). This section introduces their functions as well as their advantages and disadvantages (Fig. 2.12).

2.3.2.1 Submerged Filter

Submerged filtration is the most common type. The filter medium is submerged under the water, and the water flows either upward or downward in the biofilter tank. Aeration equipment at the bottom of the unit is required to replenish the oxygen consumed by the nitrification processes in the biofilter. Backwashing is required to wash the accumulated organic matter out of the filter medium (Fujita et al. 1994). Thus, the submerged filter is easy to install at the beginning; however, from the viewpoint of long-term operation, maintenance of the filter medium is labor intensive and time consuming.

2.3.2.2 Trickling Filter

Trickling filtration in CRAS is widely used in Western countries (Otte and Rosental 1979; Bovendeur et al. 1987; Eding et al. 2006). The medium is set in the biofilter container without submerging in the water (Kikuchi 1999). First, water is pumped up to the top of biofilter tank and is showered on the top of the filter medium. The water then flows through the biofilter medium, where the ammonium nitrogen is

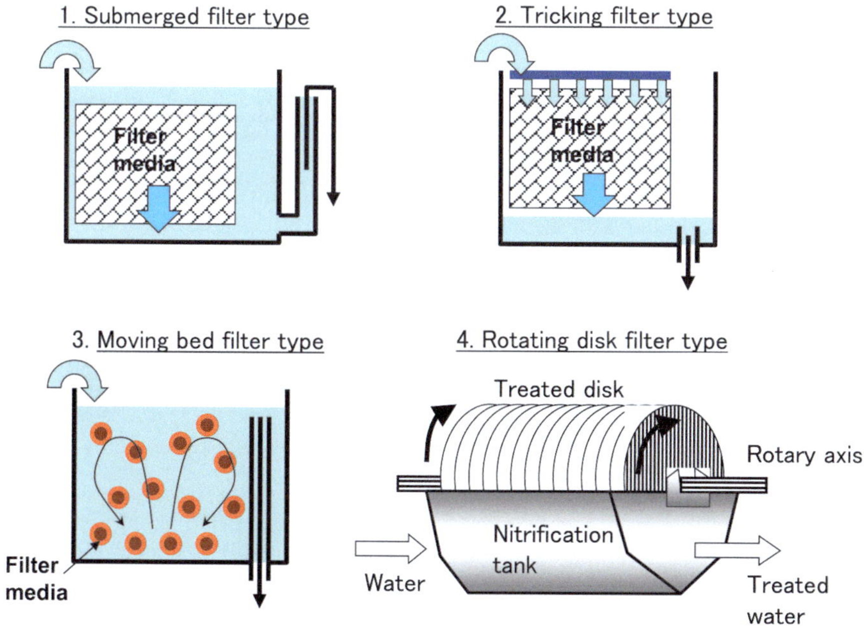

Fig. 2.12 Schematic diagram of main types of biofilter for CRAS

nitrified by nitrification bacteria. The filtered water is retrieved from the bottom of the filter container. The advantages of this non-submerged medium are the efficient supply of oxygen, which results in a high nitrification ability and less blockage of the medium. The one disadvantage is that water does not cover the entire surface of the medium at once, requiring a relatively large filter container to obtain maximal filtration.

2.3.2.3 Moving Bed Filter

A moving bed filter uses a medium with a specific gravity slightly heavier than that of water. The medium thus is moved easily in the water by current and aeration. Two types of moving bed filters are available, sealed and open type. The moving bed filter uses small beads that have a diameter of several millimeters (Fujita et al. 1994; Timmons and Losordo 1994). The medium must move continually to maintain the health of the filter, as the discharged load affects the activity of the nitrification bacteria that reside in the biofilm on the medium surface. The exfoliation of thick biofilms can often cause sudden decrease in the nitrification ability of the filter.

2.3.2.4 Rotating Disk Filter

A rotating disk filter is composed of a light disk filter, a rotary shaft, a drive motor, and a half-cylinder-conditioned reservoir tank stocked with water. Many such disks are fixed vertically and are arranged with equal distances on the center rotary shaft. Approximately 40% of the disk surface is submerged in water, and the disk rotates at a slow speed (Kikuchi 1999; Yoshino et al. 1999). Nitrification in this unit occurs by the contact of water with the nitrification bacteria on the surface of the rotating filter disk; this process is enhanced by exposing the disk to air that contains a lot of oxygen. Therefore, this unit has extremely high and efficient nitrification ability (Kikuchi 1999; Yoshino et al. 1999; Rijin 1996). However, a high load of organic matter can result in the growth of a thick biofilm on the disk, so precautions must be taken to prevent the weight of the biofilm from breaking the rotating shaft.

2.3.2.5 Intermittent Filter

The intermittent-type biofilter consists of bubble drain pipes and siphon controller unit that drains water intermittently (Fig. 2.13). A pump fills the unit with water, submerging the medium; water in the biofilter tank is then siphoned out, exposing the medium to air. Thus, the medium in this type of filter is intermittently submerged in water. This filter type has efficient nitrification ability (Yamamoto 2013). The raising and lowering of the water level in this system prevent both the blockage of the medium surface and the sedimentation of organic matter at the bottom of the biofilter tank. Therefore, this biofilter is maintenance-free and can be kept under suitable conditions for a long time of period (usually 1–2 years but depends on the rearing conditions).

We examined the nitrification ability of several types of biofilter in repeated tests using the same mass of filtration medium pre-grown with the nitrification bacteria under the same conditions. The nitrification ability of the intermittent biofilter and trickling filters was 1.5–1.7 times and 1.1–1.2 times higher than that of the

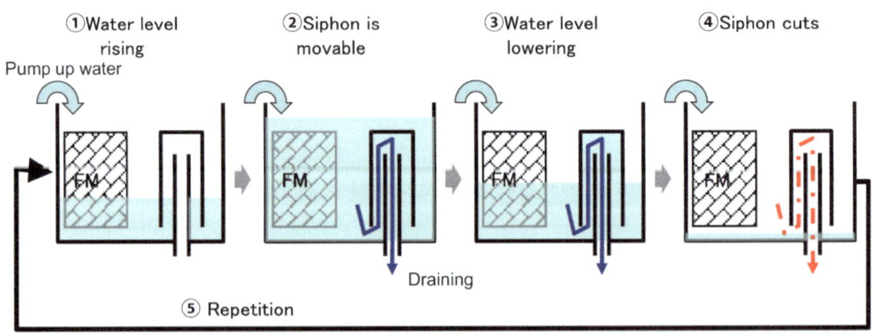

Fig. 2.13 Schematic diagram of action of intermittent-type filtration in biofilter tank. *FM, Filter materials for nitrification

submerged filter, respectively (Fig. 2.14) (Yamamoto 2013). This observation suggests that the intermittent biofilter can save 40% more space with the same nitrification ability. However, the strong downflow that occurs from draining pipe is not favorable for seed production and hinders larval swimming performance. On the other hand, in grow-out cultures, the intermittent strong current is favorable since it can be used for purging accumulated organic matter from the culturing tank.

2.3.2.6 Floating Bead Bioclarifiers

This system uses expandable plastic granular biofilter materials (Malone and Beecher 2000), and nitrification is effectively performed by bacteria attached to the surface of the granular filter (Timmons et al. 2006). At the same time, it works as a physical clarification unit that can remove organic matter (Chen et al. 1993). Since the beads possess lower specific gravity than water, they float in the upper area of the treatment unit. Suspended matter is trapped in the bottom layer of the beads. This suspended matter can be removed by backwashing using automatic bubbling treatment. One important characteristic of this system is that a very small amount of wastewater is discharged from the system. Therefore, it is quite useful for marine fish farming (Saito 2014). Recently, this filtration system has become quite popular in CRAS.

2.3.2.7 Immobilization Filter

The immobilization filter used for the nitrification was studied in seawater clarification at the Abiko Research Laboratory in the Central Research Institute of Electric Power Industry in the 1990s (Uemoto et al. 1991, 1993, 1994). In order to develop nitrification bacteria immobilized filter for use in CRAS, bacteria

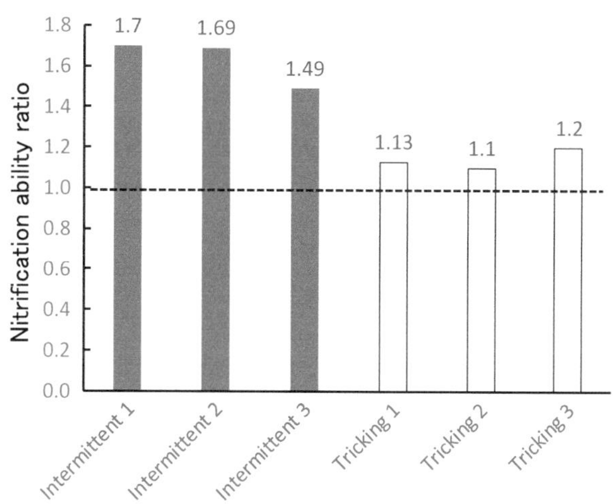

Fig. 2.14 Comparison of nitrification ability according to biofilter type: intermittent filtration or trickling filtration as compared to submerged filtration using the same volume of biofilter medium. *Nitrification ability ratio = ability of intermittent or trickling filtration/submerged filtration

isolated from the sludge were immobilized onto a the media which were 2.6–3.5 mm in a diameter and contained 10% (w/v) polyvinyl alcohol (Uemoto et al. 1991). The polymers used for the immobilization of marine nitrifying bacteria were photo-crosslinked resin (PVA-SbQ) and polyethylene glycol. The nitrification ability of the immobilized bacteria was stably maintained at high level for 6 months during the experimental period (Uemoto et al. 1993). Fish-rearing experiments on the Japanese flounder, *Paralichthys olivaceus*, were conducted for 180 days using a 2.5 kL culturing tank and 5 L of immobilized filter. The study revealed that the filter had no negative influence on the growth or survival rates of the fish (Uemoto et al. 1994). These results indicated that immobilized filters are particularly useful, as they have high nitrification activity within a very compact system.

2.3.3 *Influential Factors of Nitrification*

Since the 1950s, much research in Japan has clarified the effect of numerous environmental factors on nitrification by biofilters. The research group including Dr. Saeki, Dr. Hirayama, and others published several important papers (Saeki 1958; Hirayama 1965a, b) and are recognized as the pioneers of this field. The following section introduces the factors that affect nitrification.

2.3.3.1 Temperature

The optimum temperature of the biofilter is typically 30–35 °C (Kawai et al. 1965), and the nitrification ability decreases as temperature declines below 30 °C (Fig. 2.15). The nitrification ability becomes extremely low when the water temperature is below 10 °C and declines further as the temperature drops below 5 °C. The nitrification ability at 25 °C was 1.4–2.0 times higher than that at 15 °C (Kawai et al. 1965; Yamamoto 2013) (Fig. 2.16). In general, the temperature suitable for rearing most marine fishes ranges from 10 to 30 °C. Within this temperature range, the higher the water temperature in the biofilter, the higher the nitrification ability.

2.3.3.2 Dissolved Oxygen

Dissolved oxygen is used by nitrification bacteria in the oxidation of ammonium nitrogen during nitrification. The nitrification activity and the survival of nitrification bacteria depend on the level of dissolved oxygen. Low concentration of dissolved oxygen can be a limiting factor for nitrification. Anaerobic conditions lead to the death or inactivation of nitrification bacteria. At dissolved oxygen concentrations below 5–6 mg/L, the higher the concentration of dissolved oxygen leads to higher nitrification ability of the biofilter. At dissolved oxygen concentration below

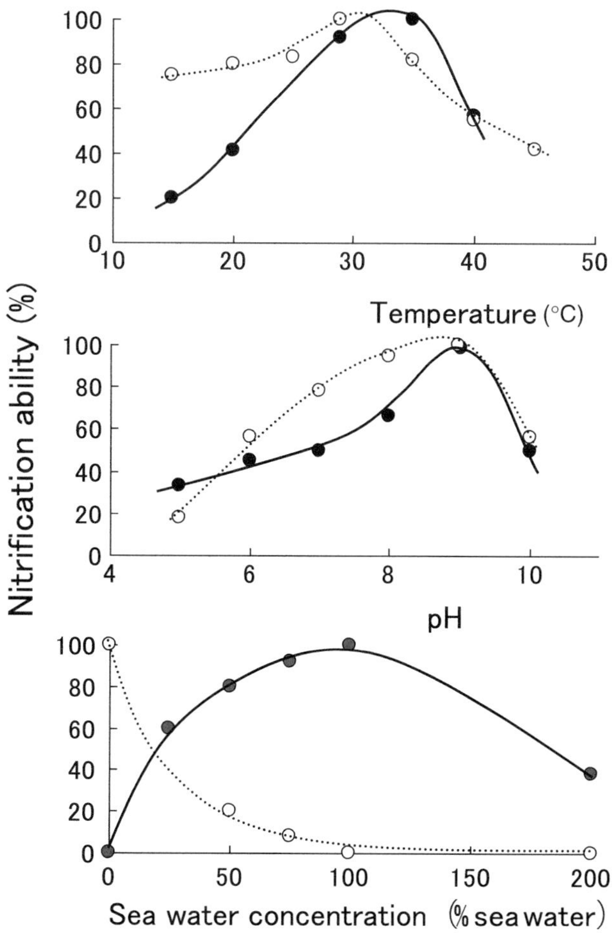

Fig. 2.15 Comparison of nitrification ability of nitrification bacteria originated in seawater and freshwater according to temperature, pH, and seawater concentration. ●, nitrification bacteria originated in seawater; ○, nitrification bacteria originated in freshwater (Kawai et al. 1965)

2–3 mg/L, the nitrification ability quickly decreases. The nitrification ability reaches a plateau at oxygen concentrations above 5–6 mg/L.

The difference in the dissolved oxygen concentration before and after nitrification treatment can be used as an indicator of the nitrification ability of the biofilter because the amount of the nitrification is nearly proportional to the amount of dissolved oxygen consumed by the nitrification bacteria (Fujita et al. 1994). Since biofiltration consumes a large amount of oxygen, it is necessary to supply oxygen to the water by several methods such as aeration or addition of pure oxygen in the biofilter tank or in the reservoir tank after biofiltration treatment.

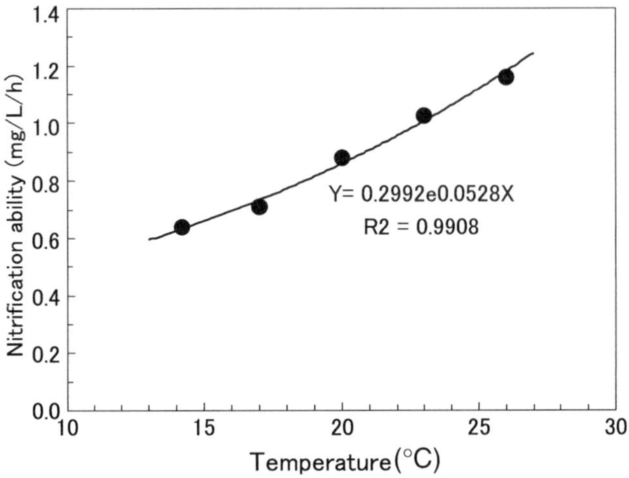

Fig. 2.16 Relationship between nitrification ability and temperature under usual rearing condition of temperate fish

2.3.3.3 pH

The nitrification process consumes alkali, thereby decreasing the pH of the water. Previous reports suggest that the optimal nitrification occurs at pH 7–9 (Fig. 2.15). At pH < 6, nitrification rapidly decreases. Although reports indicate that nitrification in fish culture tanks is not influenced by pH > 6, the recommended pH is near neutral (pH 7) (Kawai et al. 1965). To understand the effect of pH on water quality, it is necessary to consider how the fraction of dissociated and the non-dissociated forms of ammonium nitrogen changes with pH, as the non-dissociated forms are more toxic. At pH 7–9, the fraction of non-dissociated ammonium nitrogen increased with increasing pH, making the water more toxic to fish (Emerson et al. 1975).

2.3.3.4 Salinity

Nitrification bacteria are distributed in both seawater and freshwater, but the species differ depending on the salinity of the habitat. Salinity affects the nitrification activity of all types of bacteria. Nitrification bacteria found in 100% seawater exhibit the highest nitrification ability in seawater, and their nitrification ability disappears in freshwater (Fig. 2.15). The opposite is observed for nitrification bacteria found in freshwater (Kawai et al. 1965). Therefore, seawater must be used when backwashing filtration medium used for culturing seawater fish but not freshwater to prevent their death or deactivation of the bacteria during backwashing. However, recent studies showed slightly different results with different medium and experimental locations (Gonzalez-Silva et al. 2016), suggesting that more precise studies will be needed to fully understand the effects of salinity on nitrification activity.

2.3.3.5 Ammonium Nitrogen and Nitrite Nitrogen

High concentrations of ammonium nitrogen and nitrite nitrogen are reported to affect the nitrification ability of nitrification bacteria. Nitrification bacteria are cultured to expand their numbers in the filter by the process of "maturation," as described below. During this process, a small amount of ammonium nitrogen (approximately 10 mg/L) must be added because extremely low concentrations of ammonium nitrogen and nitrite nitrogen may potentially hinder the nitrification ability of the bacteria (Fujita et al. 1994).

2.3.3.6 Trace Element Ion

Trace ions of Mg, Mo, Fe, Ca, Cu, Na, and PO_4 in the water are essential to the growth of nitrification bacteria. In CRAS, the concentration of these ions should be periodically monitored if the water exchange rate is low because these essential ions can be accumulated too much in the system.

2.3.3.7 Other Items

Heavy metals and organic chlorine also affect nitrification. In addition, the antibiotics used against bacterial pathogens often affect the survival and nitrification activity of the bacteria. One study reports nearly 50% reduction in nitrification ability by addition of antibiotics to the CRAS water (Sugita et al. 2000).

2.3.4 *Biofilter Media*

2.3.4.1 Maturation of Biofilter Medium

The functions of the biofilter unit include the removal of ammonium nitrogen by nitrification bacteria and the maintenance of the biofilm that is their habitat. The process of fixing and growing of nitrification bacteria in the biofilter medium is generally called "maturation" (Kawai et al. 1964). Because nitrification bacteria grow slowly, 2–3 months are required to attain to full maturation of the biofilter medium. In many cases, the addition of an inorganic nitrogen source such as ammonium chloride as food is necessary during maturation (Kawai et al. 1964). This can be done simply by adding fertilizer regularly. The other method for maturation is using fish-rearing water, which contains inorganic and organic nitrogen.

The growth of the nitrification bacteria on the biofilter occurs through six stages, as follows: (i) the addition of ammonium nitrogen (NH_3^+, NH_4^-), a food supply for ammonium oxidation bacteria, (ii) the proliferation of ammonium oxidation bacteria, (iii) the accumulation of nitrite nitrogen (NO_2^-) produced by ammonium oxidation bacteria, (iv) the proliferation of nitrite oxidation bacteria, (v) the accumulation of nitrate nitrogen (NO_3^-) produced by nitrite oxidation bacteria, and (vi) the dissolution of nitrite nitrogen (NO_2^-), signaling that maturity has been reached (Fig. 2.17) (Kawai et al. 1964).

2.3.4.2 Characteristics and Selection of Biofilter Medium

Many different types of media are used in biofiltration units. The essential features of biofilter medium for use in CRAS are as follows: (i) suitable for biofilm attachment, (ii) large surface area and higher void ratio, (iii) chemical and biological stability without deterioration of quality, (iv) sufficient physical strength, (v) high complementation of suspended matter, (vi) maintains uniform velocity by constant void size and particle size, (vii) non-elution of harmful matter, (viii) low cost and stable supply, and (ix) easy to transport and assemble (Yamamoto 2013). However, no medium will meet all of these requirements because some of these conditions are mutually exclusive. Therefore, the user must decide which medium to use according to the specific goal and the specific CRAS design, including the biofiltration system and its maintenance (Yamamoto 2013).

One study investigating CRAS media reported that the larger the specific surface area of biofilter medium, the greater the survival of microorganism and the higher the nitrification ability (Kikuchi et al. 1994). Our study showed the same tendency

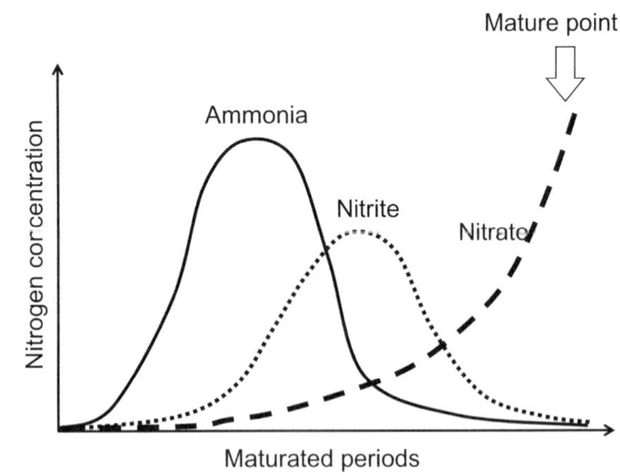

Fig. 2.17 Schematic diagram of maturation point of biofilter medium in biofilter tank, according to the three forms of nitrogen (ammonia, nitrite, and nitrate)

in comparing experiments with biofilter media made from the same materials but with the different structures and surface areas. The nitrification ability clearly differs between differing materials such as PVC (polyvinyl chloride) and coral (Fig. 2.18) (Yamamoto 2013). However, in the case of aquaculture discharged with a heavy organic load, the surface of biofilter medium is covered by biofilms, and no difference was observed in nitrification ability between a medium with a large surface area (e.g., ceramics stone) and that without (Kikuchi et al. 1994). Moreover if biofilms become too thick on the biofilter medium, the biofilm often exfoliates from the medium. To maintain a high nitrification ability, the periodic removal of blocking organic matter from the biofilter medium is necessary; alternatively, a biofilter with a high void rate can be used (Kikuchi et al. 1994). Recently, plastic floating and moving beads have become quite popular biofilter materials for CRAS. In seed production, the amount of organic waste is lower than that discharged by grow-out. The CRAS designed in the Yashima Laboratory has a foam separation unit located upstream of the biofilter unit. The efficient removal of suspended organic matter by the foam separation unit prevents the biofilm and organic matter from covering the surface of the biofilter medium (Yamamoto 2013).

Our study of biofilter media for seed production in CRAS suggested that the small-sized coral (diameter, 3–5 mm) was the most effective at nitrification. These results showed that the media with a greater surface area, such as coral, charcoal, or ceramic stone, exhibit higher nitrification ability. Further, coral can supply alkali by

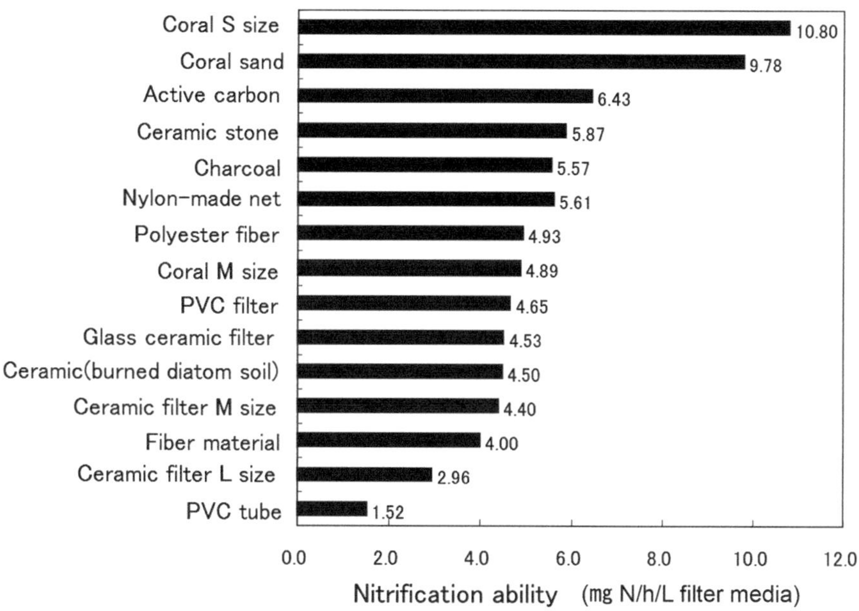

Fig. 2.18 Comparison of nitrification ability in several biofilter media by nitrification test

eluting of calcium carbonate, thereby adjusting the pH of the water (Yamamoto 2013). The nitrification ability of used fishing nets was found to be relatively high, indicating their use as a low-cost alternative biofiltration medium (Yamamoto 2013).

2.4 Disinfection Treatment

Seed production always involves the risk of infectious diseases. In the culturing of marine fish and shellfish, disinfection of the water is necessary to maintain their health. To prevent disease, seed production facilities have developed variety of treatments to disinfect the water. This section introduces the key points for disinfection in CRAS.

2.4.1 Methods of Disinfection

Water disinfection is generally conducted using the methods originally used in the waterworks. The disinfection of freshwater in the waterworks mainly relies on treatments with chlorine, UV, ozone, and ultrafiltration. On the other hand, saltwater disinfection treatments use UV, ozone, copper ions, and electrolytic techniques. Since each method has advantages and disadvantages, an appropriate method should be chosen with the specific CRAS and its purpose in mind.

2.4.1.1 Ultraviolet Ray Disinfection (UV Disinfection)

UV disinfection is used widely in fisheries. The UV disinfection unit is equipped with an ultraviolet ray lamp that disinfects by the following mechanism. An ultraviolet ray of wavelength 254 nm damages the nuclei in cells of microorganisms in the water, resulting in the death of the microorganisms. Of the various commercially available UV lamps, those with a wavelength of 254 nm are the most effective for disinfection, as the wavelength of maximum UV absorption for DNA is near 260 nm (Fig. 2.19) (Gates 1930).

This method is safe and simple; however, treatments that are too long produce excessive amount of oxidants in the treated water. Therefore, it is necessary to monitor for the negative effects of oxidants. UV disinfection of water containing large amounts of suspended matter is often incomplete because the suspended matter creates shadows. In the case of red sea bream seed production in CRAS, UV kills 99.9% of the bacteria in the water (Fig. 2.20) (Yamamoto 2013). The development of a UV-light-emitting diode (LED) is progressing. When the problems regarding the production cost and life span are solved, UV-LED will be widely used in this field.

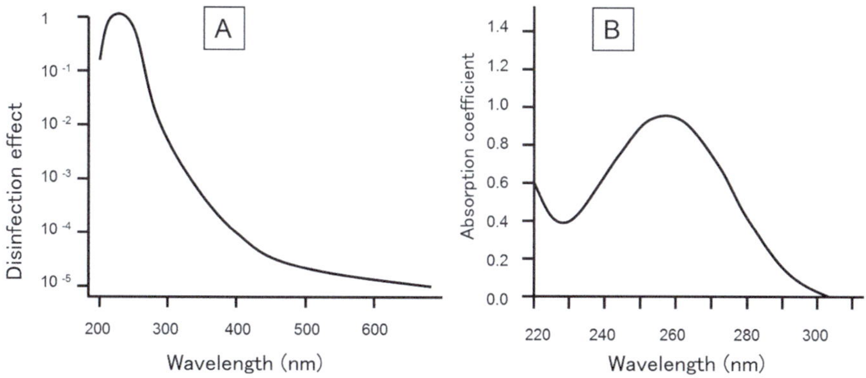

Fig. 2.19 Disinfection effect of UV spectrum for colon bacillus (**a**) and spectrum of absorption coefficient for nucleic acid (**b**). (Gates 1930)

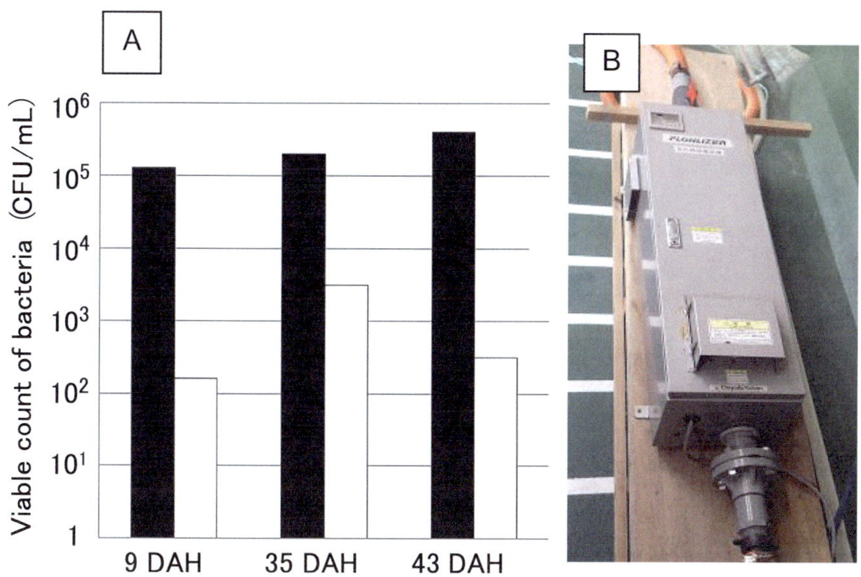

Fig. 2.20 Change in count of viable bacteria in water before and after treatment by UV disinfection unit (**a**) and outer illumination type of UV unit (**b**). *DAH, days after hatching; ■, before treatment; □, after treatment

2.4.1.2 Ozone Disinfection

The mechanism of ozone disinfection in freshwater is different from that in seawater. In freshwater, ozone remains as dissolved ozone after treatment, and the dissolved ozone directly acts as a strong oxidant on cell membranes and the nucleic acids of microorganisms, further enhancing the disinfectant effect. In contrast, ozone treatment of seawater leads to the oxidation of chlorine ion and bromide ions, producing oxidants such as hypochlorous acid and hypobromous acid. These dissolved oxidants remain in the treated seawater and destroy pathogenic bacteria and viruses.

The removal of dissolved ozone and oxidants after disinfection before using the water to rear fish is important because of their toxic effects on fish. Dissolved ozone can be easily removed from freshwater by strong aeration, but the ozone discharged into the atmosphere also must be removed for human safety. The removal of oxidants from seawater is more difficult; active carbon treatment is used to absorb them. This method does not remove all of the oxidants, and renewal of the active carbon is necessary every 1–2 years.

2.4.1.3 Electrolytic Disinfection

Electrolytic disinfection (Fig. 2.21) is used only for seawater. This simple treatment works as follows: two electrodes are placed into the seawater in the treatment tank, and the direct electrical current produces OH^- radicals on the electrodes. These OH^- radicals then oxidize chlorine and bromide ions as in the ozone treatment described above. The resulting strong oxidants disinfect the seawater through their action on microorganisms. This system also requires ozone removal before using the water for fish rearing; however, the ozone does not diffuse into the atmosphere as in the ozone disinfection mentioned above. Therefore, this method is widely used among many seed production facilities using flow-through systems (Fig. 2.22). The disadvantages of this method are as follows: (1) the accumulated "scale" caused by precipitation of Ca and Mg on the surface of the cathode electrode bar must be removed, and (2) the hydrogen gas produced during the electrolytic treatment must be collected to prevent it from catching fire. This system is used to disinfect inlet water only and is not suitable for use with recirculating water as in CRAS.

2.4.1.4 Copper Ion

The disinfection activity of trace metal ions such as silver and copper is well known and is used as a simple and safe disinfection method for fish-rearing water. Disinfection by the silver ion is widely used in freshwater but not in seawater, as the silver ion attaches to the chlorine ion in seawater, producing silver chloride deposits. However, disinfection by copper ion is commonly used in aquaculture and aquariums using freshwater and seawater.

Copper is an essential trace element for normal metabolism in organisms, but intake of copper leads to metal poisoning. Copper sensitivity is higher in invertebrates than that of vertebrates. Crustaceans are especially susceptible to even low concentrations of copper ion. Fish have a relatively high tolerance to copper ion, and it is commonly used to prevent parasitic infections and ichthyophthiriosis (white spot disease). Several types of copper ion generators are on the market

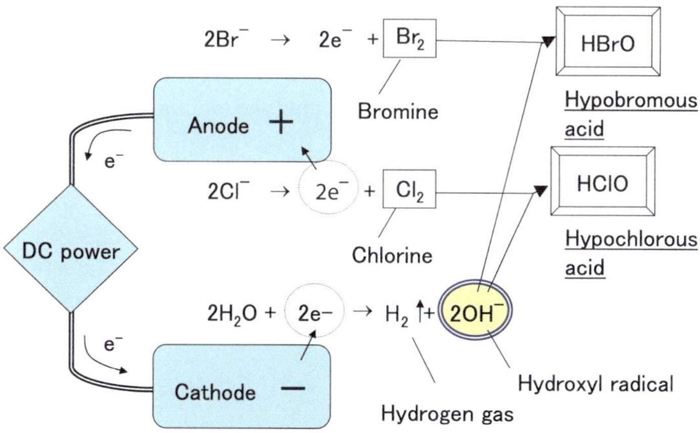

Fig. 2.21 Schematic diagram of oxidant generation by reaction of the hydroxyl radical using electrolysis unit. Oxidants: hypochlorous acid and hypobromous acid

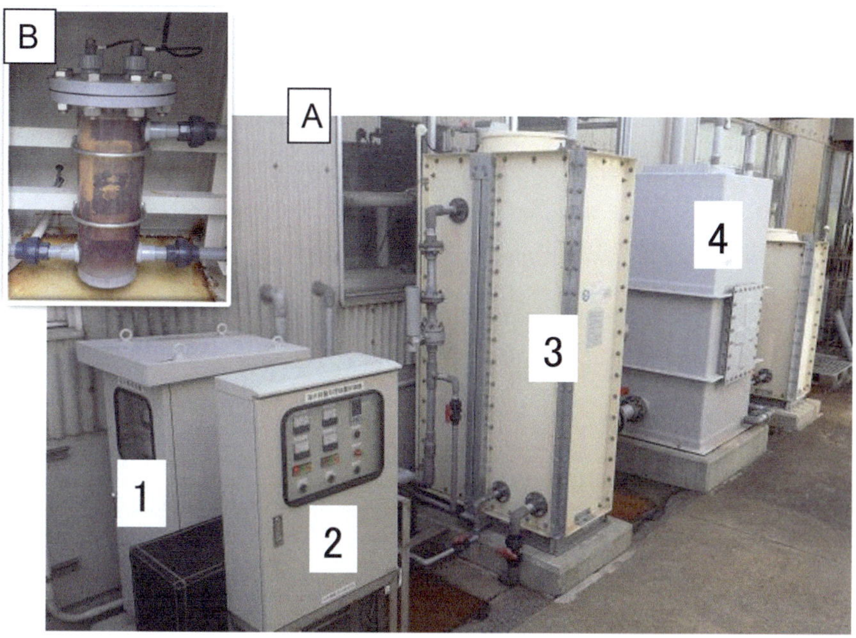

Fig. 2.22 Full view of electrolysis disinfection unit for marine fish culture using CRAS (**a**): *1*, electrolysis box; *2*, control box; *3*, reservoir tank; *4*, reactor. Electrolysis unit in its box (**b**)

(Fig. 2.23). A simple method to supply copper ion involves the immersion of copper fibers, known commercially as "copper wool" into the water.

The mechanism of disinfection by copper ion occurs by the following two processes: (1) the dissolved copper ions enter the microorganisms and bind to enzymes, inhibiting their activity; (2) active oxygen produced during the catalysis of copper ion can decompose organic matter in microorganisms.

The copper ion dose that is effective against colon bacillus is 40 ppb. Although details regarding the effect of copper ions on fish pathogens are still unclear, the suggested copper ion concentration is about 20–100 ppb for use in aquaculture or seed production (Morita 2017). Since high concentration of the copper ion is toxic to fish, precise control of the concentration is indispensable.

2.4.1.5 Other Methods

Other methods for the disinfection or removal of pathogens include ultrafiltration membrane treatment and photocatalysis. Technical advancements in ultrafiltration membrane treatment in Japan have reached the highest level in the world. The practical application of this method has spread throughout the industry. The 1–200 nm pore size of the ultrafiltration membrane blocks viruses and bacteria, so that the treated water is close to being microbe-free. Obviously no toxic components are produced because this treatment is a physical filtration. Therefore, this method may be a safer option. However, at present, the high cost of this method limits its practical applications in CRAS.

Photocatalysis is an effective technology for disinfection of the atmosphere. Using titanium oxide and sunlight, photocatalysis is a safe technology under

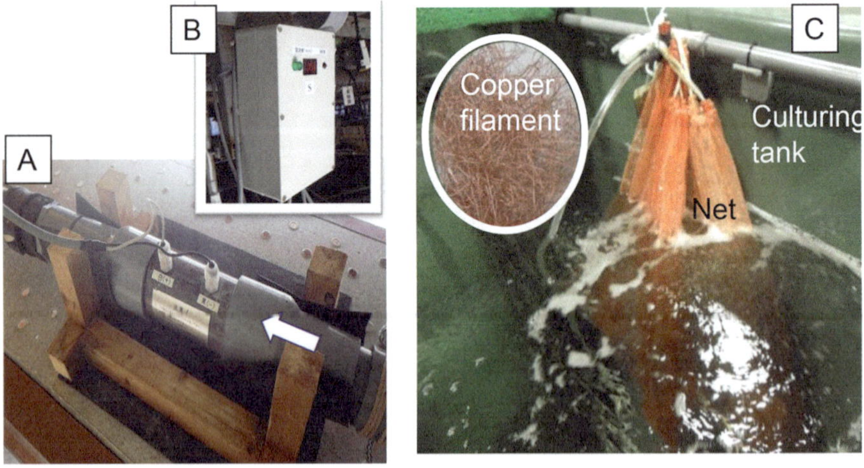

Fig. 2.23 Copper ion generator settled in pipeline (arrow shows water current) (**a**), control unit (**b**), and copper filament in the net in culturing tank of CRAS (**c**)

atmospheric conditions. However, the use of this method in water still presents a lot of problems, particularly with regard to fundamental technology and cost. Therefore, it will take time to develop suitable applications of this technology to CRAS.

2.5 Oxygen Supply System

The oxygen supply system is extremely important in CRAS. The cost of oxygen in Japan is two to three times higher than in most of the Western countries. Therefore, aeration systems using ordinary air stones and a pipe system in the culturing tank is popular in Japan (Kikuchi 1999). However, for grow-out stage cultures with a fish density $\geq$ 100 kg/kL, sufficient oxygen cannot be provided by ordinary aeration alone. Oxygen supply equipment that delivers liquid or pure oxygen formed by an oxygen generator can dissolve oxygen effectively in the water (Fig. 2.24). The cone-shaped oxygen supply equipment, called an oxygen cone, is effective and widely used in Europe and America (Kikuchi 1999; Timmons and Losordo 1994). Recently, the micro- and nano-bubble generators have been developed (Takahashi 2005; Yamamoto 2009). These devices produce many minute micro- and nano-bubbles in the water. Providing a large surface area, these bubbles dissolve oxygen with extremely high efficiency. Thus, these systems have received great deal of attention.

Simplified oxygen supply devices utilize adopted cavitation technology for the gas-liquid mixing pump. Oxygen gas is used for the cavitation, and numerous minute microbubbles are produced in the water. If the oxygen gas is supplied with air instead of pure oxygen, the gas pressure in the water must be closely monitored. Using the gas-dissolving method of cavitation under a high pressure, the water can become oversaturated with nitrogen gas, causing gas emboli in the cultured fish (Shimo et al. 2004).

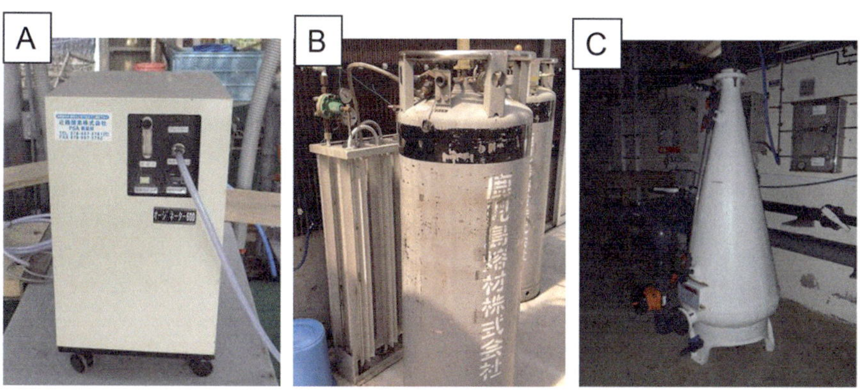

Fig. 2.24 Oxygen supply units: (**a**) oxygen generator using air, (**b**) oxygen control unit using liquid oxygen, (**c**) oxygen cone

2.6 Removal of Carbon Dioxide

In Europe and America, most CRAS utilize the pure oxygen for oxygen supply instead of aeration. This is because a lot of carbon dioxide is produced by the fish reared in high density and also by bacteria. The accumulated carbon dioxide in the water has negative effects such as causing retarded growth and acidification of the water (Itazawa and Takeda 1979). Thus, monitoring the accumulation of carbon dioxide in the water is necessary. The setting of multistage deaerators and strong aeration in the reservoir tank are effective for removing the carbon dioxide. The foam separation unit can also remove carbon dioxide from the water. As mentioned above, the foam separation unit can supply oxygen in CRAS. In some CRAS, the foam separation unit is set as the final treatment because of the multiple advantages described here.

2.7 Denitrification

During the biofiltration process, ammonium nitrogen is nitrified to form nitrate nitrogen, which accumulates in the water. The nitrate nitrogen has relatively low toxicity. Therefore, the growth and survival rate of fish is not affected by even high nitrate concentrations. However, nitrate concentrations of several hundred mg/L are reported to cause inactive feeding behavior in the Japanese flounder, *Paralichthys olivaceus*, cultivated in CRAS (Honda et al. 1993). For long-term fish culturing in CRAS, nitrate nitrogen must be removed by denitrification. This procedure utilizes the reductive reaction function of denitrification bacteria, which converts nitrate nitrogen into nitrogen gas. The denitrification equipment contains filtration medium that carries denitrification bacteria, and the water is recirculated slowly through the device. By keeping the equipment under anaerobic conditions, the denitrification bacteria can proliferate within the bacterial flora and denitrify the water continuously. This process requires a carbon source for denitrification. Typically, methanol or biodegradable plastic can be used as a carbon source. Much research on denitrification during sewage treatment has been conducted, resulting in technical advancements in denitrification equipment and a greater understanding of denitrification bacteria in Japan (Watanabe et al. 1989, 1991). A small-scale study examined the efficiency of the denitrifying system using a fixed bacterial reactor for culturing Japanese flounder in CRAS and evaluated the denitrifying activity of a fluidized-bed system using immobilized marine denitrifying bacteria (Watanabe et al. 1993). As a result, the denitrifying activity was shown to be about three times better than that of the biofilm fixed-bed system.

"Self-cleaning inherent gas denitrification-reactor" has been also developed, and it was proven to be quite effective for denitrification for pike perch farming (Müller-Belecke et al. 2013). This kind of novel system will be also applied to CRAS in the near future. Since the technologies have been mainly developed for freshwater sewage treatment, technical improvement, development of equipment, and researches on denitrification bacteria are essential for its application to CRAS, especially in seawater systems. Recently, a system that integrates aerobic and anaerobic microbial processes, including a novel combination of denitrification, anaerobic ammonium oxidation (anammox), and methanogenesis, has developed to eliminate toxic inorganic nitrogen compounds and organic solids. The effectiveness of this system was proved by rearing gilthead sea bream in CRAS (Tal et al. 2009). Therefore, such a novel technology will play important roles for future development of biofilter having denitrification activity.

2.8 Wastewater Treatment

The organic matter trapped by the physical filtration unit, including feces, remaining feed, and bacterial flock, must be discharged from the CRAS. This waste must be appropriately treated before to protect the environment. In Europe and America, the cost for such treatment is an issue for the management of CRAS. After the coagulation-sedimentation treatment, the solid sediments of organic matter are taken out of the sediment tank, molded into a cake form, and then dehydrated. The dried solid sediments are disposed as industrial waste. The supernatant liquid is stored in the reservoir pond, where the water evaporates naturally and the nitrogen and phosphorus in the water are absorbed by plants and phytoplankton. The water left after the treatment is drained when its quality level drops below the emission control standard. For a large facility, a reservoir pond and vast space are necessary for wastewater treatment. Therefore, technological improvements in wastewater treatment are urgently needed.

2.9 Conclusions

The water clarification methods used in CRAS to disinfect and remove organic matter and inorganic nitrogen must be quick to use and safe for fish and human. Same goes for disinfection. Numerous methods and equipment are now used in the treatments described here. When building a CRAS, it is important to select and combine appropriate equipment depending on the specific purpose of the unit.

Water from the culturing tank is treated as it flows through several water clarification devices and is then returned to the culturing tank. In CRAS, inadequate water clarification equipment will cause the water quality to decline. If this situation occurs, one solution for recovering water quality is to use a combination of different types of water exchanging methods. However, if the water exchange rate is increased, the influence of external risks also increases (e.g., infectious disease). Therefore, the performance of the disinfection unit must be improved to eliminate pathogenic microorganisms from the inlet water. Needless to say, a constant exchange of water will defeat the advantages of CRAS, such as preserving the environment, avoiding external risks, and minimizing costs to maintain suitable temperature. Thus, there is a great need for the development a fully closed system that is eco-friendly, poses no risk of pathogen infection, and is less expensive.

References

Bovendeur J, Eding EH, Henken AM (1987) Design and performance of a water recirculation system for high-density culture of the African catfish, *Clarias gariepinus*. Aquaculture 63:329–353

Chen S, Timmons MB, Aneshansley DJ, Bisogni JJ Jr (1993) Suspended solids characteristic from recirculating aquaculture systems and design implications. Aquaculture 112:143–155

Ebeling JM, Sibrell PL, Ogden SR, Summerfelt ST (2003) Evaluation of chemical coagulation–flocculation aids for the removal of suspended solids and phosphorus from intensive recirculating aquaculture effluent discharge. Aquac Eng 29:23–42

Ebeling JM, Rishel KL, Sibrell PL (2005) Screening and evaluation of polymers as flocculation aids for the treatment of aquacultural effluents. Aquac Eng 33:235–249

Eding EH, Kamstra A, Verreth JAJ, Huisman EA, Klapwijk A (2006) Design and operation of nitrifying trickling filters in recirculating aquaculture - a review. Aquac Eng 34:234–260

Emerson K, Russo RC, Lund RE, Thurston RV (1975) Aqueous ammonia equilibrium calculations: effect of pH and temperature. J Fish Res Board Can 32:2379–2383

Ermukdakul T, Chawaloesphonsiya N, Boonchayaahant B, pungraswi W, Painmanakul P (2013) Different approaches for the separation of suspended solids in aquaculture system. Water Environ Technol 11(2):59–70

Fujita K, Yamamoto K, Takizawa S (1994) Classification of filtration, rapid-filtration, bio-filtration, membrane-filtration. Gihodo, Tokyo, p 301. (In Japanese)

Gates FL (1930) A study of the bacterial action of ultra violet light III. The absorption of ultra violet light by bacteria. Genetic. Physiology 14(1):31–42

Gonzalez-Silva BM, Jonassenl KR, Bakke I, Østgaard K, Vadstein O (2016) Nitrification at different salinities, biofilm community composition and physiological plasticity. Water Res 95:48–58

Gutierrez-Wing MT, Malone RF (2006) Biological filters in aquaculture, trends and research directions for freshwater and marine applications. Aquac Eng 34:163–171

Hall POJ, Holby O, Kollberg S, Sarnuelsson MO (1992) Chemical fluxes and mass balances in a marine fish cage farm IV. Nitrogen. Mar Ecol Prog Ser 89:81–91

Hirayama K (1965a) Studies on water control by filtration through sand bed in a marine aquarium with closed circulating system-I. Oxygen consumption during filtration as an index in evaluating the degree of purification of breeding water. Nippon Suisan Gakkaishi 31:977–982. (In Japanese with English abstract)

Hirayama K (1965b) Studies on water control by filtration through sand bed in a marine aquarium with closed circulating system-II. Relation of filtering velocity and depth of sand layer to purification of breeding water. Nippon Suisan Gakkaishi 23(11):983–990. (In Japanese with English abstract)

Honda H, Watanabe Y, Kikuchi K, Iwata N, Takeda S, Uemoto H, Furuta T, Kiyono M (1993) High density rearing of Japanese flounder, Paralichthys olivaceus with a closed saltwater recirculation system equipped with denitrification unit. Suisanzoshoku 41:19–26

Huguenin JE, Colt J (1989) Design and operating guide for aquaculture saltwater system. Elsevire, Amsterdam. 262pp

Itazawa Y, Takeda T (1979) An estimation of the minimum level of dissolved oxygen in water required for normal life of fish. II. An experiment with carp under carbon dioxide accumulation. Nippon Suisan Gakkaishi 45:323–327. (In Japanese with English abstract)

Kawai A, Yoshida Y, Kimata M (1964) Biochemical studies on the bacteria in the aquarium with a circulating system-I. Changes of the qualities of breeding water and bacterial population of the aquarium during fish cultivation. Nippon Suisan Gakkaishi 30:55–62. (In Japanese with English abstract)

Kawai A, Yoshida Y, Kimata M (1965) Biochemical studies on the bacteria in the aquarium with a circulating system-II. Nitrifying activity of the filter-sand. Nippon Suisan Gakkaishi 31:65–71. (In Japanese with English abstract)

Kikuchi K (1995) Nitrogen excretion rate of Japanese flounder – a criterion for designing closed recirculating fish culture systems. Israeli J. Aquaculture 47:112–118

Kikuchi K (1999) Introduction of recirculating aquaculture systems and their prospects. In: Hino A, Maruyama T, Kurokura H (eds) Reduction of environmental emissions from aquaculture. Kouseisha-kouseikaku, Tokyo, pp 64–76. (In Japanese)

Kikuchi K, Honda H, Kiyono M (1994) Ammonia oxidation in marine biological filters with plastic filter media. Fish Sci 60(2):133–136

Kitamura R, Ishii K, Maeda I, Kozaki T, Iwabuchi K, Saito T (2016) Evaluation of bacterial communities by bacteriome analysis targeting 16S rRNA genes and quantitative analysis of ammonia monooxygenase gene in different types of compost. J Biosci Bioeng 121:57–65

Kumara VJR, Sukumarana V, Achuthana C, Josepha V, Philipb R, Singha ISB (2013) Molecular characterization of the nitrifying bacterial consortia employed for the activation of bioreactors used in brackish and marine aquaculture systems. Int Biodeter Biodegr 78:74–81

Lee DE, Lee J, Kim YM, Myeong JI, Kim KH (2016) Uncultured bacterial diversity in a seawater recirculating aquaculture system revealed by 16S rRNA gene amplicon sequencing. J Microbiol 54:296–304

Losordo TM, Hobbs AO, Delong DP (2000) The design and operational characteristics of the CP&L/EPRI fish barn: a demonstration of recirculating aquaculture technology. Aquac Eng 22:3–16

Malone RF, Beecher LE (2000) Use of floating bead filters to recondition recirculating waters in warmwater aquaculture production systems. Aquac Eng 22:57–73

Maruyama T, Suzuki Y, Sato D, Kanda T, Michishita T (1999) Performance of a closed recirculating system with foam-separation and nitrification units for intensive culture of Japanese flounder. Nippon Suisan Gakkaishi 65(5):818–825. (In Japanese with English abstract)

Morita T (2017) Prevention of parasite and sterilization of system after fish rearing, instance of latest technology and industrialization in recirculation aquaculture system. In: Yamamoto Y, Morita T (eds) RAS study meeting. Midorishobou, Tokyo, pp 105–114. (In Japanese)

Müller-Beleckea A, Zienerta S, Thürmera C, Kaufholda S, Sprangerb U (2013) The "self cleaning inherent gas denitrification-reactor" for nitrate elimination in RAS for pike perch (*Sander lucioperca*) production. Aquac Eng 57:18–23

Otte G, Rosental H (1979) Management of a closed brackish water system for high density fish culture by biological and chemical water treatment. Aquaculture 18:169–181

Rijin JV (1996) The potential for integrated biological treatment system in recirculating fish culture – a review. Aquaculture 139:181–201

Rishel KL, Ebeling JM (2006) Screening and evaluation of alum and polymer combinations as coagulation/flocculation aids to treat effluents from intensive aquaculture systems. J World Aquac Soc 37:191–199

Saeki A (1958) Studies on fish culture in the aquarium of closed-circulating System. Its fundamental theory and standard plan. Nippon Suisan Gakkaishi 23(11):684–695. (In Japanese)

Saito S (2014) Recirculating Aquaculture System(RAS) (Review) I. System structure and technological components. Sci Rep Hokkaido Fish Res Inst 86:81–102

Sakami T, Ando T, Morita T, Yamamoto Y (2012) Phylogenetic diversity of ammonia-oxidizing archaea and bacteria in biofilters of recirculating aquaculture systems. Mar Genomics 7:27–32

Schreier HJ, Mirzoyan N, Saito K (2010) Microbial diversity of biological filters in recirculating aquaculture systems. Curr Opin Biotechnol 21:318–325

Shimo S, Akimoto Y, Takahama H (2004) Review of the water quality effects on marine organisms. Rep Mar Ecol Res Inst 6:1–159

Sugita H, Makuta Y, Mizuno H, Yoshihara K (2000) Effects of antibacterial substances on the nitrifying activity of filter materials in a model system. Suisanzosyoku 48(1):131–134. (In Japanese with English abstract)

Suzuki Y, Maruyama T (1999) Performance of a closed recirculating system with foam separation, nitrification and denitrification units for intensive culture of Japanese eel. In: Hino A, Maruyama T, Kurokura H (eds) Reduction of environmental emissions from aquaculture. Kouseisha-kouseikaku, Tokyo, pp 98–115. (In Japanese)

Suzuki Y, Maruyama T (2000) Removal of suspended matter from an aqueous solution by foam separation with fish mucus. J Jpn Soci Water Environ 23(3):181–186. (In Japanese with English abstract)

Suzuki Y, Maruyama T, Seki H (1996) Absorption behavior of organic substance onto bubble surface in nonforming bubble separation, J. Chemical Engineering Japan 29(5):794–798

Suzuki Y, Maruyama T, Takemoto S, Oda R (1999) Performance of a closed recirculating system with foam separation, nitrification and denitrification units for the intensive culture of eel. J Jpn Soc Water Environ 22(11):896–903. (In Japanese with English abstract)

Suzuki Y, Maruyama T, Sato D, Kanda T, Michishita T (2000) Water quality and mass balances in a culture of Japanese flounder using a closed recirculating system with foam-separation and nitrification units. Nippon Suisan Gakkaishi 66(1):1–9. (In Japanese with English abstract)

Tabata K, Ishibashi N (1984) Long term toxicity of anionic and nonionic polyacrylamide flocculants to fish. Jpn J Water Pollut Res 7(6):361–368. (In Japanese with English abstract)

Takahashi M (2005) ζ potential of microbubbles in aqueous solutions, electrical properties of the gas–water interface. J Phys Chem B 109(46):21858–21864

Tal Y, Schreier HJ, Sowers KR, Stubblefield JD, Place AR, Zohar Y (2009) Environmentally sustainable land-based marine aquaculture. Aquaculture 286:28–35

Timmons MB, Ebeling JM (2013) Biofilters, recirculating aquaculture, 3rd edn. Cayuga Aqua Ventures, Ithaca, pp 254–273

Timmons MB, Losordo TM (1994) Aquaculture water reuse systems: engineering design and management. Elsevier, Amsterdam. 333pp

Timmons MB, Holder JL, Ebeling JM (2006) Application of microbead biological filters. Aquac Eng 34:332–343

Twarowska JG, Westerman PW, Losordo TM (1997) Water treatment and waste characterization evaluation of an intensive recirculating fish production system. Aquac Eng 16:133–147

Uemoto H, Kikuchi K, Kiyono M (1991) Development of water purification technologies for intensive fish culture, 6. Saltwater treatment with immobilized marine nitrifying bacteria. Central Research Institute of Electric Power Industry Research Reports, U90056: 26 pp. (In Japanese with English abstract)

Uemoto H, Kikuchi K, Kiyono M (1993) Development of water purification technologies for intensive fish culture, 8. Mass cultivation and immobilization of marine nitrifying bacterium isolated from closed recirculation system. Central Research Institute of Electric Power Industry Research Reports, U90056: 21 pp. (In Japanese with English abstract)

Uemoto H, Watanabe Y, Kikuchi K (1994) Development of water purification technologies for intensive fish culture, 10. Rearing of Japanese flounder with closed saltwater recirculation system equipped with immobilized bacteria. Central Research Institute of Electric Power Industry Research Reports, U93048: 22 pp. (In Japanese with English abstract)

Watanabe Y, Kikuchi K, Uemoto H, Takeda S, Kiyono M (1989) Development of water purification technologies for intensive fish culture, 3; Selection of marine denitrifying bacteria by

their denitrification patter. Central Research Institute of Electric Power Industry Research Reports, U89035: 16 pp. (In Japanese with English abstract)
Watanabe Y, Kikuchi K, Takeda S, Kiyono M (1991) Development of water purification technologies for intensive fish culture, 7. Removal of nitrate accumulating in cultural saltwater and some factors affecting the denitrification of marine denitrifying bacteria. Central Research Institute of Electric Power Industry Research Reports, U91002: 22 pp. (In Japanese with English abstract)
Watanabe Y, Kikuchi K, Uemoto H, Kiyono M (1993) Development of water purification technologies for intensive fish culture, 9. High cell density mass cultivation of marine denitrifying bacteria and efficient denitrifying system by using immobilized bacteria. Central Research Institute of Electric Power Industry Research Reports, U93023: 22 pp. (In Japanese with English abstract)
Wheaton F, Hochheimer J, Kaiser GE (1991) Fixed film nitrification filters for aquaculture. In: Brune DE, Tomaso JR (eds) Aquaculture and water quality. World Aquaculture Society, Baton Rouge, pp 272–303
Yamamoto Y (2009) Application of micro-babble for fisheries and aquaculture. Mater Integration 22(5):24–29. (In Japanese)
Yamamoto Y (2013) Research of development for closed recirculation system for seed production in marine fishes. Doctoral dissertation of graduated school of marine science and technology, Tokyo University of Marine Science and Technology. 206 pp. (In Japanese)
Yamamoto Y, Miyata T, Yoseda K (2011) Existing research of closed recirculating aquaculture in Europe. J Fish Technol 3(2):153–156. (In Japanese)
Yoshino H, Gruenberg DE, Watanabe I, Miyajima K, Satoh O (1999) Changes in water quality and performance of a closed recirculating saltwater aquaculture system for rearing Pejerrey, *Odonthestes bonariensis*. Suisanzoshoku 47(2):289–297. (In Japanese with English abstract)

Part III
Tackle of Recirculating Aquaculture in Fishes

Chapter 3
Eel *Anguilla japonica*: Toward Zero Emission

Yoshihiro Suzuki and Toshiro Maruyama

Abstract The development of a closed recirculating aquaculture system that does not discharge effluents would reduce a large amount of pollutant load on aquatic bodies. In this study, eel were reared in a closed recirculating system, which consisted of a rearing tank, a foam separation unit, a nitrification unit, and a denitrification unit. The foam separation unit has an inhalation-type aerator and supplies air bubbles to the rearing water. The gross eel weight increased 2.8–24 kg, and the magnification was 8.6 times. Then, the feed conversion ratio [(growth wet weight)/(feed dry weigh) × 100] was 67%, which is equal to that obtained by culture farms. The maximum fish density in the rearing tank became 44 kg-fish/m^3, which is about seven times higher than that of conventional eel culture farms in Japan. The survival rate under the congested experimental conditions was 91%. The foam separation unit maintained oxygen saturation in the rearing water above approximately 87%. Furthermore, fine colloidal substances were absorbed on the stable foam formed from eel mucus and were removed from the rearing water by foam separation. Ammonia oxidation and the removal of suspended solids were accomplished rapidly and simultaneously in the nitrification unit. The ammonia concentration and turbidity were almost kept at less than 1.0 mg-N/L and 10 units, respectively. When the denitrification process was operated, nitrate that accumulated in the rearing water was reduced to 10 mg-N/L. The sludge was easily recovered from the nitrification and denitrification tanks, and the components were found suitable as compost. Based on these results, the intensive aquaculture of freshwater fish such as eel can be achieved using a closed recirculating system without emission.

Keywords Aquaculture system • Recirculating • Eel • Foam separation • Zero emission

Y. Suzuki (✉) • T. Maruyama
Department of Civil and Environmental Engineering, Faculty of Engineering, University of Miyazaki, Miyazaki 889-2192, Japan
e-mail: suzuki@civil.miyazaki-u.ac.jp

T. Takeuchi (ed.), *Application of Recirculating Aquaculture Systems in Japan*, Fisheries Science Series, https://doi.org/10.1007/978-4-431-56585-7_3

3.1 Introduction

Worldwide, the development of a recycling system that brings environmental load close to zero is required today in most, if not all, industries. The aquaculture industry is likewise urged to convert to a new system that introduces the concept of zero emission, and this requires the development of a closed recirculating system using innovative technology for fish production. The technology of a recirculating aquaculture system with a high fish density has been developed (Bovendeur et al. 1987; Heinsbroek and Kamsta 1990; van Rijn and Rivera 1990; van Rijn 1996; Geiner and Timmous 1998), and the remarkably high productivity and energy efficiency of such a system have become possible (Blancheton 2000). In many cases, however, some closed recirculating systems require the drainage of a certain percentage of rearing water into the environment. The conventional eel culture in Japan is also water flow system with control of the exchange rate of the rearing water. If rearing water is drained without appropriate processing, the high nitrogen and phosphorus loads contained in the drainage can pollute the environment, even if the frequency of change is very low. In addition, the sludge and water used to wash filter media are also directly discharged in many cases. The characteristics of an economical and environmentally friendly zero-emission system are as follows: water use is minimized, drainage water is purified to the same level as raw water, and sludge is further utilized as fertilizer. The following contaminants produced during fish culture should be made as close to zero as possible: degradable organic substances, nitrogen, and phosphorus. Currently, zero-emission aquaculture system in an actual aquaculture farm has not yet been developed. However, if intensive fish culture in a perfectly closed recirculating system becomes technically possible, the development of zero-emission systems can be realized.

We have developed the zero-emission system composed of foam separation, nitrification, and denitrification units, with the combined aim of increased fish production and reduced nutrient load in aquatic environments (Maruyama et al. 1998; Suzuki et al. 2003). The advantage of this system is that it is equipped with an effective foam separation unit as part of its main purification process. Oxygen supply, removal of suspended substances, and deaeration can be achieved simultaneously by the foam separation process (Maruyama et al. 1991, 1996).

An ideal aquaculture system is one which purifies the rearing water while obtaining high biomass productivity. In this study, eel productivity, function of each water treatment process, sludge recovery, and load reduction were examined.

3.2 Experimental Condition

3.2.1 System Description

A closed recirculating system with foam separation, nitrification, and denitrification units is shown in Fig. 3.1. This system consisted of a fish-rearing tank (0.5 m^3, water volume 0.43 m^3, water surface area 1.0 m^2), a foam separation tank (0.25 m^3) equipped with an inhalation-type aerator (200 V, 0.2 kw), a nitrification tank (0.16 m^3), and a denitrification tank (0.21 m^3). The total amount of water in this system was 1.05 m^3. A heater (100 V, 1 kw) and a pH control pump (Iwaki Co., EH/W-PH, 5% sodium hydrogen carbonate solution) for adjusting the conditions of the rearing water (25 °C and pH 7.5) were set in the foam separation tank, and a water conditioner was set on the recirculating pipe. First, tap water was introduced to the system, and one cycle was carried out for 15 min at 56 L/min. The rearing water was transported to the foam separation tank by a circulating pump, and oxygen supply and foam separation processing were simultaneously carried out using this unit. The rearing water was then introduced into the nitrification tank with an upflow style, and the treated rearing water was returned to the rearing tank. Water temperature, pH, electric conductivity (EC), and dissolved oxygen concentration of the rearing water were continuously monitored using a water quality monitoring system (Horiba Co., WP-100).

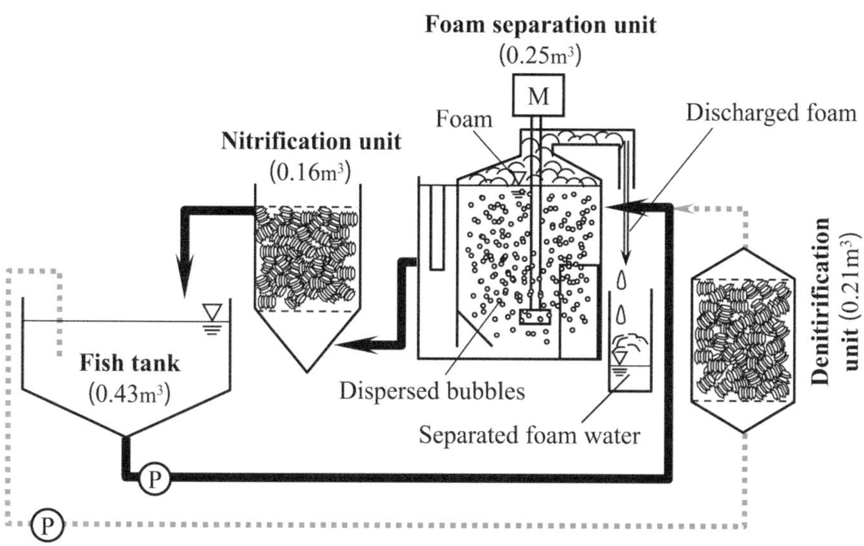

Fig. 3.1 Schematic diagram (not to scale) of the closed recirculating system with foam separation, nitrification, and denitrification units (Modified from Suzuki et al. 2003)

The main core of this system was the foam separation unit (Fig. 3.2), which was equipped with an inhalation-type aerator (Fig. 3.3, Plesca Co., Japan). By rotating the impeller, negative pressure is generated at the back of the impeller, and air is drawn from water in the shaft tube that connects to the outside environment (Fig. 3.3). Air is sheared with a blade immersed in water, and numerous bubbles are extensively dispersed in water. Surface-active materials in the rearing water adsorb on bubbles, and the bubbles are carried to the water surface. Then, foam generates on the water surface. The foam generated continuously is spontaneously removed from the foam duct placed at the upper part of the tank equipped with air exhaust (Fig. 3.2). Furthermore, air bubbles were vigorously mixed in water, and oxygen was efficiently dissolved in the rearing water until it passes through the foam separation unit.

A cylindrical medium made of polyethylene (Furukawa Electrician Industry Co., 14 mm diameter, 11 mm inside diameter, 14 mm length, 0.93 specific gravity) was used to fill the nitrification tank up to the 0.16 m^3 (surface area 93 m^2) mark. Nitrifying bacteria were immobilized onto the medium prior to the fish-rearing experiment.

In the denitrification process, a portion of the rearing water was made to flow into the denitrification tank using another line via a circulating pump. The same medium as that used in the nitrification process was used as the denitrification medium. The methanol dose tube was established at the midpoint of the inflow line to the denitrification tank, and methanol was continuously injected by a metering pump (Iwaki Co., EH-B15) with an appropriate amount of methanol corresponding to three times the concentration of nitrate nitrogen in the rearing water (Suzuki and Maruyama 1999). Then, the mixture of rearing water and methanol was introduced into the denitrification tank. Methanol injection was adjusted taking into account the nitrate concentration in the rearing water every week. The treated water that passed through the denitrification tank was returned to the foam separation tank.

The above system is considered an almost perfectly closed system because water is added only to replace that which is lost to evaporation and foam generation.

3.2.2 Fish Rearing

In an aquaculture pond, the size of an eel differs greatly with the cause of the growth rate. An important point is sorting out and breeding for every size of a fish in order to produce the eel efficiently in the actual eel farm. Since Japanese consumers like the fatty meat of eel, it is common to add fish oil to assorted feed for rearing eel. Therefore, in the rearing experiment as closing to the actual eel farm, rearing eel were sorted out, and the fish oil was added to the feed. Larval eel *A. japonica* (total gross weight 2.8 kg, 217 tails, about 13 g/tail) were placed in the rearing tank. In the initial part of the experiment, the proportion of total eel weight to water quantity in the rearing tank was 0.8%. Throughout the rearing experiment, the eel were fed a commercial diet (Chubu Shiryo Co., Japan, 48% protein) daily except Sundays. In

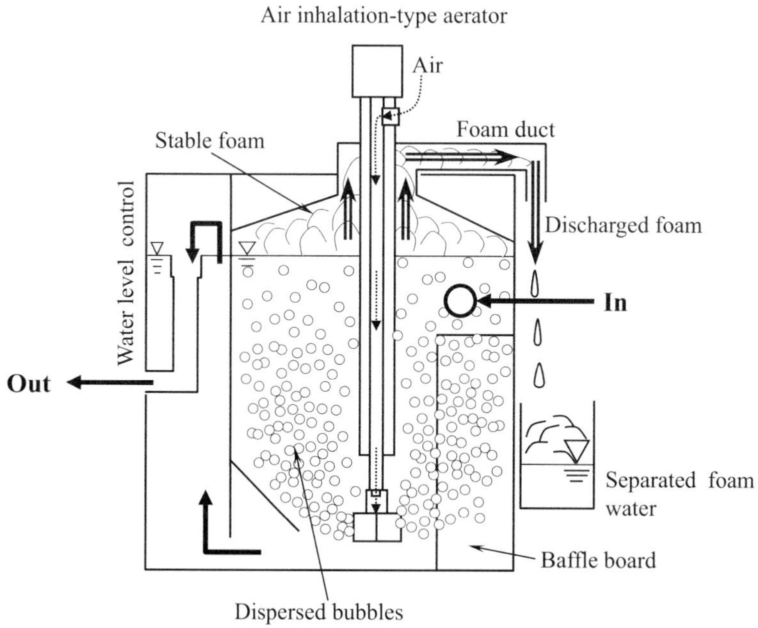

Fig. 3.2 Schematic diagram of the foam separation unit (Modified from Suzuki et al. 2003)

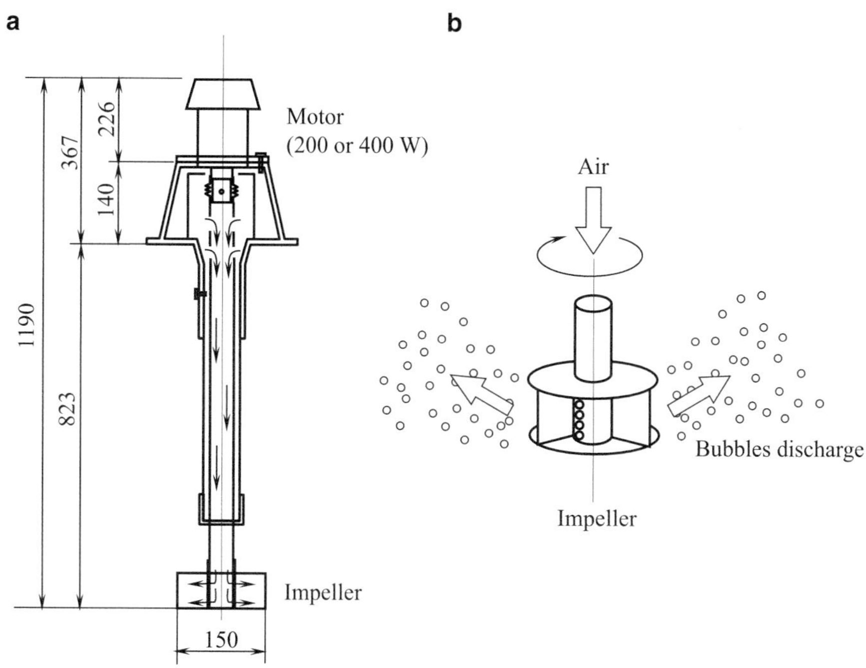

Fig. 3.3 Schematic diagram of an air inhalation-type aerator (Modified from Suzuki et al. 2003): (**a**) the whole shape (unit: mm) and (**b**) the impeller shape

the initial stage, 100 g of the feed with 5% of fish oil was given once daily every 10 a.m. This became twice daily (10 a.m. and 6 p.m., feed 200 g) when the baiting became active. Eel were cultured for 9 months (262 days). The sorting periods were beginning to 4 months (Period-1), 4–7 months (Period-2), and 7–9 months (Period-3).

3.2.3 Procedure for Sludge Recovery

After the water supply to the nitrification tank was stopped by a bypass pipe, the sludge that accumulated in the nitrification tank was drained from the bottom and then transferred to another tank. All medium was also removed from the denitrification tank and washed with sludge water. After washing, the media were returned to the nitrification tank, and the sludge was allowed to settle for 2 h. The supernatant was returned to the system, and fish rearing was continued in the usual way. The concentrated sludge was frozen until analyses. At the end of the experiment, the denitrification tank was also washed in the same way.

3.2.4 Analytical Methods

To determine the quality of rearing water, a sample was collected daily (except Sundays) from the rearing tank before feeding. Dissolved oxygen (DO), turbidity as kaolin standard (Mitsubishi Kagaku Co., SEP-PT-706D), total organic carbon (TOC, Shimadzu Co., TOC-5000), color as platinum-cobalt standard, absorbance at 260 nm (E260, Shimadzu Co., UV-2200), total ammonia nitrogen (TAN, HACH Co., DR-2000), nitrate (NO_3-N, HACH Co., DR-2000), nitrite (NO_2-N, HACH, DR-2000), total nitrogen (T-N), phosphate (PO_4-P), and total phosphorus (T-P) were analyzed. The standard platinum-cobalt method of measuring color was used, in which the unit of color is produced by 1 mg-Pt/L in the form of chloroplatinate ion. The collapsed-foam water samples were also obtained, and TOC, color, E260, suspended solids (SS), T-N, and T-P were analyzed. The analytical methods followed that of the Japanese Industrial Standard (JIS K 0102) or HACH Co. analytical manual.

Carbon and nitrogen in the solid samples, such as feed, fish tissue, and dried sludge, were analyzed using a CHN coder (analyzed by Shimadzu Techno-Research Co., Japan). Phosphorus in solid matter was decomposed in a mixture of perchloric acid and nitric acid and analyzed in the same way as T-P in rearing water.

3.3 Eel Culture in Closed Recirculating System

3.3.1 *Survival and Growth of Eel*

The survival rate, eel growth, and feed quantity during the experimental period (262 days) are shown in Fig. 3.4. Throughout the rearing period, the eel fed actively, and their total weight increased over time. Histograms showing the number of eel at given sizes at each period are shown in Fig. 3.5. The individuals which were about 13 g/tail at the beginning grew up over 100 g/tail in 4 months of Period-1. The eel that attained commercial size (over 120 g/tail) were harvested, and rearing was continued. The percentage of the commercial number to a total number of individuals was 18%. The group of 80–100 g/tail of average size grew satisfactorily, and the percentage of commercial-sized individuals was reached to 80% in Period-2. Finally at the end of Period-3, 27 of 36 individuals, which remained in the fish tank, grew over the commercial size. The gross weight and the production weight of commercial eel throughout the rearing test are shown in Fig. 3.6. The magnification of growth weight and the feed efficiency in Period-1 were 5.9 and 79.4%, respectively. The productivity of eel fell in Period-2, and the magnification of growth weight and the feed efficiency were 1.8 and 53.5%, respectively. In Period-3 as the last period, the productivity improved again, and the magnification of growth weight and the feed efficiency were 2.4 and 89.1%, respectively.

3.3.2 *Fish-Rearing Density and Productivity*

The total feed consumed by the end of the rearing experiment was 33.4 kg, and the increase in eel weight and magnification were 24 kg and 8.6 times. Then, the feed conversion ratio was 67%, which is equal to that obtained by culture farms in Japan.

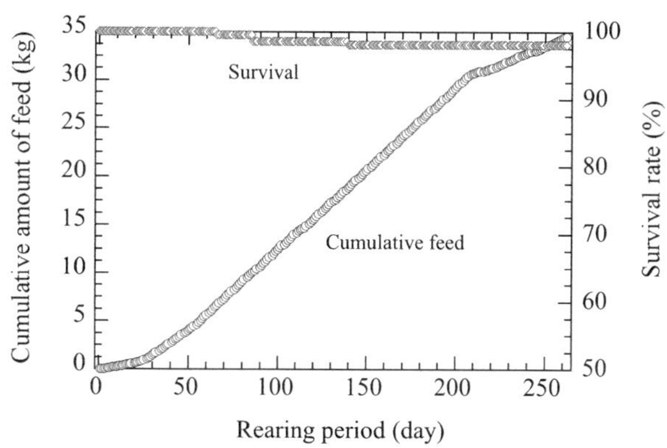

Fig. 3.4 Cumulative amount of feed and survival rate during the rearing period (Modified from Suzuki et al. 2004)

Fig. 3.5 The histogram of rearing eel during the rearing period (Modified from Suzuki et al. 2004)

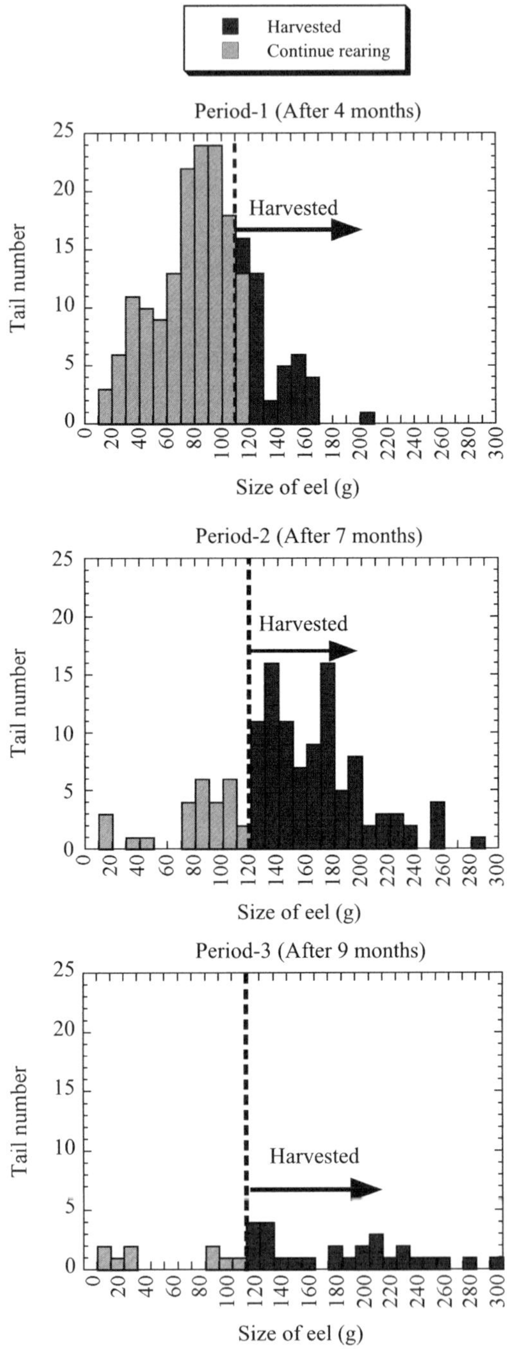

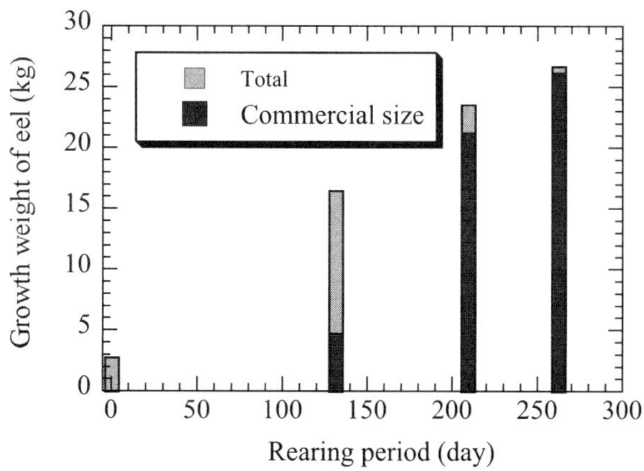

Fig. 3.6 Growth weight of eel during the rearing period (Modified from Suzuki et al. 2004)

The maximum fish density in the rearing tank (water volume, 0.43 m^3; water surface, 1.0 m^2) became 44 kg-fish/m^3 (4.4%) which is about seven times higher than that of conventional eel culture farms in Japan (about 0.6%). The production weight for 9 months was the rearing experiment to 24 kg-fish/m^3 in the whole recirculating system. In addition, the total volume of using water for rearing was 2100 L (system, 1050 L; methanol solution, 1020 L; added tap water, 30 L), and then the amount of water to produce 1 kg of eel was only 90 L. It has been reported that though the maximum fish density was reached to 80 kg-fish/m^3, the amount of using water for eel production was more than 700 L (Liao et al. 2002). Therefore, the closed recirculating system was extremely effective for eel production as based on the fish density and water consumption.

3.3.3 Quality of Rearing Water

Dissolved Oxygen The changes in DO of the rearing water are shown in Fig. 3.7. The oxygen saturation percentage was kept above approximately 87% throughout the experimental period. The maximum and minimum DO concentrations were 7.6 mg/L and 4.8 mg/L, respectively, with a mean of 6 mg/L. The DO concentration decreased to 4.8 mg/L because water temperature increased, thereby lowering the DO saturation value, when the water conditioner momentarily encountered problems. While this system does not provide an extraneous source of oxygen except for the aerator in the foam separation tank, a high DO concentration was properly maintained in the rearing water.

Turbidity, Color, and TOC The changes in turbidity, color, and TOC of the rearing water are shown in Fig. 3.8. The turbidity of the rearing water was maintained in the range of 1–10 units, whereby almost no suspended substances could be observed. The turbidity standard for tap water in Japan is two units. However, the rearing water turned yellowish brown as the experiment progressed, and color and TOC

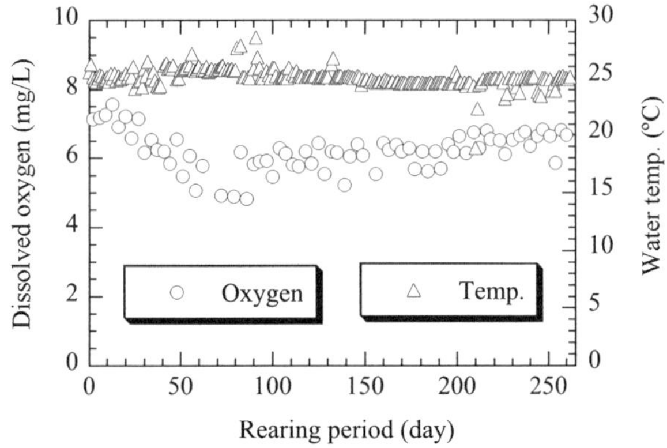

Fig. 3.7 Dissolved oxygen concentration and water temp. in the rearing water (Modified from Suzuki et al. 2004)

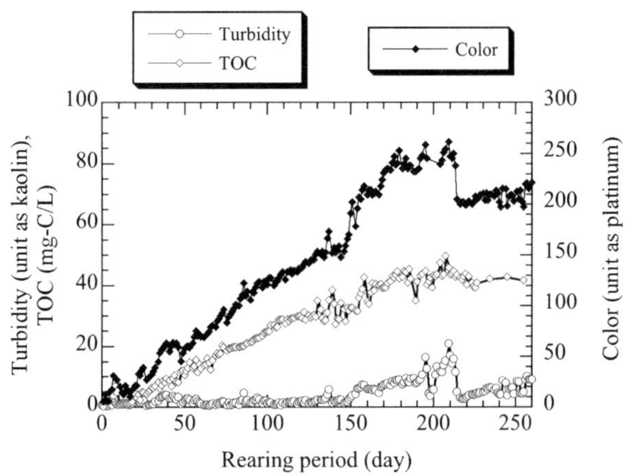

Fig. 3.8 Turbidity, color, and TOC in the rearing water during the rearing period (Modified from Suzuki et al. 2004)

gradually increased. The color determined by the platinum-cobalt method is useful in measuring the color of potable water and water whose color is due to naturally occurring materials. The color units reached 250 degrees by the end of the Period-2. The change in TOC was similar to the change in the color. The color units of the rearing water showed a very high correlation with TOC (Fig. 3.9). Therefore, the yellowish-brown material that accumulated seemed to be slightly decomposed organic substances such as humin (Tambo and Kamei 1998). In the Period-3 when the loading with rearing decreased, turbidity, color, and TOC of the rearing water decreased rapidly. The water treatment process installed in the rearing system has the capability to remove the substances originated in turbidity, color, and TOC. In this experiment, the accumulation of the yellowish-brown material did not interfere with eel growth. Moreover, it is easy to remove the brown material via activated carbon adsorption (Suzuki et al. 2000) once the water color interferes with fish feeding or observation.

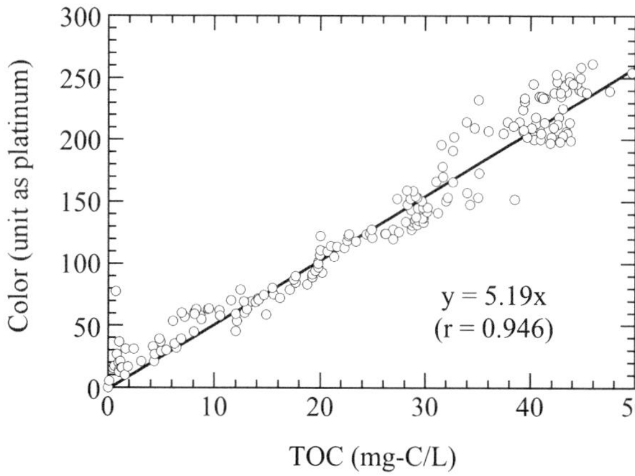

Fig. 3.9 Relationship between TOC concentration and color in the rearing water (Modified from Suzuki et al. 2004)

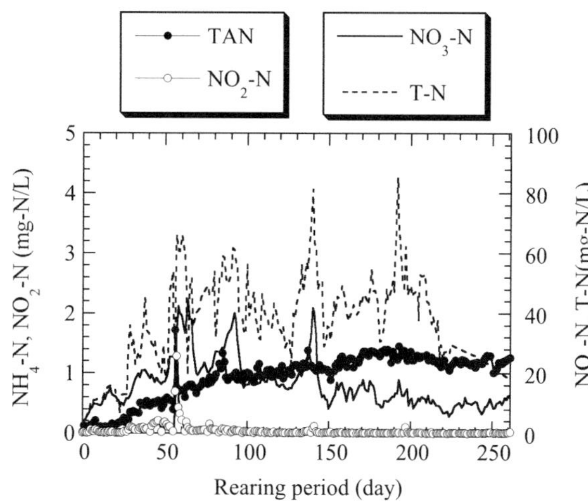

Fig. 3.10 Concentration of nitrogen compounds in the rearing water during the rearing period (Modified from Suzuki et al. 2004)

Nitrogen The changes in the concentrations of TAN, NO_2-N, NO_3-N, and T-N in the rearing water are shown in Fig. 3.10. While there was a sudden increase in TAN and NO_2-N concentrations when the amount of feeds was increased to 200 g/day, the concentrations immediately returned to a low level. The system's nitrification process functions well. The mean concentrations of TAN and NO_2-N within the rearing period were 0.87 mg-N/L and 0.052 mg-N/L, respectively. The TAN concentration was kept low throughout the study period. The NO_2-N concentration was maintained at a very low level for 270 days. In addition, the concentration of NO_3-N formed via TAN oxidation was maintained in the range of 2.5–44.5 mg-N/L by denitrification process. The NO_3-N increased rapidly as a pulse was caused by the pump trouble for adding methanol. Maintenance and management of methanol injection is of primary importance in denitrification process. Organic nitrogen accounted for the difference in concentration between T-N and NO_3-N. The amount of injected methanol was higher at about 20% than the stoichiometric methanol quantity needs for denitrification reaction, so that the quantity of residual methanol would be small. Actually, the interference effect of eel growth could not be observed with denitrification. It seemed that residual methanol that passed through the denitrification tank was biodegraded under aerobic condition until circulation in the system.

Phosphorus The changes in phosphate and T-P in the rearing water are shown in Fig. 3.11. Most of the phosphorous in the rearing water was phosphate because it accumulated only in the absence of a phosphate removal process in this system. The color units reached 110 mg-P/L by the end of the rearing experiment. It was necessary to remove phosphorous from waste water after rearing for zero emission. Therefore, the removal of phosphorous from rearing water was examined by coagulation using a ferric coagulant. The optimum condition determined by a jar test, coagulant dosage and pH range were 100 mg-Fe/L and 4–7, respectively. The concentration of phosphorous in the treated rearing water was less than 1 mg-P/L. As the result, after eel production, phosphate in the rearing water should be removed using appropriate processing methods such as coagulation using the ferric coagulant.

Evaluation of Water Quality The conventional criteria of water qualities for rearing eel in Japan are as follows: DO, more than 2–3 mg/L; TAN, less than 10 mg-N/L; NO_2-N, less than 10 mg-N/L; and NO_3-N, less than 300–400 mg-N/L. In Europe, the water quality standards for rearing eel are a little severe rather than Japan, and the standards are as follows: suspended solids, 25 mg/L; TAN, less than 6 mg-N/L; NO_2-N, less than 2 mg-N/L; and NO_3-N, less than 200 mg-N/L. As compared with these criteria and standards, each parameter of rearing water in this system was observed at the very low concentration. When we use the closed recirculating system with foam separation, nitrification, and denitrification processes for rearing eel, each process always functions appropriately throughout the eel production and can maintain a good water quality control, even if the fish density is markedly high and fish oil is added to the feed.

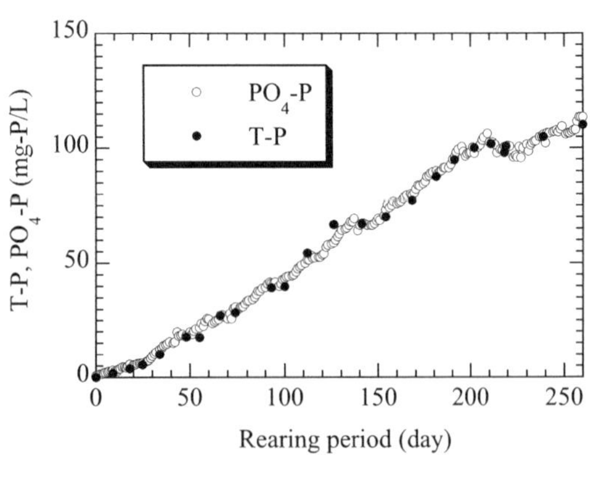

Fig. 3.11 Concentrations of phosphorus compounds in the rearing water during the rearing period (Modified from Suzuki et al. 2004)

3.3.4 Characteristics of Foam Separation Process

Foam Generation The amount of foam water discharged and the changes in turbidity and color in the foam water samples are shown in Fig. 3.12. The discharged foam water was observed on the 74th day of the study period, and foam water was continuously generated to the end of the experiment. The EC in rearing water increased with rearing eel, and when about 1000 μS/cm was exceeded, the foam was generated. The total amount of discharged foam water was 43 L (10% of the total volume in this system), and the quantity of water discharged per day was 301 ± 234 mL (mean ± SD, $n = 143$). It has been reported that foam generation of fish mucus is dependent on the concentrations of mucus and coexisting solvent ions (Suzuki and Maruyama 2000). In addition, it has been reported that the substance that contributed to foam generation is mucus glycoprotein secreted from the skin of eel cultured in the rearing tank (Suzuki et al. 2003). Therefore, foam generation was considered to be related to the concentration of surface-active materials, such as fish mucus, and EC in rearing water.

Removal of Suspended Solids and Color Components The suspended solids were significantly concentrated in the separated foam water, and the turbidity of the separated foam water was two to three orders of magnitude higher than that of the rearing water. The turbidity in foam water changed irregularly and varied from 5 to 1000 mg/L, making it necessary to remove SS from the system by a foam separation process. The mean concentrations of turbidity within the rearing period were 198 ± 20 units. Moreover, a brown material was significantly concentrated in the foam water. The color unit of the foam water was reached to 6000. The mean concentrations of color within the rearing period were 1896 ± 18 units. The foam

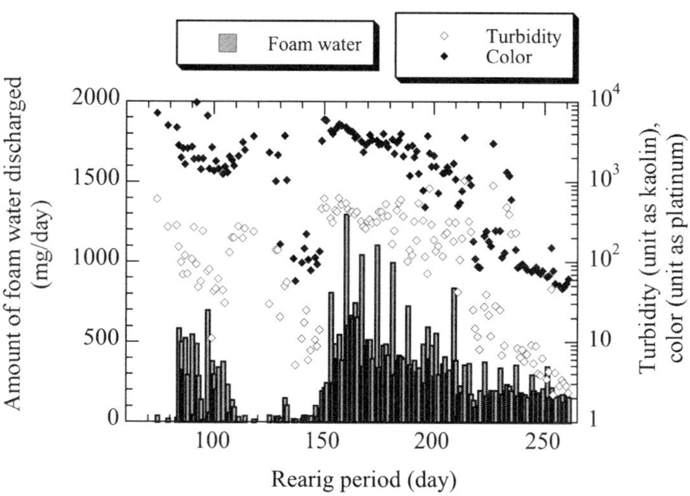

Fig. 3.12 The amount of foam water discharged and the changes in turbidity and color in the foam water samples during the rearing period (Modified from Suzuki et al. 2004)

separation process was able to remove the color components, which are difficult to remove by biological treatment or physical filtration. In addition, the mean concentrations of TOC as a combined parameter for organic substances load were 600 ± 477 mg-C/L (mean ± SD, $n = 101$). While an analysis of the effect of bacterial removal was not undertaken in this study, it has been reported that bacteria are concentrated and suspended in foam water (Maruyama et al. 1991, 1996; Suzuki et al. 2000).

Removal of Nitrogen and Phosphorus The T-N concentration, 190 ± 150 mg-N/L (mean ± SD, $n = 101$), in the foam water was several times higher than that in the rearing water. Since most of the nitrogen components were NO_3-N, highly efficient removal was not attained by foam separation. The T-P concentration in the foam water, which ranged from 93 ± 68 mg-P/L (mean ± SD, $n = 99$), was also higher than that in the rearing water. The removal efficiency of phosphorus by foam separation was higher than that of nitrogen.

3.3.5 Recovery and Utilization of Sludge

Suspended solids accumulated in the nitrification tank as the rearing period progressed. Therefore, in order to prevent plugging of the nitrification tank, cleaning of the medium and recovery of the sludge were carried out on each experiment period. By the end of the study, the sludge that accumulated in the nitrification and denitrification tanks was also recovered. More than 90% of the total sludge accumulated in the nitrification tank. The polypropylene cylindrical medium was very light, such that taking them out from the nitrification tank was very easy. The sludge immediately separated from the media after washing and then settled down. The sedimentation velocity of the sludge was rapid, and the sludge settled down in 10 min. This proved that the sludge could be quickly separated into solid and liquid components.

The components of the recovered sludge from the nitrification tank on the end were analyzed, and the C, N, and P contents as dry weight were 17%, 3%, and 12%, respectively, and the calculated C/N ratio was 6. The C/N ratio and N content satisfied the organic fertilizer composition recommended by the Japan Sewage Works Association (1994). Moreover, the phosphorus content of the sludge was also very high, making it possible to use the recovered sludge as a good compost material.

3.3.6 Mass Balances

The total amount of feed was 34,524 g-dry weight during the rearing experiment for 262 days. The N and P contents in the feed were 8.1 and 3.1%, respectively. The eel production was 2669 g-dry weight. The N and P contents in the fish body were 8.5

and 0.91%, respectively. The N and P contents in rearing water and sludge were obtained with the concentrations and volumes at the time of the experimental end. The total N and P contents in the feed were considered as 100%, and the mass balances of this system are shown in Fig. 3.13.

In the case of total nitrogen, 33% was utilized for eel growth, 0.3% was accumulated in the rearing water as nitrate and organic nitrogen, 0.3% was removed by foam separation, and 4.4% was accumulated in the nitrification and denitrification tanks as sediment (Fig. 3.13a). Regarding mass balances in the culture, the assimilation of nitrogen in the fish body varied from 25 to 35% of the total nitrogen input without regarding the difference in fish species (Folke and Kautsky 1989; Hall et al. 1992; Maruyama and Suzuki 1998; Skjølstrup et al. 1998; Suzuki and Maruyama 1999). These results agree well with this study. Although the nitrogen content in the sludge was very small for the total nitrogen input, the close values obtained in this study were reported in other culture systems. The nitrogen contents in the sludge were 6% in the recirculation system for rainbow trout (Skjølstrup et al. 1998) and 5% in the flowing system for carp (Maruyama and Suzuki 1998). Almost all the nitrogen that must be treated in this system was present as a dissolved fraction. In this study, the remaining 62% of nitrogen in the system was removed as nitrogen gas by denitrification. Denitrification could have removed the residual nitrate in the rearing water if the operation was continued for a few days after the fish was harvested. Since the recovered sludge can be utilized as fertilizer, the zero emission of nitrogen is almost possible in this system.

In the case of phosphorus, 10% was utilized for eel growth, 12% was accumulated in the rearing water, 0.4% was removed by foam separation, and 61% was accumulated in the nitrification and denitrification tanks as sediment (Fig. 3.13b). Because of analytical error, the total percentage was lacking to 100%. The percentage of phosphorus in the eel was lower than that of nitrogen. The decrease in the phosphorus content in the fish (17–28% of the total phosphorus input) was observed in other previous reports (Folke and Kautsky 1989; Holby and Hall 1991; Maruyama and Suzuki 1998; Suzuki and Maruyama 1999; Suzuki et al. 2003). On the other hand, the proportion of phosphorus in the sludge remarkably increased further than that of nitrogen. Phosphorus was loaded as suspended substances and accumulated in the sludge in the nitrification tank. Sludge recovery and effective sludge utilization are important in order to achieve the zero emission of phosphorus.

3.3.7 Economical Efficiency

The running cost for eel production in the rearing experiment became 2300 yen/kg-fish which is about two to three times higher than that of conventional eel culture farms in Japan. The percentage that the power consumption as the aerator and the recirculating pump to the total running cost was more than 70%. The electric power capacity of the pump is excessive to the amount of circulating water flow rate, and the miniaturization of the aerator is also possible in the system. Reduction of power consumption is a subject of this system. At present, water reservation is the most

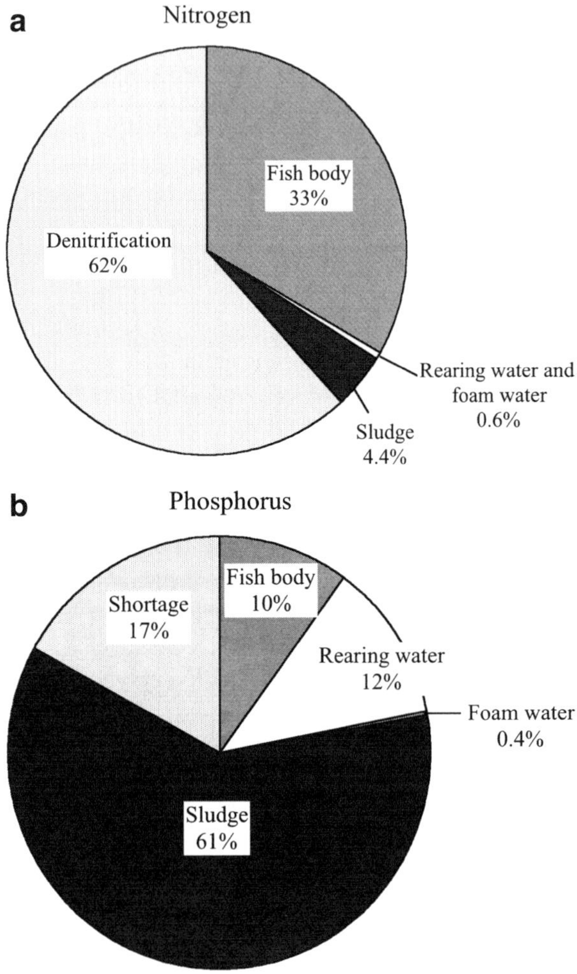

Fig. 3.13 Mass balances of nitrogen (**a**) and phosphorus (**b**) in the closed recirculating system (Modified from Suzuki et al. 2004)

important and serious problem for the actual inland aquaculture. However, the amount of water to produce 1 kg of eel was less than 100 L in this system. Since using water for fish production is drastically reducible, it is the principal merit of the closed recirculating system that unitization of tap water, which is the perfectly safe for pathogens, is also possible. In particular, this system could be effective as a stockyard of the mature eel for spawning under specific pathogen-free condition.

3.4 Conclusions

Our proposed system achieved intensive eel culture in a perfectly closed cycle for 9 months. Eel growth was satisfactory, and gross weight multiplied seven times during the study period, with a survival rate of 98%. In order to produce 1 kg of eel,

90 L of water was used. Oxygen was efficiently supplied to the rearing water by a foam separation unit, and oxygen saturation was maintained at 87% throughout the experiment. Simultaneously, the foam separation process removed the brown colloidal substances generated by fish mucus. The nitrification tank removed suspended solids and likewise rapidly nitrified TAN. In addition, denitrification effectively removed NO_3-N, and the concentration of NO_3-N was maintained less than 45 mg-N/L by denitrification process. More than 90% of the total nitrogen in the system was removed by denitrification. Sludge was easily recovered from the nitrification and denitrification tanks and proved to be suitable as compost material.

Since a recirculating system with draining has been built in conventional eel culture farms, it is easy to install foam separation, nitrification, and denitrification process to the actual eel rearing pond. This system has a high application potential as a novel aquaculture technology that aims at zero emission.

References

Blancheton JP (2000) Developments in recirculation systems for Mediterranean fish species. Aquac Eng 22:17–31

Bovendeur J, Eding EH, Henken AM (1987) Design and performance of water recirculation system for high-density culture of the African catfish, Clarias gariepinus (Burchell 1822). Aquaculture 63:329–353

Folke C, Kautsky N (1989) The role of ecosystems for a sustainable development of aquaculture. Ambio 18:234–243

Geiner AD, Timmous MB (1998) Evaluation of the nitrification rates of microbead and trickling filter in an intensive recirculating tilapia production facility. Aquac Eng 18:189–200

Hall PO, Holby O, Kollberg S, Samuelsson M (1992) Chemical fluxes and mass balances in a marine fish cage farm. IV Nitrogen Mar Ecol Prog Ser 89:81–91

Heinsbroek LTH, Kamsta A (1990) Design and performance of water recirculation systems for eel culture. Aquac Eng 9:187–207

Holby O, Hall PO (1991) Chemical fluxes and mass balances in a marine fish cage farm. II Phosphor Mar Ecol Prog Ser 70:263–272

Japan Sewage Works Association (1994) Quality standard for special fertilizers such as compost. Japan Sewage Works Association, Tokyo. 97pp. (In Japanese)

Liao IC, Hsu Y, Lee WC (2002) Technical innovations in eel culture systems. Rev Fish Sci 10:33–450

Maruyama T, Suzuki Y (1998) The present stage of effluent control in Japan and pollutant load from fish culture to environment.-possibility of intensive recirculating fish culture systems. Nippon Suisan Gakkaishi 64:216–226. (In Japanese with English abstract)

Maruyama T, Okuzumi M, Saheki A, Shimamura S (1991) The purification effect of the foam separating system in living fish transportation and preservation. Nippon Suisan Gakkaishi 57:219–225. (In Japanese with English abstract)

Maruyama T, Okuzumi M, Satoh Y (1996) The purification of rearing seawater of Japanese flounder with the closed foam separation-filtration system. Nippon Suisan Gakkaishi 62:578–585. (In Japanese with English abstract)

Maruyama T, Suzuki Y, Sato D, Kanda T, Michisita T (1998) Performance of a closed recirculating system with foam-separation and nitrification units for intensive culture of Japanese flounder. Nippon Suisan Gakkaishi 65:818–825. (In Japanese with English abstract)

Skjølstrup J, Nielsen PH, Frier FO, McLean E (1998) Performance characteristics of fluidized bed biofilters in a novel laboratory scale recirculation system for rainbow trout: nitrification rates, oxygen consumption and sludge collection. Aquac Eng 18:265–276

Suzuki Y, Maruyama T (1999) Performance of a closed recirculating system with foam separation, nitrification and denitrification units for the intensive culture of Japanese eel. In: Hino A, Maruyama T, Kurokura H (eds) Reduction of environmental emissions from aquaculture. Koseisha-koseikaku, Tokyo, pp 98–115. (In Japanese)

Suzuki Y, Maruyama T (2000) Removal of suspended matter from aqueous solution by foam separation with fish mucus. J Jpn Soc Water Environ 23:181–186. (In Japanese with English abstract)

Suzuki Y, Maruyama T, Sato D, Kanda T, Michisita T (2000) Water quality and mass balances in a culture of Japanese flounder using a closed recirculation system with foam-separation and nitrification units. Nippon Suisan Gakkaishi 66:1–9. (In Japanese with English abstract)

Suzuki Y, Maruyama T, Numata H, Sato H, Asakawa M (2003) Performance of a closed recirculating system with foam separation, nitrification and denitrification units for the intensive culture of eel: towards zero emission. Aquac Eng 29:165–182

Suzuki Y, Maruyama T, Tanigichi H, Kametani T, Yoshida T (2004) Development of a closed recirculating system for intensive culture with the combined aim of high production and nutrient load reduction. J Water Waste 46:758–768. (In Japanese)

Tambo N, Kamei T (1998) Coagulation and flocculation on water quality matrix. Water Sci Technol 37:31–41

van Rijn J (1996) The potential for integrated biological treatment systems in recirculating fish culture—a review. Aquaculture 139:181–201

van Rijn J, Rivera G (1990) Aerobic and anaerobic biofiltration in aquaculture unit - nitrate accumulation as a result of nitrification and denitrification. Aquac Eng 9:217–234

Chapter 4
Pejerrey *Odontesthes bonariensis*

Hiroyuki Yoshino

Abstract Pejerrey *Odontesthes bonariensis* was reared in the recirculation system for 1 year. The system consisted of two particle-trap equipped octagonal tanks (0.75 m^3) for the rapid collection of feces and uneaten feed, a mechanical filter, a UV unit, a biological filter which provided nitrification, and a hollow-fiber oxygen injection system. Total system volume was about 2.1m^3. The flow rate was fixed at 2.1 m^3/h. Three hundred and eighty-three pejerreyes (avg. wt. 1.6 g) were stocked in the system with salinity of 7‰, and temperature was maintained at 24 °C. After 1 year, survival rate was 92%. An average of 8.5 L of saltwater and quantity of evaporation water was supplemented to the system daily. At maximum load, just before expiry (avg. wt. 109.1 g), water quality was very good with NH_4-N at 0.2 mg/L and NO_2-N at 0.1 mg/L. During the experiment, nitrates peaked at 900 mg/L, but dropped to 150 mg/L after a denitrification system was installed. The effects of the water velocity on the growth of the fish were investigated by rearing in the two tanks in which the velocity differentiated. As a result, pejerreyes were reared with swimming exercise under a rotational flow at comparatively slow water velocity for a long period. A decrease in variance in growth rate was observed. It was suggested that this rearing condition, which is relatively low water velocity condition (under 1 BL/s), could be a contributing factor to not only the flesh quality of cultured fishes but their efficient production and improvement as commodities.

Keywords Pejerrey • Aquaculture • Closed recirculating • Denitrification • Swimming • Exercise • Growth • Octagonal tank

4.1 Introduction

Although it is not difficult to realize running water pond culture in seawater or freshwater, the culture in brackish water, salinity of 7‰, hardly realizes due to the limited availability of the rearing water. Closed recirculating aquaculture, however,

H. Yoshino (✉)
Hokkaido Industrial Technology Center, Hakodate, Hokkaido 041-0801, Japan
e-mail: yoshino@techakodate.or.jp

T. Takeuchi (ed.), *Application of Recirculating Aquaculture Systems in Japan*, Fisheries Science Series, https://doi.org/10.1007/978-4-431-56585-7_4

can make it possible because this system requires artificial seawater, which will be used for a long time through purification. Few cases have been reported referring to closed recirculating aquaculture in brackish water, while many cases on seawater or freshwater aquaculture have been reported such as the culture of eels in freshwater with the partial running water system (Knösche 1994) and the culture of Japanese flounder *Paralichthys olivaceus* with a closed seawater recirculation system (Honda et al. 1993).

Pejerrey inhibits around South America and transplanted from Argentina to Japan in 1966. After the transplantation, public research institutes have taken the leading role in research, and the culture of pejerrey has been commercialized. The rearing condition is reported to be the best in brackish water. The fish has white meat and is cooked into the variety of dishes such as sashimi (sliced raw fish), the grilled, tempura, and the fried. The fish in the market varies in size between 80 g and 400 g and in price between 1500 Japanese yen and 3800 yen per kilogram, which can be said as relatively high priced.

In this study, the author developed the closed recirculating aquaculture system which does not require continuous water exchange nor regular partial water exchange and can operate for longer period with adding some water. Then the author conducted a 1-year rearing experiment of pejerrey (*Odontesthes bonariensis*) in the closed recirculating aquaculture system in the salinity of 7‰ rearing water. Next, the practicality of the system and the performance of purification are reported with testing the water quality changes (Yoshino et al. 1999). The effects of the water velocity on the growth of the fish were investigated by rearing in the two tanks in which the velocity differentiated.

4.2 The Experimental System

The features of this system are the following: preventing uneaten feed and excrement from cumulating in the system by enough physical filtration, denitrifying through the built-in denitrification unit, water recirculation except adding water to compensate the loss due to evaporation, and disposal of uneaten feed and excrement.

4.2.1 The System Summary

The system design was stated below:

1. Pejerrey was designated as experimental fish and the salinity of the rearing water is 7‰.
2. Two culture tanks (0.75 m^3) were contained in a prefabricated house (width 7.2 m × length 2.5 m × height 2.28 m) and used.

3. Juvenile pejerrey had been reared for 1 year without continuous or intensive water changes.
4. Water quality targets were under 1 mg-N/L of both NH_4-N and NO_2-N.
5. After confirming that natural denitrification did not occur because of accumulation of uneaten feed and excrement, a denitrification unit was installed.

The basic water treatment followed the rules below. The rearing water that was out of the culture tank went through physical water filtration to remove solid matter and then biological filtration to remove dissolved ammonium. Then oxygen was being dissolved into the water and run into the tank. The schematic depiction of this system is shown in Fig. 4.1. This system is equipped mainly with two octagonal culture tanks with particle trap for the rapid collection of uneaten feed and excrement, a drum screen filter, a foam fractionation unit, a UV sterilization unit, a rotating biological contactor (RBC), a fluidized bed filter, a denitrification unit, an oxygen injection system, and a circulation pump.

4.2.2 Details of the System

The Culture Tanks Two octagonal culture tanks (0.75 m^3) are equipped with a particle trap for the rapid collection of uneaten feed and excrement. The culture tank and the particle trap are shown in Figs. 4.2 and 4.3, respectively. The rearing water emitted from the culture tanks is divided into two channels: one running into a drum screen filter from the center of the tank and another to sludge collector from the central bottom of the tank. The uneaten feed and excrement settled on the bottom is emitted to the sludge collector and sunk, and clean water flows to the drum screen filter. The amount of the precipitate is between 3 and 4 liters every day.

Physical Water Filtration Unit A drum screen filter and a foam fractionation unit were employed to filter solid matter greater than 45 μm and those smaller than 45 μm, respectively.

1. Drum Screen Filter (Water Volume 0.1 m^3)

   The drum screen filter is shown in Fig. 4.4. A plain weave net woven of 40 μm nylon yarn was used in the filter, and sieve opening was 45 μm. The surface of the filter was sucked by a vacuum pump to prevent from clogging up. Sucked water flowed into a settling tank (volume, 0.02 m^3; flow rate, 0.03 m^3/h) to settle particles, and then the water flowed into circulation through the rotating biological contactor.
2. Foam Fractionation Unit

   Figure 4.5 depicts the foam fractionation unit. It is self-priming (air flow rate 5 m^3/h), and gas-liquid ratio and gas-liquid contact time were set at 2.4 and 3.8 min, respectively. The foam fractionation unit had been used for 305 days from the beginning of the culture, but terminated on the 306th day because stable foams were not recognized and no suspended material was removed.

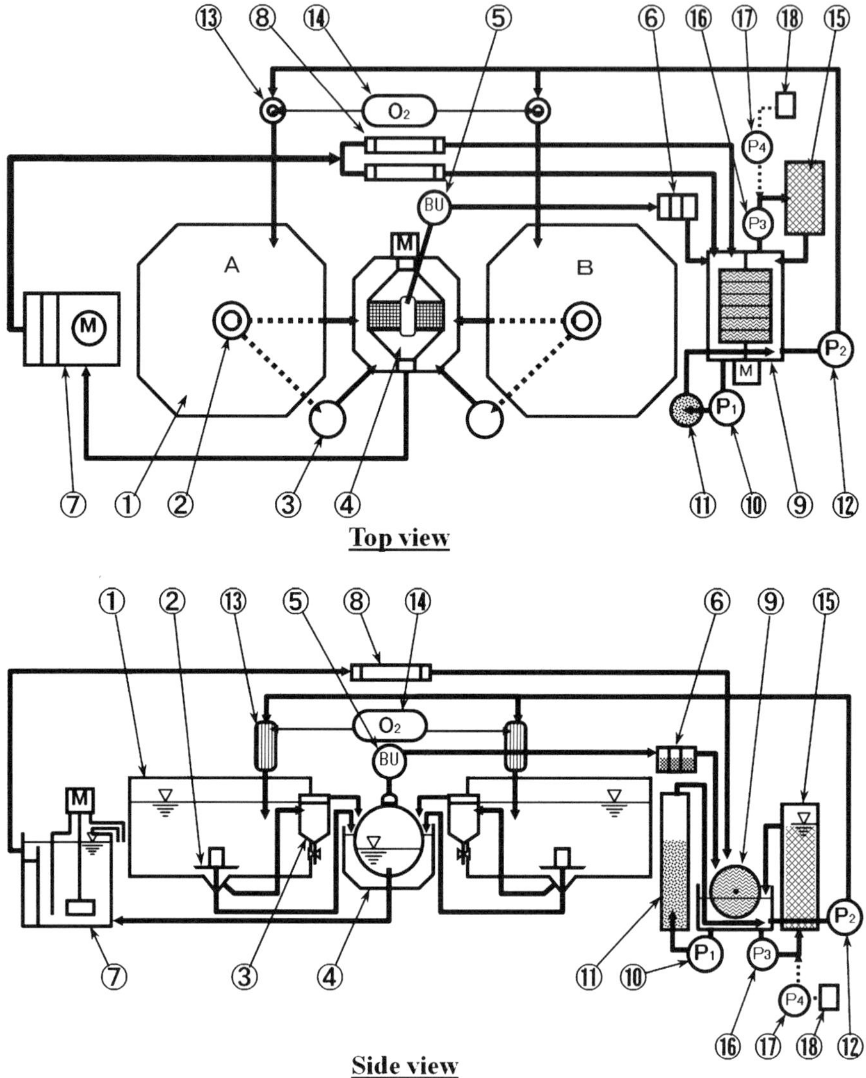

Fig. 4.1 A schematic view of closed recirculating aquaculture system. ① Culture tank, ② particle trap, ③ sludge collector, ④ drum screen filter, ⑤ vacuum pump, ⑥ settling tank, ⑦ foam separation unit, ⑧ UV sterilization unit, ⑨ rotating biological contactor (RBC), ⑩ pump, ⑪ fluidized bed filter, ⑫ circulation pump, ⑬ hollow-fiber oxygen injection system, ⑭ pure oxygen, ⑮ denitrification unit, ⑯ diaphragm pump, ⑰ peristaltic pump, and ⑱ methanol tank (Reproduced from Yoshino et al. 1999)

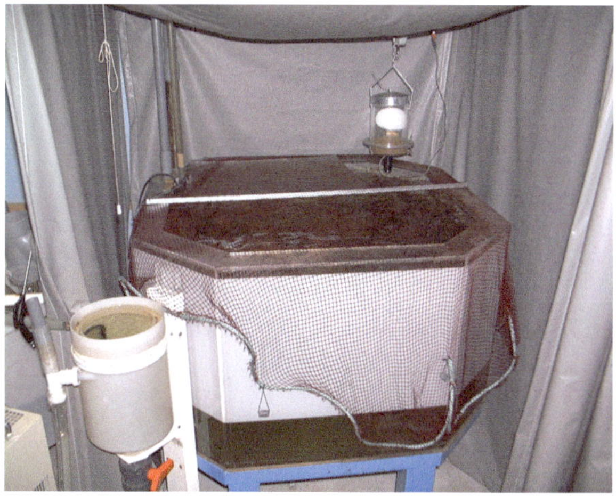

Fig. 4.2 Culture tank and sludge collector (front left)

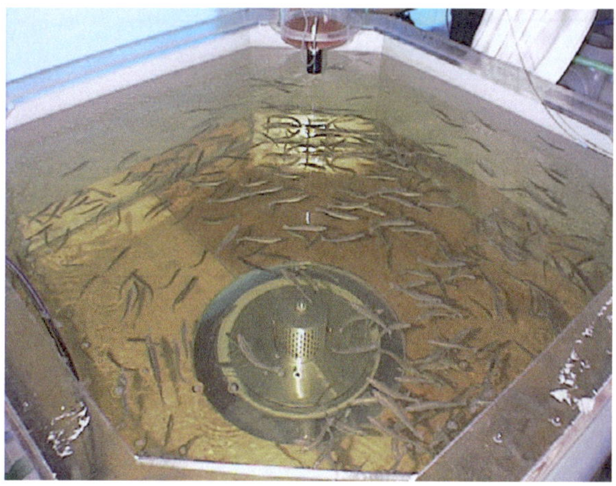

Fig. 4.3 Particle trap (center of tank bottom)

Biological Water Filtration Unit A biological water filtration unit consisted of a rotating biological contactor and a fluidized bed filter as a nitrification unit and a submerged filter as a denitrification unit.

1. Nitrification Unit

   It is the first attempt in aquaculture to use interlocked fiber as a medium of RBC, and we confirmed the practicality of a fluidized bed filter of sand, which has an advantage in the installation. FritzZyme No. 9 (Fritz Industries Inc.), as solution containing marine nitrifying bacteria, was added to the RBC and the fluidized bed filter 45 days prior to the initiation of the culture, and they had been matured.

Fig. 4.4 Drum screen filter

Fig. 4.5 Foam separation unit

(a) RBC

Figure 4.6 depicts the RBC. Interlocked fiber (specific surface area 410 m^2/m^3) was employed as medium of the RBC. The other conditions were set as the following: The volume of the medium was 0.11 m^3; the submersion ratio was between 35 and 40%; the rotational velocity was 7 rpm (peripheral velocity 0.18 m/s); the apparent water velocity was 0.5 cm/s; the average residence time was 2.9 min.

(b) Fluidized bed filter (volume 0.06 m^3, pump output 0.2 kw, flow rate 2.8 m^3/h)

Figure 4.7 depicts the fluidized bed filter. Sand was used as the medium (specific surface area 8000 m^2/m^3), and the volume of the sand was 0.01 m^3. The flow rate, the average water velocity, and the average residence time were set at 2.8 m^3/h, 1.3 m/min, and 1.3 min, respectively.

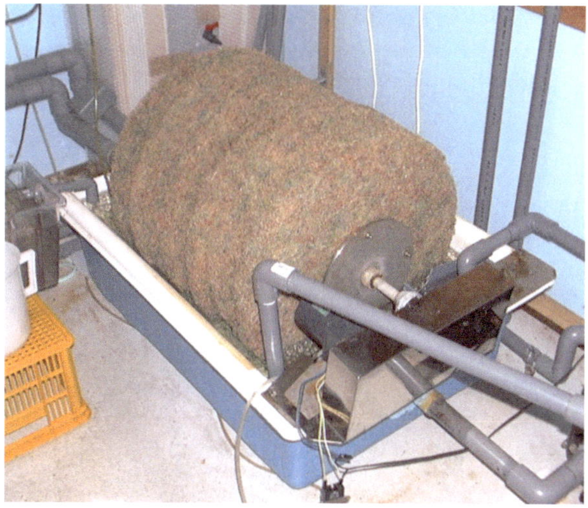

Fig. 4.6 Rotating biological contactor (RBC)

Fig. 4.7 Fluidized bed filter

2. Denitrification Unit (Water Volume 0.12 m^3)

   The denitrification unit was a submerged filter of which medium was fiber (specific surface area 740 m^2/m^3). The volume was 0.12 m^3, and the flow rate and the average residence time were set at, respectively, 0.16 m^3/day and 18 hours. One hundred liters of the rearing water taken from the tank on the 174th day and 50 grams of glucose were added to the denitrification unit, and then 2 liters of denitrifier culture solution was added before 6 days of the preliminary operation of the denitrification unit itself. A decline of the concentration of NO_3-N and an increase of alkalinity were confirmed on the 178th day, and the circulation of the rearing water was begun on the 180th day after the installation of the denitrification unit. The marine denitrifying bacteria *Alcaligenes* sp. Ab-A-1 strain (Watanabe et al. 1991) was inoculated as denitrification bacteria.

Oxygen Injection System Hollow fibers are allocated in the tubes (diameter, 50 mm; length, 800 mm), which dissolve oxygen through their membrane. They are located in front of each tank. The amount of dissolved oxygen in each tank was monitored by a dissolved oxygen analyzer in each tank, and the amount of dissolved oxygen was controlled to be saturated by a computer.

Ultraviolet Sterilizer The rearing water emitted from the foam fractionation unit was separated into two routes, and ultraviolet sterilizers were installed on both routes. A low-pressure mercury lump (wavelength 254 nm, 40 w) was employed, and the ultraviolet radiation dose was 9.8×10^4 μW·sec/cm^2 (1.1 m^3/h) at first and 5.9×10^4 μW·sec/cm^2 (1.1 m^3/h) when the output was decreased by 60%, which is expected to be a year later (nominal capacities).

Others Installed were a circulation pump (output 0.2 kw, flow rate 2.1 m^3/h), an electronic immersion heater (in the foam fractionation unit, 2 kw), an automatic feeder, and a security alarm system.

4.3 Experiment Materials and Method

4.3.1 Experiment Materials

System Only one culture tank had been employed till the 36th day, and the total quantity of water was 1.35 m^3, while the circulating water amount was 1.26 m^3/h. Another tank was added on the 37th day, and the total quantity of water was set at 2.1 m^3, while the circulating water amount was 2.1 m^3/h.

Rearing Water Tap water was dechlorinated, and then artificial seawater formulation was used to make the water a salinity of 7‰. No water exchange was conducted during the experimental term; however tap water was added without any treatment to compensate the loss by evaporation and by drainage while ejecting the uneaten feed and excrement. The salinity was measured twice a week, and the artificial seawater formulation was added, if necessary, to keep the salinity between 7 and 8‰. pH was measured once a day, and either 5% (w/v) of sodium hydrogen carbonate solution or 5% (v/v) of hydrochloric acid was added to keep pH in a range of 7–8.2.

Experimental Fish Used were 383 pejerreyes (average weight 1.57 g) which were preliminarily reared in the still water tank of the salinity of 7‰ for 105 days from the incubation. No selection had been conducted since the incubation. Housed in the

Table 4.1 Schedule of feed manufacturer, code, size, and protein content used during this study

Day	Feed size (mm)	Protein content (%)[a]
1–83	1.2–2	Min. 50
84–209	2	62
210–267	3	62
268–365	3.1	Min. 48

Modified from Yoshino et al. (1999)
[a]Manufacturer indication

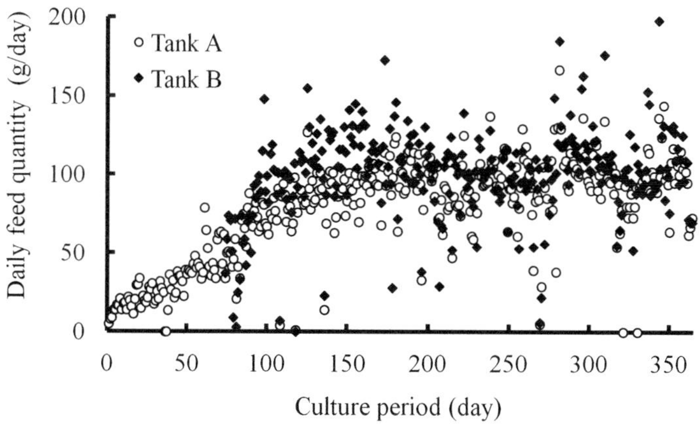

Fig. 4.8 Daily feed quantity supplied to tanks A and B (Reproduced from Yoshino et al. 1999)

section A tank were all 383 pejerreyes (the volume of water 0.713 m^3) at the beginning of the rearing, and on the 73rd day, 192 pejerreyes were moved into the section B tank (the volume of water 0.739 m^3) and reared separately.

Feed A feed was mainly an assorted feed for flatfish, and the size of feed was changed in line with growth and provided by an automatic feeder. Table 4.1 depicts the used feed. Feeding was provided at 10–20-minute intervals from 4:00 to 22:00 every day, and the amount of feeding was measured by a load cell. The amount was adjusted to leave few feed uneaten. The amount of daily feed was described in Fig. 4.8.

4.4 Result and Discussion

4.4.1 Rearing Result

Figure 4.9 depicts pejerreyes at the end of rearing, and Table 4.2 shows the rearing result. Since published data in regard to the growth of pejerrey is very few, it is

Fig. 4.9 Pejerreyes at the cultured end

Table 4.2 Culture conditions for 1 year

		Tank A	Tank B
Initial	Body weight	1.57 ± 0.54[a]g ($n = 50$)	
	Number of fish	383	
73 day	Body weight	14.2 ± 4.7[a]g ($n = 80$)	
	Number of fish	191	192
Final	Body weight	109.0 ± 41.9[a]g ($n = 175$)	109.2 ± 33.4[a]g ($n = 178$)
	Total length	22.9 ± 2.5[a]cm ($n = 175$)	22.8 ± 2.2[a]cm ($n = 178$)
	Number of fish	175	178
	Survival rate	91.6%	92.7%
	Density	2.7%	2.7%
	Feed conversion ratio	1.49	1.52

Reproduced from Yoshino et al. (1999)
[a]Mean ± S.D.

difficult to compare with other rearing methods. However, it is reasonable to state that pejerreyes were reared normally because the rearing density was 2.7% and the survival rate was 92% at the end of the rearing. Details will be given in 5.2.4 with information about the effect of the flow velocity.

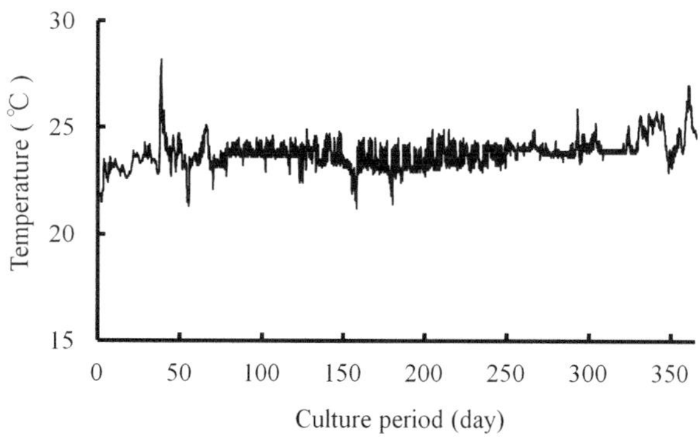

Fig. 4.10 Changes of water temperature in the culture tank (Reproduced from Yoshino et al. 1999)

4.4.2 Water Quality

Water Temperature Figure 4.10 depicts a change in water temperature during the rearing experiment. The average room temperature ± standard deviation (the number of samples, range) was 20.2 ± 3.5 °C (n = 4372, 8.5–29.2 °C). Water temperature changed in line with room temperature, and an average was 23.8 ± 0.7 °C (n = 4372, 21.2–28.2 °C), which indicates that it was almost kept at the target water temperature (24.0 °C).

pH and Alkalinity Nitrification well occurs in a range of pH 7.4–8.4 in general (Kitao 1997), and it is possible that nitrification hardly occurs under pH 7.0 due to inactivity of nitrifying bacteria. In this experiment, pH was kept above 7.0 by adding 5% (w/v) of sodium hydrogen carbonate solution if necessary. Figure 4.11 shows changes of pH and alkalinity during the experiment. pH at the beginning of the rearing was 7.8, but it rose to 8.18 due to addition of sodium hydrogen carbonate solution. pH was kept in a range of 7–7.7 from the 50th day to the 310th day by controlling the amount of the solution added. Because pH of the inflow water from the denitrification unit got higher (pH 8.6–8.8) and pH of the rearing water rose from the 310th day, 5% (v/v) of hydrochloric acid was added to prevent pH from rising from the 340th day. Through the procedures above, the average pH was 7.60 ± 0.26 (n = 265, 7.00–8.18), and the average alkalinity was 107.5 ± 80 mg-$CaCO_3$/L (n = 274, 52–363.5). The cumulative amount of sodium hydrogen carbonate added was 14.1 kg and that of hydrochloric acid was 2.47 L.

Saturation of Dissolved Oxygen Figure 4.12 shows a change of saturation of dissolved oxygen in the culture tank. An average percent saturation of dissolved oxygen was 103.5 ± 3.0% (n = 365, 95.7–135.1%), and it was almost maintained at target percentage (100%).

Salinity Figure 4.13 shows a change of the salinity of the rearing water. The salinities at the beginning of the rearing and on the 40th day were 7.7‰ and 8.6‰, respectively. However, it dropped gradually and hovered around 6.8–7.8‰

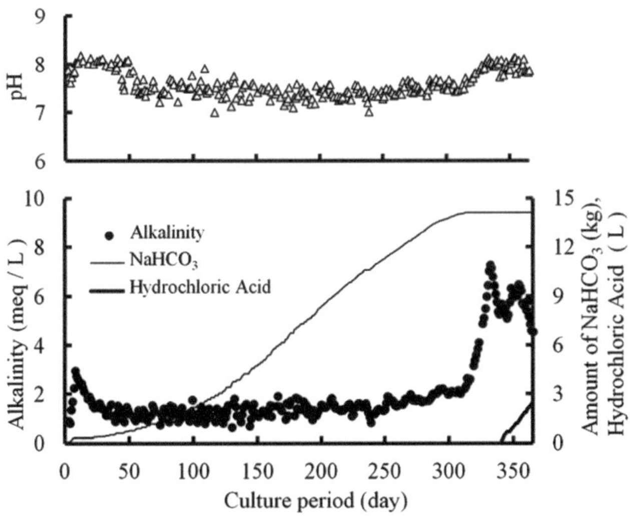

Fig. 4.11 Daily changes in pH and alkalinity in the culture water and cumulative amounts of $NaHCO_3$ and hydrochloric acid added to the system (Reproduced from Yoshino et al. 1999)

Fig. 4.12 Changes in dissolved oxygen saturation level in the culture tank (Reproduced from Yoshino et al. 1999)

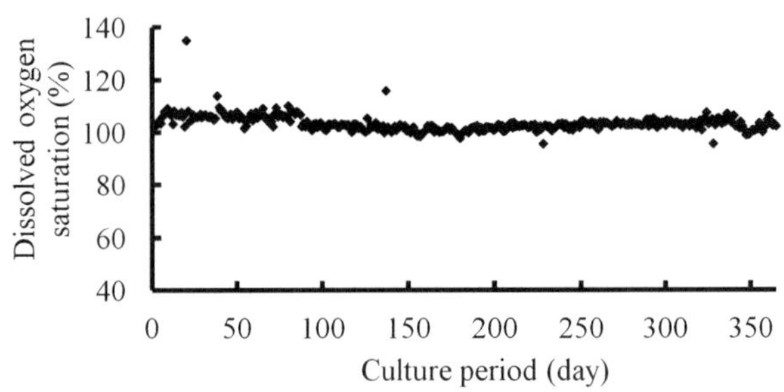

Fig. 4.13 Changes in salinity in the culture water (Reproduced from Yoshino et al. 1999)

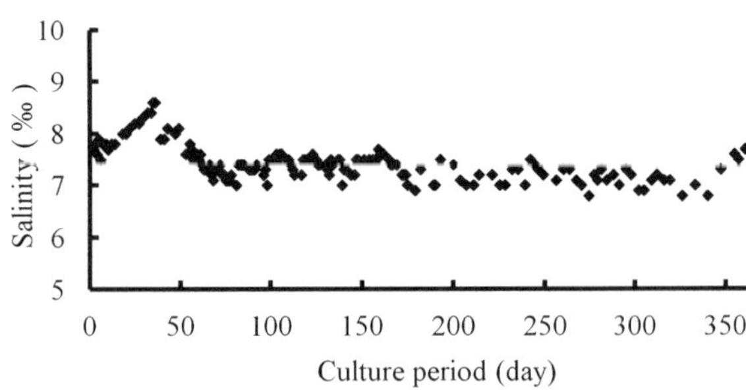

after the 60th day till the end. The salinity rose till the 40th day because artificial seawater of 7–12‰ salinity was supplied. However, the salinity decreased from the 41st day due to the supplement of tap water. This is why the salinity was adjusted by adding artificial seawater formulation from the 60th day. By this adjustment, average salinity during the experiment was 7.4 ± 0.3‰ (n = 162, 6.8–8.6‰).

Amount of Water Supplied Although we did not conduct any intentional water change in this system, we supplied water to compensate the loss due to evaporation and due to drainage while ejecting the uneaten feed and excrement. We needed to estimate the amount of compensation because water was automatically and directly supplied into the RBC to keep the water level constant, and the amount of water supplied was not measured. While it is difficult to estimate the amount of evaporation, the amount of water supplied corresponding to the drainage was estimated, on the contrary. Estimated from the amount of artificial seawater formulation that was added to adjust the salinity, the amount is assumed 2763 L and 8.5 L/day on average because the amount of artificial seawater formulation added after the 41st day was 20.25 kg and an average salinity from the 41st day to 365th day was 7.3 ± 0.2‰ (n = 140, 6.8–8.1‰). Thus, the total amount of water supplied is 8.5 L/day on average plus the amount of evaporation.

NH_4-N, NO_2-N Changes in the concentration of NH_4-N and NO_2-N during the experiment are shown in Fig. 4.14. The concentration of NH_4-N had hovered in a range of 0.05–0.15 mg-N/L for the first 150 days of the experiment and in a range of 0.15–0.3 mg-N/L from the 150th day. The average concentration of NH_4-N during the experiment was 0.18 ± 0.07 mg-N/L (n = 358, ND(<0.05 mg-N/L)- 0.36 mg-N/L), and it was kept stably low around the year.

The concentration of NO_2-N started to rise from the beginning of the rearing, and it reached 0.77 mg-N/L on the 21st day, and then it started decreasing and reached 0.17 mg-N/L on the 60th day. It had hovered in a range of 0.15–0.3 mg-N/L during the 60th to 130th day and in a range of 0.06–0.2 mg-N/L from the 130th day. The average concentration of NO_2-N during the rearing was 0.18 ± 0.12 mg-N/L (n = 358, ND(< 0.01 mg-N/L)- 0.77 mg-N/L). It was assumed that the delay of maturation of nitrite-oxidizing bacteria raised the concentration of NO_2-N. Nitrifying bacteria were planted, and ammonium chloride and sodium nitrite were added 45 days prior to the initiation of the experiment, and maturation was promoted. However, it took about 100 days to mature nitrite-oxidizing bacteria as a result. Factors which kept stable after the maturation were the stability of water temperature (23.8 ± 0.7 °C) and that pH, alkalinity, salinity, and dissolved oxygen were kept in the best range to activate nitrifying bacteria.

NH_3-N The concentration of NH_3-N was calculated from the concentration of NH_4-N, pH, salinity, and water temperature (US Environmental Protection Agency 1989) and is shown in Fig. 4.15. Un-ionized ammonia are strongly poisonous to fish among ammonia in water (NH_3, NH_4^+), and the proportion of un-ionized ammonia among ammonia in water depends mainly on pH, salinity, and water temperature. pH and water temperature increase in proportion to the proportion of the un-ionized

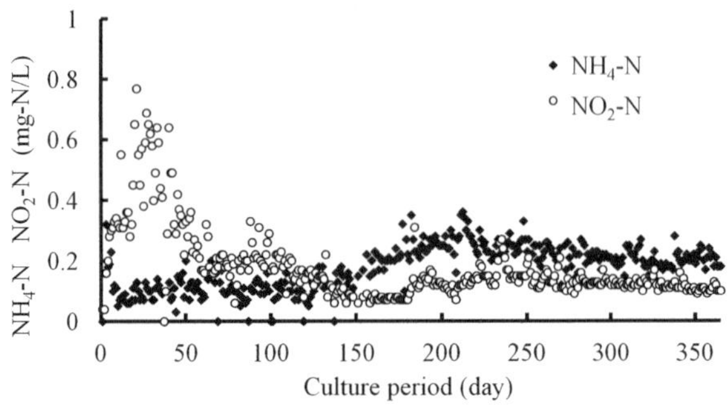

Fig. 4.14 Changes in concentrations of NH_4-N and NO_2-N in the culture water (Reproduced from Yoshino et al. 1999)

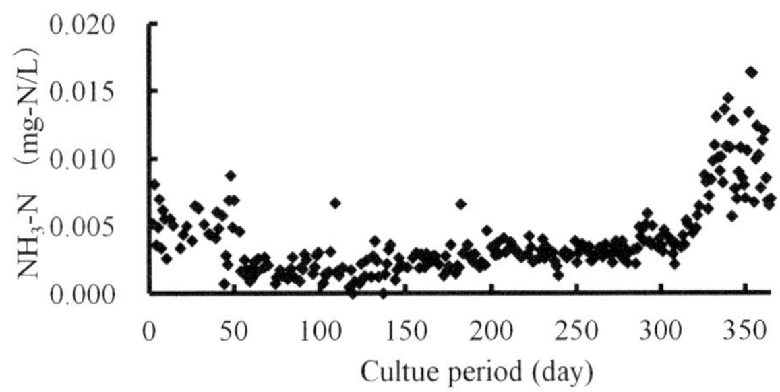

Fig. 4.15 Changes in concentrations of calculated NH_3-N in the culture water (Reproduced from Yoshino et al. 1999)

ammonia (NH_3) (Nomura 1980). The average concentration of NH_3-N during the rearing was 0.004 ± 0.003 mg-N/L ($n = 264$, ND (< 0.001 mg-N/L) - 0.014 mg-N/L). The reason that the concentration of NH_3-N was high till the 50th day and after the 310th day while that of NH_4-N is relatively stable is that the proportion of un-ionized ammonia among ammonia in water increased due to the increase of pH. The increasing trend from the 310th day stopped on the 340th day because a pH increase was intervened by pH adjustment, and this implies the effect and the importance of pH control.

The toxicity of un-ionized ammonia to pejerrey is unclear. However, given the result obtained from the 9–12-month rearing that the maximum safe concentration of un-ionized ammonia to the rainbow trout was 0.0125 ppm (Smith and Piper 1975), 0.004 mg-N/L on average was low enough though the concentration exceeded 0.0125 mg-N/L for a certain time due to the delay of pH control.

NO_3-N Figure 4.16 shows a change in the concentration of NO_3-N in the rearing water. The concentration gradually increased till the 190th day. The concentration dipped on the 36th day because the total amount of water increased by adding another culture tank. The denitrification unit was installed on the 180th day. It was only after the 280th day that denitrification significantly started, which is why the concentration of NO_3-N reached 900 mg-N/L before then. The concentration started to decrease linearly from the 300th day, and it dropped to 150 mg-N/L at the end of the experiment. Figure 4.17 depicts the chronological changes in the

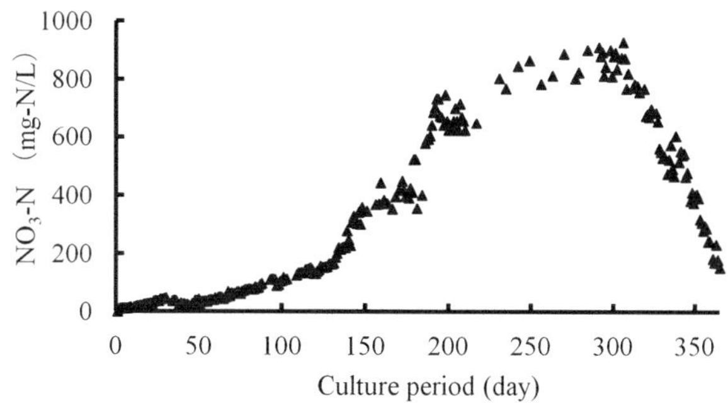

Fig. 4.16 Changes in NO_3-N concentrations in the culture water (Reproduced from Yoshino et al. 1999)

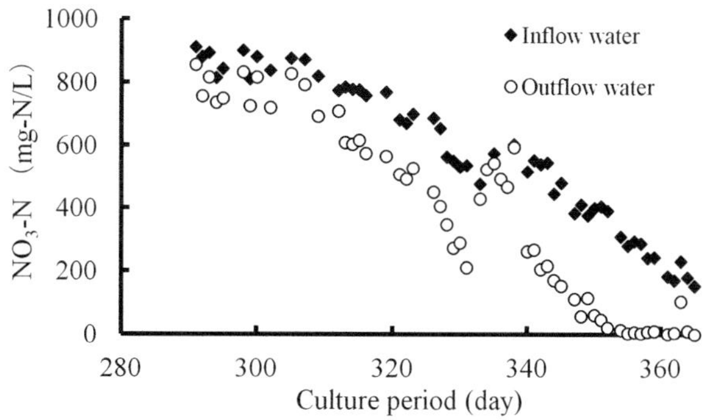

Fig. 4.17 Changes in NO_3-N concentrations of inflow and outflow water in denitrification filter (Reproduced from Yoshino et al. 1999)

concentrations of NO_3-N of the inflow and outflow of the denitrification unit from the 290th day. The effectiveness of the denitrification unit was confirmed by the significant decrease of the concentration of NO_3-N of the outflow compared to the inflow, and few NO_3-N was detected (ND $<$ 0.5 mg-N/L) after the 353rd day. When there is an anaerobic place in the closed recirculating aquaculture system and natural seawater is used, denitrification could happen because denitrifying bacteria in natural seawater might breed. The accumulation of nitric acid till the 180th day was checked. It is, however, assumed that no denitrification happened until the installation of the denitrification unit because (1) tap water was used in this experiment and marine denitrifying bacteria may have not existed, (2) the system was kept in the aerobic condition, and it was difficult for anaerobic denitrifying bacteria to breed due to the physical filtration. An active denitrification such as an installation of a denitrification unit is needed under the circumstances where the whole system is kept in the aerobic condition or few denitrifying bacteria exist in a long-term closed recirculating aquaculture accompanied with nitrification of a biological filtration.

Denitrification was accelerated rapidly from the 300th day and brought about the decline in the concentration of NO_3-N for 750 mg-N/L in 65 days as well as huge changes in pH and alkalinity, which led to the pH control. Although it has been already known that alkalinity rises in line with denitrification (Maruyama et al. 1996), it is possible to downsize the denitrification unit and decrease the usage of pH control chemicals if the denitrification unit is installed at the beginning of the rearing.

4.4.3 System Efficiency

Excrement Disposal Speed of Each Unit The amount of nitrogen in an average weight of the consumed feed and the amount of excrement and ammonia treated per hour in each unit are shown in Table 4.3. The amount of nitrogen treatment was calculated from the NH_4-N concentration difference between inflow and outflow of each system. An average weight of the consumed feed per day is 188.5 ± 11.7 g ($n = 4175$, 5–203.3 g), and an average weight per hour and intake of nitrogen were 10.5 g and 744 ± 46 mg-N/h, respectively, since hours of feeding was 18 hours a day. Given the amount of nitrogen consumed was 100%, the percentage of caption as solids and removal as ammonium nitrogen was the following: 12% as solids such as excrement, 28% at the nitrification unit, and 7% in the pipe.

Purification Mechanism in Closed Recirculating Aquaculture System Main nitrogenous products emitted into water by fish are ammonia, urea, and excrement which contains undigested feed. Urea is rapidly resolved into ammonia and carbon dioxide in water (Deguchi 1980). Forty-seven percent of nitrogen intake was identified as excrement or ammonia. Thus, it is assumed that the rest was not emitted and almost all was taken into the fish body. We will discuss the purification mechanism in terms of the amount of nitrogen collected and the amount of nitrification of ammonia at each unit. Seventy-four percent of the total emitted nitrogen was nitrified as ammonia, in which 60% of the total emitted nitrogen was nitrified in the nitrification unit (38% was in the RBC and 22% was in the fluidized bed filter) and 14% was in the other units. Nitrifying bacteria attached to the surface of pipes other than nitrification units seemed to have nitrified. After ammonia were nitrified

Table 4.3 The amount of nitrogen in an average weight of the consumed feed and the amount of excrement and ammonia treated per hour in each unit in the closed recirculating aquaculture system for pejerrey[a]

	Feed	Feces		Ammonia		
Parameter	Consumed nitrogen[b]	Sludge collector	Drum filter	RBC	Fluidized bed filter	Others
Nitrogen (mg-N/h)[c]	744 ± 46	66 ± 36	23 ± 9	133 ± 52	77 ± 42	50 ± 38
(%)	100	9	3	18	10	7

Reproduced from Yoshino et al. (1999)
[a]Total body weight, 38.5 kg; body weight, 109.1 ± 37.8 g (mean ± S.D., $n = 353$)
[b]Feed consumption rates, 10.5 ± 0.65 g/h (mean ± S.D., $n = 4$); nitrogen content in feed, 7.1%
[c]Mean ± S.D., $n = 4$

to nitric acid, they were deoxidized into nitrogen and emitted out of the system. Furthermore, 26% of the total emitted nitrogen was collected in a solid form (excrement) and emitted out of the system. Nineteen percent of the total emitted nitrogen was collected at a particle trap for the rapid collection of uneaten feed and excrement and 7% in the drum screen filter. It should be noted that the figures above do not count uneaten feed, and the numbers would increase if uneaten feed appears. Solids such as uneaten feed and excrement themselves do not worsen the water quality immediately, but proteins contained in them change into an amino acid and ammonia by bacteria (Kawai 1980), and they will lead to a cause of water contamination. Thus, it is assumed that early removal of uneaten feed and excrement contributes to lessen the load of the nitrification units.

In this system, uneaten feed and excrement are removed from the culture tanks in 1–2 min after the appearance without their getting out of shape and are emitted out of the system after settling in the sludge collector. Solid particles which are greater than 45 μm and did not settle in the sludge collector are collected by the drum screen filter. For the reasons above, inside of the system was kept clean and no precipitates without any cleaning for a year, and blockade of the biological filtration system did not occur. Generally speaking, the biological filtration system without preprocessing to remove solid particles needs to backwash regularly because blockade of filter media lessens the capacity of nitrification, which is the instability factor of the system because the condition of nitrifying bacteria differs before and after the backwashing (Yoshino et al. 1999). This implies the importance of physical filtration such as early removal of uneaten feed and excrement by the particle trap to keep the system operating stably for the long run.

The effect of disinfection in this system is described below. Feed such as a pellet contains 10^5 CFU/g of bacteria, and excrement contains 10^{7-8} CFU/g of bacteria in general. Bacteria are continuously released into the rearing water with uneaten feed and excrement. In this system, as mentioned above, uneaten feed and excrement were collected immediately by the particle trap and emitted out of the system, and solid particles left in the rearing water were collected by the drum screen filter and the foam fractionation unit. Bacteria left in the rearing water were disinfected by the UV sterilization unit. The number of live bacteria was not counted in this experiment. The prior studies report that the number of live bacteria in the culture tank is halved by physical filtration such as foam fractionation unit and is reduced to 1/10–1/100 by UV sterilization unit. The number of bacteria is also reported to increase 10–100 times in the culture tank (Yoshimizu and Kasai 2002). The effect of disinfection was greater than 99.9% by irradiating more than 1.0×10^4 μW·sec/cm^2 of UV light after settling uneaten feed and excrement in the drainage of the rearing water by coagulation process at National Salmon Resources Center Shiribetsu Factory (Yoshimizu 1998). Since it is assumed that UV intensity was kept above 4.9×10^4 μW·sec/cm^2 by conducting the physical treatment mentioned above in this system, it is also assumed that the live bacteria in the rearing water were few enough for rearing. No diseases were observed in this experiment.

4.5 Rearing in Two Kinds of Water Velocity

A water current can be created in a culture tank by changing the way of pouring water. A rotating flow can be created particularly in circular and polygonal tanks. Designed was the system in which solid particles such as uneaten feed and excrement are gathered in the center and emitted by using this water current (Cobb and Titcomb 1930; Surber 1936; Larmoyeux et al. 1973). It has another effect to rapidly uniform water quality in the tank (Timmons et al. 1998). A lot of prior studies that examined the effect of swimming exercises on Salmonidae were conducted, and it is reported that swimming exercises in the proper speed could improve the growth rate and *FCR* (Christiansen and Jobling 1990; Houlihan and Laurent 1987; Jørgensen and Jobling 1993). Besides the studies on Salmonidae, there are reports that swimming exercises harden the flesh of the red sea bream *Pagrus major*, which is indicated from the experimental results of 12-day swimming exercise (Tachibana et al. 1988) and that 12 cm ayu *Plecoglossus altivelis* best grew in the 35 cm/s and 45 cm/s flow rate, and the feed efficiency was also the best in those conditions, which is demonstrated in the 60-day rearing in the various flow rates (4–45 cm/s) (Nakagawa et al. 1991).

Most of the prior experiments on the effect of flow rate are short-term (about 60 days) rearing experiments in which a uniform current is created in the circulation tank. However, under the actual aquaculture circumstances, it is difficult to rear in a uniform current because the current is usually a rotating current in circular and polygonal tanks. Furthermore, the flow rate has to be sustainable for the long term because rearing often lasts a year. A water current is usually created by water pressure of inflow water, and in this case it is hardly possible to create a fast current. An expensive submarine pump, which adds undesirable cost, must be used to obtain a fast current. However, already utilized is the method to create a water current fast enough to gather uneaten feed and excrement into the center of the tank only by a rotating flow of inflow water in a circular tank.

This experiment examines velocity distribution in the tanks, feeding rates, and growth of the fish in comparison by using two culture tanks in which the water velocity differs.

4.5.1 Method

Culture Tanks Details of the culture tanks (Fig. 4.2) embedded in the closed circulating aquaculture system (Fig. 4.1) are shown in Fig. 4.18. The culture tanks are two octagonal culture tanks (0.75 m^3) equipped with a particle trap for the rapid collection of uneaten feed and excrement, and different velocities can be set. Open inlets were three 7 mm diameter holes in the 32 mm diameter vinyl chloride pipe, which was vertically inserted at the edge of the tanks, and outlets were located at the lower center of the tanks. Rotating flows in the tanks created by

Fig. 4.18 A schematic view of the experimental tank and measurement point of water velocity. (a), (b), (c), and (d) are measurement point of water velocity (275 mm in depth)

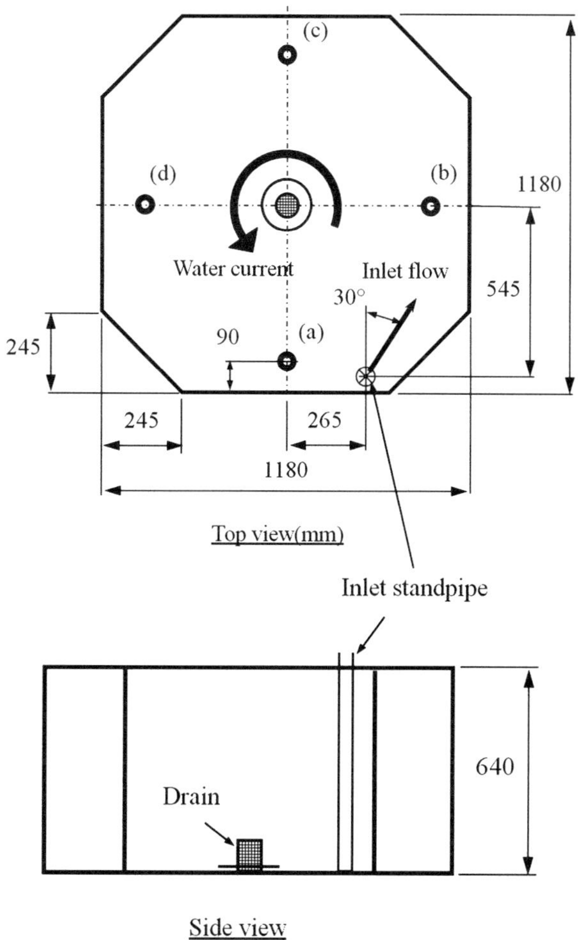

this inflow water gathered solid particles such as uneaten feed and excrement into the lower center of the tanks, and they were emitted to sludge collectors and settled.

Since both tanks were connected to the same purification system and constituted the closed recirculating aquaculture system, the same quality of rearing water was provided. The volume of inflow into the tanks was controlled by both the inverter, which controlled the flow rate of circulation pump, and the valve, which was installed in the middle of the pipe connected to each tank.

Current Velocity Current velocities were set relatively low so that uneaten feed and excrement can be emitted by the water flow of the circulating water without any special equipment. Figure 4.19 depicts the water velocities set in each tank with a term. Reared were 383 pejerreyes till the 72nd day in section A, and 192 pejerreyes were randomly moved to section B on the 73rd day. Average current velocity was set constant at 8 cm/s in section A from the 1st day to 365th day and in section B

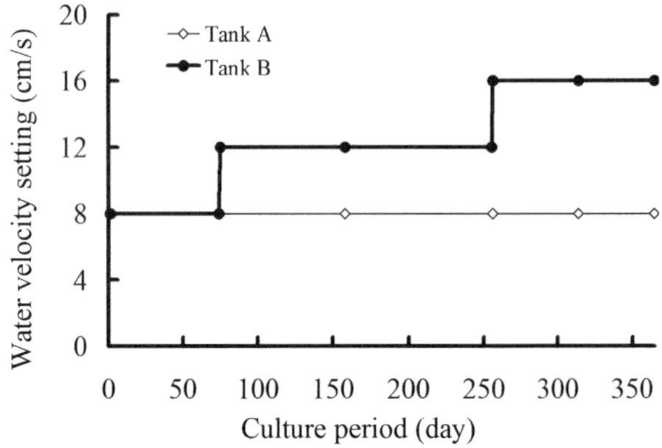

Fig. 4.19 Water velocity setting and period for each tank

12 cm/s (75th–255th day) and 16 cm/s (256th–365th day). In section B, the current velocity was set higher in line with the growth of pejerreyes. Measurement was conducted with three-dimensional electromagnetic current meter, which was adjusted so that a mean current velocity of four points at a depth of 27.5 cm and 9 cm away from the wall of the tank (Fig. 4.18) was obtained.

Measurement of the Weight of Feed and Uneaten Feed Feed weight was measured by the load cell every day. The weight of uneaten feed was measured eight times in the interval of 7–14 days between the 306th and 62nd day by the following method. The sediments emitted into the sludge collector for 24 hours were filtered through a screen mesh of 1.5 mm to collect uneaten feed. After the collection of uneaten feed, wet weight and the amount of moisture with the feed were measured in the uneaten feed, and the dry weight was calculated. Average daily food intake rates (*FI*, %/day) of the fish were calculated using the following formula:

$$FI = \left(1 - \frac{W_{\mathrm{duf}}}{W_{\mathrm{mf}}\left(1 - \frac{M}{100}\right)}\right) \times 100 \tag{4.1}$$

in which W_{duf} is the dry weight of uneaten feed, W_{mf} is the weight of feed, and M is the moisture with the feed at feeding.

The feed conversion rate (*FCR*) was calculated using the following formula:

$$FCR = \frac{F_{\mathrm{w}}}{W_{\mathrm{f}} - w_{\mathrm{i}}} \tag{4.2}$$

in which F_{w} is the total weight of feed and W_{i} and W_{f} are the total weights of the fish at the start and the end of the experiment, respectively.

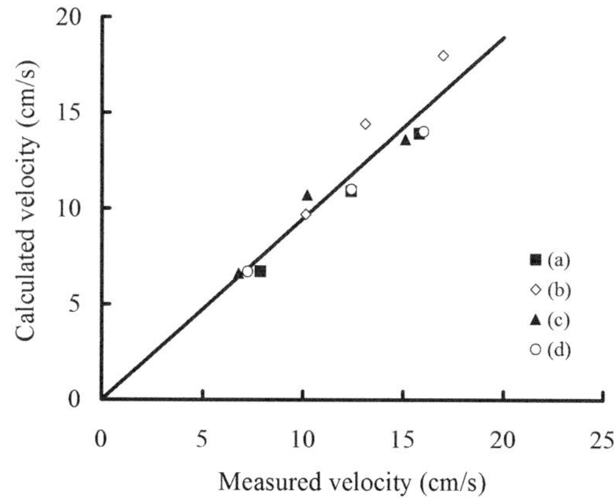

Fig. 4.20 Comparison of the calculated velocity by the fluid analysis and the measured velocity at a measurement point (Fig. 4.18, (a) (b) (c) (d), 275 mm in depth) at each velocity set ($y = 0.95\,x$, $n = 12$, $r = 0.95$)

4.5.2 *Result and Discussion*

Water Velocity in the Tank Figure 4.20 depicts the comparison of the calculated velocity by the fluid analysis and the measured velocity at measurement points (Fig. 4.18a, b, c, d) at each velocity set. The calculated water velocity was about 5% lower than the measured velocity, but the correlation was very high ($r = 0.95$). The distribution of water velocity was estimated from the fluid analysis and is shown in Fig. 4.21. Water velocity in the wide range of 0–20 cm/s occurred as a maximum at the front of the inlet even in case of 8 cm/s. Water velocity was fast on the outside and slow in the center, which was similar to the observation in circular tanks. Thus, the fish was able to select a variety of water velocities in the octagonal tank, as compared to raceway designs in which velocity is uniform along the channel (Timmons et al. 1998). This was confirmed by the fluid analysis.

Swimming Conditions It was observed that pejerreyes dispersed around the whole of the tank without gathering in a specific place. Figure 4.22 depicts the swimming conditions. The swimming conditions were classified into four patterns: when the fish swam faster than current, at the same velocity as the current (orientation), independently in the current, or drifted by the current. Richkus (1975) examined the response of juvenile alewives to water currents and observed that most fish were strongly oriented to water currents and their direction was influenced by the current velocity. It was observed that most pejerreyes swam either faster than the current or at the same velocity. These conditions were the same as those in sections A and B. The fish in the tank with faster water velocity were assumed to increase their swimming momentum because all conditions were the same except for water velocity.

Food Intake Rate Table 4.4 shows the weight of feed, the dry weight of uneaten feed, and the food intake rate (*FI*) from the 306th day to 362nd day. The food intake rate in sections A and B did not differ significantly ($p > 0.1$). Concerning the water

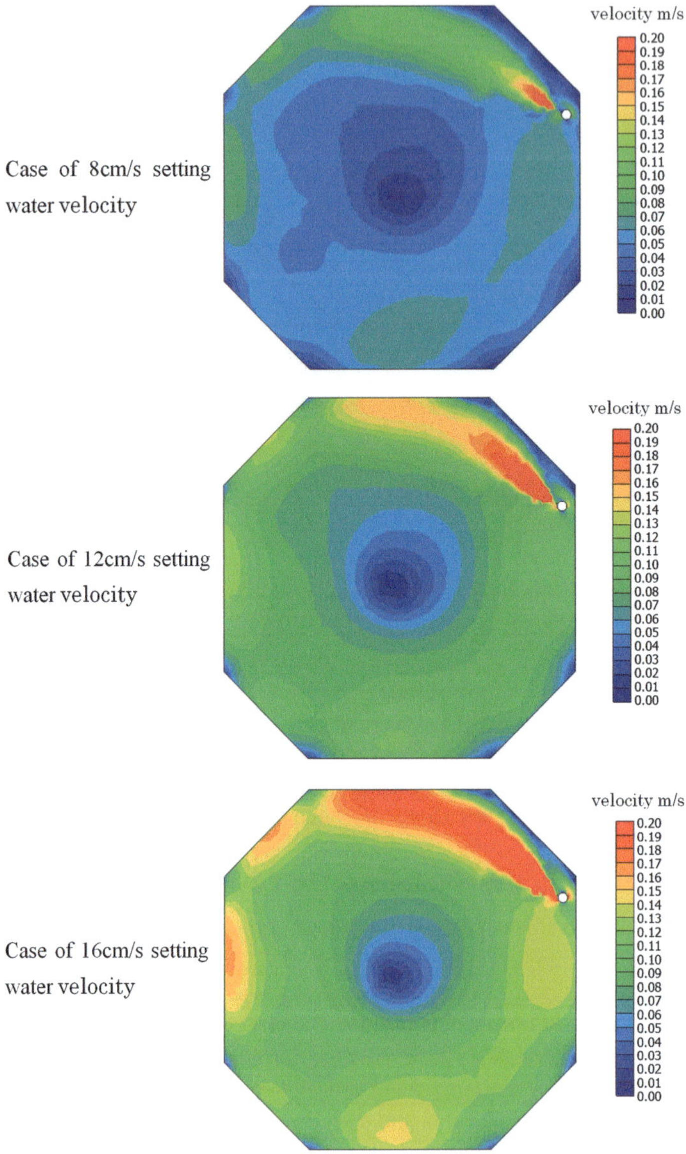

Fig. 4.21 Water velocity simulation contours at 275 mm in depth calculated by the fluid analysis

velocity set during the measurement of uneaten feed, section B (16 cm/s) was set at twice the velocity of section A (8 cm/s), which was the maximum velocity setting for the experiment. In respect to the relationship of water velocity and food intake rate, the food intake rate increased when the water velocity was moderate because

Fig. 4.22 The swimming situation of the pejerrey

Table 4.4 Feed weight, uneaten feed dry weight, and food intake rate (*FI*) per 1 day in tanks A and B (between day 306 and 362)

	Tank A	Tank B
Feeding weight (g)	94.4 ± 9.8[a]	100.9 ± 15.9[a]
Uneaten feed dry weight (g)	5.5 ± 3.9	6.2 ± 4.0
FI (%)	93.9 ± 4.4	93.7 ± 3.9

Mean ± S.D. ($n = 8$)
[a]The moisture content of the feed was 7.7%

feed spread in the entire tank (Jørgensen and Jobling 1993). Flore and Keckeis (1998) reported that food intake rate decreased when the water velocity was too fast because predation cannot be done. However, there is assumed to be no effect on the food intake of pejerrey in this range of water velocity.

Culture Result The survival rate after culture, culture density, body length, body weight, and feed conversion ratio for 1 year of culture are shown in Table 4.2. No significant differences were observed in the survival rate (A 91.6%, B 92.7%) and the culture density (the rate of the total weight of the fish to the water weight of the tank) (A 2.68%, B 2.63%). This indicates that there were no large differences in the culture environment except the water velocity. The mean body length and weight were not significantly different between sections A and B ($p > 0.05$). However, concerning the variance in body length and body weight, there were significant differences between sections A and B ($p = 0.022$, $p = 0.001$, respectively), and the standard deviation in section A was twice as large as that in section B. These facts indicate that the unevenness in the growth decreases when the water velocity is increased. The relationships between the water velocity and the body length in sections A and B were 0.4 BL/s and 0.8 BL/s, respectively.

FCR was 1.49 in section A and 1.52 in section B. *FCR* in section B was a little higher. When the *FCR* was low, it appeared that the feed had been effectively used to increase body weight; on the other hand, when there was a high volume of uneaten feed, *FCR* appeared to be high. In this experiment, however, *FI*s during the experiment are assumed as nearly the same as is mentioned above; *FCR* in section B is thought to be actually higher. This is because the fish of the high water velocity group consumed much more energy in swimming than the fish of the low water velocity group.

In this experiment, pejerreyes were reared with swimming exercise under a rotational flow at comparatively slow water velocity for a long period. A decrease in variance in growth rate was observed. It was suggested that this rearing condition, which is relatively low water velocity condition (under 1 BL/s), could be a contributing factor to not only the flesh quality of cultured fishes but their efficient production and improvement as commodities.

4.6 Conclusions

The author developed the closed recirculating aquaculture system which does not require continuous water exchange nor regular partial water exchange and can operate for longer period with adding some water. Then the author conducted a 1-year rearing experiment of pejerrey in the closed recirculating aquaculture system in the salinity of 7‰ rearing water and performed various kinds of examinations and obtained the following results:

1. The water quality was maintained in the good condition, and the rearing was done without a problem. The system was kept aerobic by physical filtration, and no natural denitrification was observed. The built-in denitrification unit lowered the amount of nitric acid, and its efficacy was presented.
2. The percentages of nitrogen treatment in each system were as follows: as dissolved, 38% in the RBC, 22% in the fluidized bed filter, and 14% in the other units and as solid matters such as excrement, 19% in the particle trap for the rapid collection of uneaten feed and excrement and 7% in the drum screen filter.
3. No significant differences were observed in the survival rate, the body length, and the body weight between the high current condition (maximum current velocity16 cm/s, 0.8 BL/s) and the low current condition (current velocity 8 cm/s, 0.4 BL/s). However, the variances of the body length and weight were significantly larger in the low current condition.

References

Christiansen JS, Jobling M (1990) The behavior and the relationship between food intake and growth of juvenile Arctic charr, Salvelinus alpinus L., subjected to sustained exercise. Can J Zool 68:2185–2191

Cobb WW, Titcomb JW (1930) A circular pond with central outlet for rearing fry and fingerlings of the salmonidae. Fish Soc 60:121–123

Deguchi Y (1980) Aquacultural environment of recirculation filtration pond. In: Nippon Suisan Gakkai (ed) Fresh water culture and culture water. Kouseisha-kouseikaku, Tokyo, pp 87-94. (In Japanese)

Flore L, Keckeis H (1998) The effect of water current on foraging behaviour of the rheophilic cyprinid *chondrostoma nasus* (L.) during ontogeny: evidence of a trade-off between energetic gain and swimming cost. Regul Rivers: Res Manage 14(1):141–154

Honda H, Watanabe Y et al (1993) High density rearing of Japanese flounder, *Paralichthys olivaceus* with a closed seawater recirculation system equipped with denitrification unit. Suisanzoshoku 41:19–26

Houlihan DF, Laurent P (1987) Effects of exercise training on the performance, growth, and protein turnover of Rainbow trout. Can J Fish Aquat Sci 44:1614–1621

Jørgensen EH, Jobling M (1993) The effects of exercise on growth, food utilization and osmoregulatory capacity of juvenile Atlantic salmon. Aquaculture 116:233–246

Kawai A (1980) Microbiological purification mechanism. In: Nippon Suisan Gakkai (ed) Fresh water culture and culture water. Kouseisha-kouseikaku, Tokyo, pp 111–122. (In Japanese)

Kitao T (1997) Aerobic biotreatment. In: Iwai S et al (eds) Wastewater and waste treatment wastewater. Kodansha, Tokyo, pp 131–192. (In Japanese)

Knosche R (1994) An effective biofilter type for eel culture in recirculating systems. Aquaculture Eng 13:71–82

Larmoyeux JD, Piper RG, Chenoweth HH (1973) Evaluation of circular tanks for salmonid production. Fish-Cult 35(5):122–131

Maruyama T et al (1996) The purification of rearing seawater of Japanese flounder with the closed foam separation-filtration system. Nippon Suisan Gakkaishi 62(4):578–585

Nakagawa H et al (1991) Effects of water velocities on lipid reserves in Ayu. Nippon Suisan Gakkaishi 57:1737–1741

Nomura M (1980) Environment of running water pond and fish production. In: Nippon Suisan Gakkai (ed) Fresh water culture and culture water. Kouseisha-kouseikaku, Tokyo, pp 64–83. (In Japanese)

Richkus WA (1975) The response of juvenile alewives to water currents in an experimental chamber. Trans Am Fish Soc 104:494–498

Smith CE, Piper RG (1975) Lesions associated with chronic exposure to ammonia. In: Ribelin WE, Migaki G (eds) The pathology of fishes. Univ of Wisconsin Press, Madison, pp 497–514

Surber EW (1936) Circular rearing pools for trout and bass. Fish Soc 21:1–14

Tachibana K et al (1988) The effect of swimming exercise on flesh texture of cultured red sea-bream. Nippon Suisan Gakkaishi 54:677–681. (In Japanese with English abstract)

Timmons MB et al (1998) Review of circular tank technology and management. Auacultural Eng 18:51–69

U.S. Environmental Protection Agency (1989) Ambient water quality criteria for ammonia (saltwater). EPA 440/5-88/004 Office of Water Regulations and Standards, Criteria and Standards Division, Washington, DC, pp 1–2

Watanabe Y et al (1991) Development of water purification technologies for intensive fish culture, 7. Removal of nitrate accumulating in cultural seawater and some factors affecting the

denitrification of marine denitrifying bacteria. Abiko Research Laboratory Rep 2. Abiko Research Laboratory, Abiko, pp 1–21
Yoshimizu M (1998) Disinfection of water and waste water for aquaculture. Kaiyo monthly extra 14:112–117. (In Japanese)
Yoshimizu M, Kasai H (2002) Disinfection of hatchery water and waste water for aquaculture. Industrial Water 523:13–26. (In Japanese)
Yoshino H et al (1999) Changes in water quality and performance of a closed recirculating seawater aquaculture system for rearing pejerrey, *Odontesthes bonariensis*. Suisanzoshoku 47(2):289–297. (In Japanese with English abstract)

Chapter 5
Japanese Flounder *Paralichthys olivaceus*

Kotaro Kikuchi

Abstract Fish production with closed recirculation systems, which makes reduced discharge of organic and inorganic wastes possible, is considered to be one of the promising approaches for sustainable development of aquaculture. We conducted research on the closed production of Japanese flounder *Paralichthys olivaceus* since 1986 and obtained information to develop closed aquaculture systems for the flounder such as nitrogenous excretion rate of fish and nitrification activity of marine biological filters for designing water treatment unit, optimum water temperature for the growth, proper stocking density and effects of water quality changes on the growth of fish for increasing production performance, and effective feed composition. Based on the experimental results as well as available knowledge, a pilot scale closed recirculation system of 10 m^3 in total water volume was developed with operation manuals specific for Japanese flounder. The system consisted of a culture tank of 4 m in diameter, settling tank, drum screen filter, submerged biological filter, heating-cooling unit, oxygen generator and supplier, blower, and UV sterilizer. In the feeding experiment, 1015 fish of 2 g initial body weight were introduced to the culture tank and fed commercial pellet diet twice daily to satiation each. Fish grew to 456 g after 259 days with good survival rate (85%) and feed efficiency (97%). Culture density at the end was 31 kg/m^2; the bottom area of the culture tank and total culture water used was 25m^3. Dissolved oxygen ranged from 90 to 130% of saturation through the rearing period. Ammonia and nitrite were maintained at less than 4 mg-N/L, and no apparent adverse effects on the feeding and growth were observed.

Keywords Japanese flounder • Closed recirculating aquaculture • Nitrogen excretion • Nitrification • Temperature • Growth

K. Kikuchi (✉)
Environmental Science Research Laboratory, Central Research Institute of Electric Power Industry, Abiko, Chiba 270-1194, Japan
e-mail: kikuchi@criepi.denken.or.jp

T. Takeuchi (ed.), *Application of Recirculating Aquaculture Systems in Japan*,
Fisheries Science Series, https://doi.org/10.1007/978-4-431-56585-7_5

5.1 Aquaculture of Japanese Flounder in Japan

Increasing productivity without serious impact on the environment is believed to be essential for sustainable development of aquaculture. Several efforts have been carried out to develop effective technologies. Fish production in closed recirculation systems, which makes reduced discharge of organic and inorganic wastes possible, is considered to be one of the promising approaches. Closed recirculating aquaculture was initially tried with carp and eel in Japan in the mid-twentieth century (Japan Aquicultural Research group 1962). Currently, based on several technological developments made by engineers and biologists during the past three decades, commercial production is being conducted with fresh water fish such as tilapia, eel, and catfish using closed recirculation technologies mainly in western countries. However, there have been few reports regarding research on recirculation systems for marine finfish that are more familiar to Japanese.

Japanese flounder *Paralichthys olivaceus* (Fig. 5.1) is the only flatfish species that has been commercially produced in Japan. The fish has one of the highest commercial values, and its market price (/kg) is 2–3 times higher than that of yellowtail *Seriola quinqueradiata*, and red sea bream *Pagrus major*, the most popular cultured finfish in Japan. Aquaculture of Japanese flounder started in the mid-1970s, and the commercial production became extensive in the early 1980s with development of fingerling production and farming techniques. Aquaculture production of Japanese flounder increased gradually and reached a peak of 8562 metric ton (MT) in 1997, nearly equal to that of the wild catch which has been constant since 1980. However, it decreased gradually thereafter to 3475 MT in 2011 (Fig. 5.2) and is much smaller than those of yellowtail and red sea bream, which yielded 146,240 MT and 61,186 MT, respectively, in 2011. There are 300 to 400 flounder farms located mostly in southern Japan, particularly Shikoku and Kyushu islands, because of good condition in water temperature. Many of the farms produce 10,000–20,000 fish on average per year.

Unlike other marine finfish species that have been produced in floating net cages near the coast, land-based (onshore) culture tanks are the prevalent culture system for Japanese flounder (Fig. 5.3). The tanks are constructed of various materials: wooden panel with vinyl sheet, concrete, plastic, and/or a combination of these. The most popular shape is a circular tank of 6–8 m in diameter, but square or octagonal tanks are also available. Water depth in a tank is 60–100 cm. Culture tanks are generally installed inside or are covered with shade cloths. Sand-filtered seawater is continually supplied to the tank with 12–24 times of water change daily. The water change rate depends on water temperature and stocking density. The production cycle changes farm by farm, because fingerlings are available throughout the year. One to 3 g fingerlings are stocked in the culture tank at 100 to 200 fish/m^2 bottom area of the culture tank in winter to early spring. Fish grow to 0.5 kg in 9–10 months and 1 kg, favorable commercial size, in 14–16 months. Culture density for 1 kg size fish is less than 15 kg/m^2. Locally available sardines and sand lance, fresh or frozen,

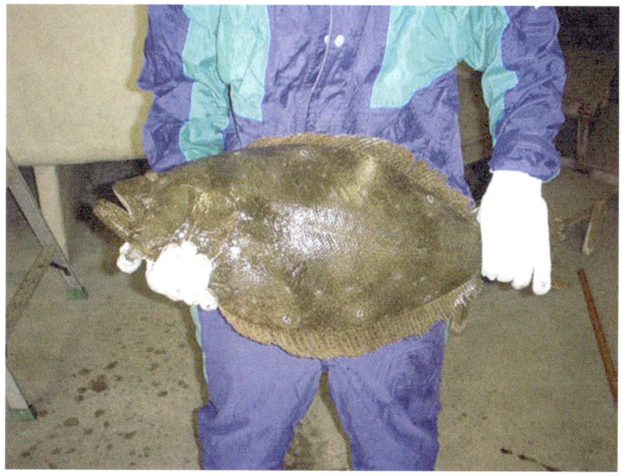

Fig. 5.1 Japanese flounder *Paralichthys olivaceus*

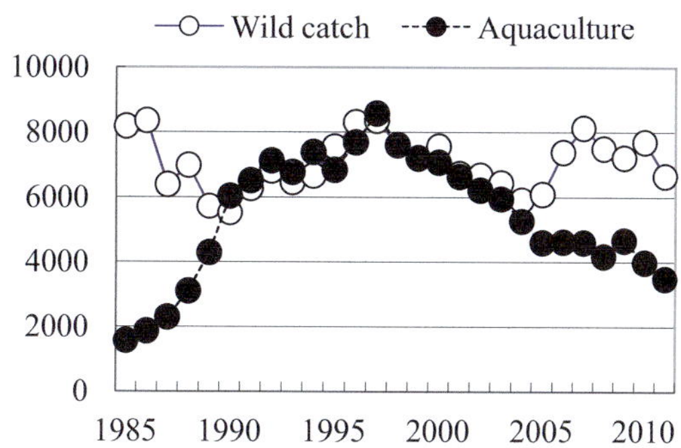

Fig. 5.2 Wild catch and aquaculture production of Japanese flounder in Japan (MT)

Fig. 5.3 A typical Japanese flounder farm at southern Japan, land-based culture tanks with running seawater

are still major source of feed; however, the use of moist or dry pellet has been increasing. There are several viral, bacterial, and parasitic diseases, such as *rhabdovirus, vibriosis, edwardsiellosis,* and white spot disease, which cause problems for commercial production (Muroga and Nakai 1990). One of the most severe diseases is *edwardsiellosis*, occurring frequently in summer and resulting in serious damage (Nakastugawa 1983; Kanai et al. 1988). Prevention of *edwardsiellosis* by vaccination has been shown (Iida and Wakabayashi 1992; Furuta et al. 1995).

5.2 Development of Closed Recirculating Aquaculture Techniques for Japanese Flounder

Central Research Institute of Electric Power Industry (CRIEPI) started a project to develop closed recirculating aquaculture production techniques for marine finfish since 1986. Japanese flounder was considered to be ideal fish for closed production because of their high market value and several production techniques established for land-based culture tanks. We planned to maximize productivity of the production system by controlling culture environment optimum for the growth of the flounder based on the biological information such as optimum water temperature for the growth, proper stocking density, nitrogenous excretion rate of fish, and effects of water quality changes on the growth.

5.2.1 Design of Water Treatment System

Treatment of fish wastes and leftover feeds is indispensable for water recycling aquaculture system to keep culture condition good for the growth of fish. Nitrogenous substances, especially ammonia, are primary concerns for all recirculation systems because ammonia has high toxicity to almost all aquatic organisms. Leftover feeds and fecal waste must be removed quickly because of negative impacts of suspended solids on the fish condition and consumption of dissolved oxygen during biodegradation. Water treatment system for closed aquaculture generally consists of two functions: mechanical filtration for solid materials and biological treatment (bacterial nitrification) for dissolved substances, and both must be designed based on the quantitative information on fish wastes. Therefore, we examined the nitrogenous excretion of Japanese flounder in terms of fish size, feeding rate and water temperature, and also nitrifying activity of biofilters.

Nitrogenous Excretion

1. *Fish size*: Major nitrogen excretes of fish are ammonia and urea, and excretion rate of both substances for starved Japanese flounder decreased with the growth of fish on unit weight basis (Fig. 5.4, Kikuchi et al. 1990). Especially, ammonia

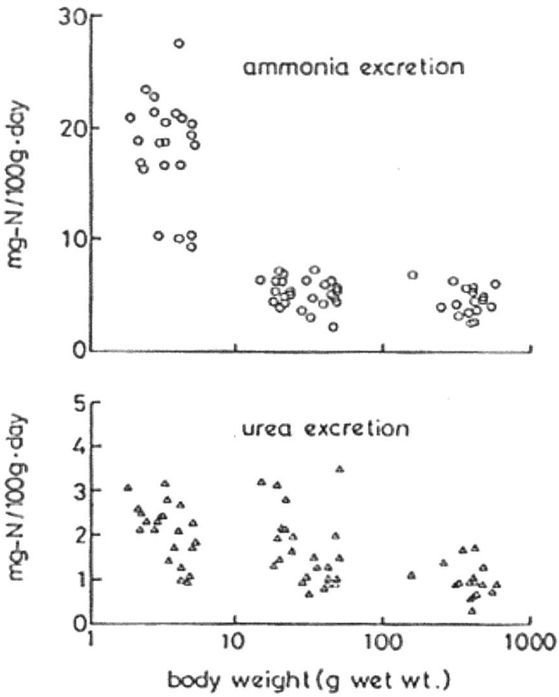

Fig. 5.4 Ammonia and urea excretion of starved Japanese flounder at 20 °C. Reproduced from Kikuchi et al. 1990

excretion rate changed considerably at 10 g body weight, and the average rate for fish at the size of 1.8–5.1 g was 18.3 mg-N/100 g body weight/day, 15–49 g fish was 5.4 mg, and 163–575 g fish was 4.8 mg. On the other hand, urea excretion rate decreased gradually with the fish growth, and the values were 2.1 mg, 1.7 mg, and 1.0 mg for respective fish size group. Fish size also affects nitrogenous excretion after feeding. Table 5.1 shows daily nitrogen excretion of 1.6–6.5 g fish (4.2 g in average, juvenile), 15–65 g (42 g, young), and 163–675 g (363 g, immature) fed commercial pellet diet at average daily ration in commercial aquaculture (Kikuchi et al. 1991, 1992b). Feeding rate per unit weight was higher for smaller size fish, and all measured excreted nitrogenous substances showed similar trends being highest for juvenile fish. There were fluctuations in the proportions of excreted nitrogen to the consumed nitrogen, and ammonia-N accounted for 21–32% of the consumed nitrogen, urea for 3–5%, and feces for 8–13%. Nearly 40–50% of consumed nitrogen may be excreted into culture water (natural environment) in flounder aquaculture when fish were fed pellet diet. When we look at the excretion per individual fish, ammonia, urea, and feces-N excretion of immature fish were much higher than those of juvenile and young flounder. The numbers of fish in the culture tank usually decrease with the growth of fish in commercial aquaculture; however, the largest waste load to water treatment system is considered to happen when

Table 5.1 Daily rates of nitrogen excretion of juvenile, young, and immature Japanese flounder

Stage	Feeding rate	Consumed nitrogen	Nitrogen excretion (mg-N/100 g fish/day)			
	(%)	(mg-N/100 g fish)	Ammonia	Urea	Feces	Total
Juvenile	2.8 ± 0.6	206.2 ± 43.3	47.8 ± 11.2	6.0 ± 1.0	27.5 ± 10.8	81.3 ± 19.5
			(23.2)	(2.9)	(13.3)	(39.4)
Young	1.3 ± 0.2	98.8 ± 14.5	20.7 ± 5.3	3.8 ± 1.6	12.8 ± 7.8	37.3 ± 10.0
			(21.0)	(3.8)	(13.0)	(37.8)
Immature	0.5 ± 0.2	40.4 ± 11.2	12.9 ± 4.3	1.8 ± 0.8	3.1 ± 1.6	17.8 ± 5.7
			(31.9)	(4.5)	(7.7)	(44.1)

Figures in the parentheses show proportions to the consumed nitrogen. Data represent means and standard deviations

fish grows to near commercial size. Therefore, the excretion rate of immature fish is more important to design water treatment system for dissolved and fecal wastes.

2. *Feeding rate*: More detailed information on nitrogenous excretion rate is required for stable operation of water treatment system. Fig. 5.5 shows hourly ammonia and urea excretion of immature flounder after feeding commercial pellet diet at 0.5, 1.0, and 1.5% of their body weight at 20 °C (Kikuchi et al. 1991). Although daily feeding rate of the flounder is approximately 0.5% with commercial pellet as mentioned above, it was not clear that fish feed 0.5% every day or 1.0% every other day. Hourly ammonia excretion rate increased just after the feeding from the starved level, reached a peak during 3–6 h after feeding, and decreased gradually thereafter. The rate during 12–24 h of fish fed 0.5% was not different from that of the starved fish; however, higher values continued for more than 36 h for 1.0 and 1.5% feeding groups. Noticeable findings of this study is that the highest rate of hourly ammonia excretion was almost the same regardless of feeding rate and was about 1.0 mg-N/100 g body weight/h. Similar trends were shown for urea excretion, and peak value was almost equal among three ration groups. Different from ammonia, urea excretion did not increase during initial 6 h after feeding and was the highest during 6–12 or 12–24 h. Peak value of urea was 10–20% of that of ammonia. Existence of upper limit in the excretion rate regardless of ration level was also reported for plaice (Jobling 1981) and Atlantic cod (Ramnarine et al. 1987).
3. *Water temperature*: Water temperature is one of the most influential factors affecting the nitrogen excretion of fish. The daily ammonia excretion rate of starved flounder increased with increasing temperature from 16 to 25 °C, and the value at 25 °C was nearly twice of that at 16 °C for juvenile, young, and immature flounder (Kikuchi et al. 1995). Although urea excretion rate tended to increase with the temperature, clear relationship was not found between temperature and the excretion rate. Increasing temperature from 16 to 25 °C also increased ammonia, urea, and feces-N excretion of 6 g size fish fed

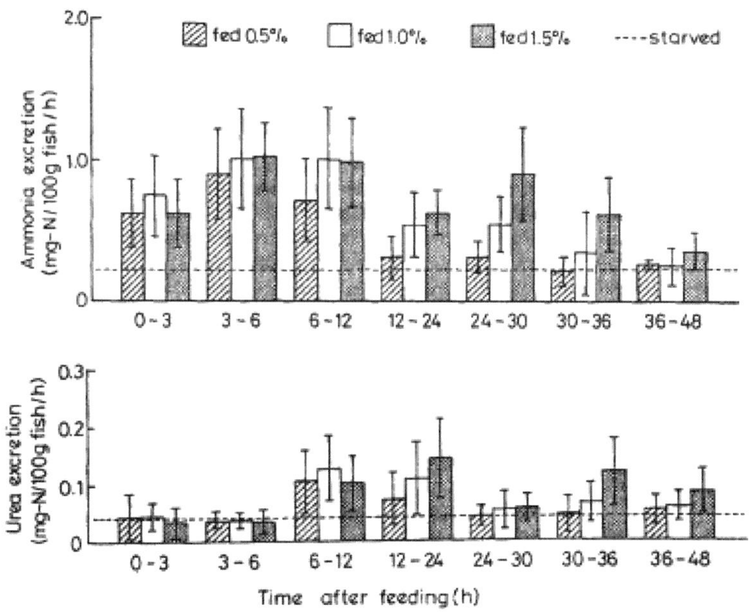

Fig. 5.5 Diurnal changes in ammonia and urea excretion of Japanese flounder fed commercial pellet at 0.5, 1.0 and 1.5% of their body weight at 20 °C. Reproduced from Kikuchi et al. 1991

commercial pellet; however, effect of water temperature on the fed fish seemed to be smaller than that on the starved fish (Kikuchi et al. 1995).

4. *Urine and feces*: Kidney and gills are major excretory organs in most fish species, and nitrogenous substances are excreted through both organs. Our preliminary study showed that volume of urine of 400 g starved Japanese flounder was about 2 mL/100 g body weight/day containing 500 mg total-N/L urine, 70 mg urea-N, and 2 mg ammonia-N. The results indicate that ammonia and urea excreted through urine were extremely smaller than those of total values even compared with starved fish (Fig. 5.4), therefore, most of ammonia and urea are considered to be excreted through gills in Japanese flounder. Few publications were found for feces composition of fish, and our preliminary study showed that flounder feces fed commercial pellet diet (54% of crude protein, 18% of lipid) contained 19% of protein, 6% of lipid, 56% of ash, and 19% of other substances including carbohydrates.

Treatment of Ammonia There are some techniques to remove dissolved ammonia such as ammonia stripping method, breakpoint chlorination, ion exchange, and biological treatment. Among these, biological method, especially bacterial nitrification, is considered to be one of the most realistic techniques for the production of marine organisms because of its little toxicity and easy operation. Not activated sludge and immobilized bacteria but fixed film filtration is most popular as bacterial nitrification, and there are several treatment types such as submerged filter, trickling filter, rotating biological contactor, fluidized bed, and beads filter. However, there was little information available for the comparison of the technologies. We chose submerged biofilter as the nitrification unit for closed aquaculture of Japanese flounder and examined nitrifying activity with artificial filter materials in 10 L volume of experimental seawater recirculation system at 20 °C (Kikuchi et al. 1994a). Earthenware ball (0.27 m^2/L, surface area/volume), honeycomb tube with three different surface areas (0.13, 0.50, 1.00), net filter material (0.35), and fibriform filter material (1.44) of 1 L volume each were tested (Fig. 5.6). The nitrification activity seemed to be higher for materials having higher specific surface area when ammonia was only added to the system as shown in Fig. 5.7. However, there was not significant relationship between nitrification rate and surface area of the filter. Daily loading of organic substance (Ehrlich meat extract) with ammonia for 3 months reduced the nitrification activity of the filter by 30–50% depending on filter material. The maximum nitrification activity after loading organic matter was obtained for the filter with net filter material and was estimated to be 11 mg-N/L filter/h or 0.55 g-N/m^2 surface area (including filter walls surface)/day. This value was similar to those reported by Rijn and Rivera (1990) and Nijhof and Bovendeur (1990). Alkalinity and pH of recycling seawater decreased with the progress of ammonia oxidation (Fig. 5.8), and ammonia oxidation was inhibited when alkalinity and pH reached to 0.5 meq/L and 6.0, respectively.

Design of Water Treatment System for Closed Recirculating Aquaculture The highest ammonia excretion rate of the flounder and nitrification activity of submerged biofilter with net filter material at 20 °C are estimated to be 24 mg-N/100 g fish/day and 550 mg-N/m^2 surface area/day, respectively. Based on these results, we need at least 45 m^2 of surface area for submerged nitrifying filter to treat ammonia derived from 100 kg flounder of 500 g size. We must consider other dissolved nitrogen and fecal nitrogen that are decomposed to ammonia during recirculation; however, sum of these is considered to be less than the value of ammonia excretion (Table 5.1). Therefore, submerged biofilter with 90 m^2 of

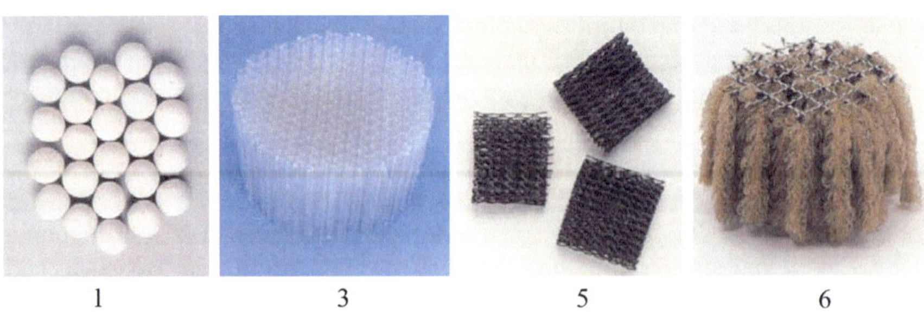

Fig. 5.6 Filter materials used to examine ammonia oxidation rate of marine biological filter. 1, earthenware balls; 3, honeycomb tube of 8 mm in cell size (0.50 m^2/L); 5, net filter material; 6, fibriform filter material. Reproduced from Kikuchi et al. 1994a

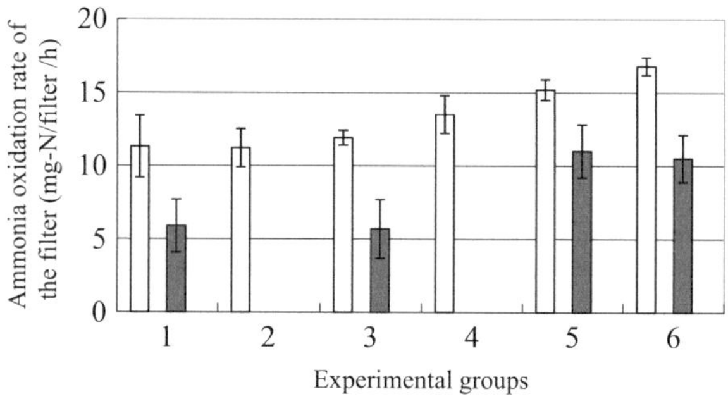

Fig. 5.7 Ammonia oxidation rate in well-conditioned biological filters with different filter material. 1, earthenware balls; 2–4, honeycomb tube having different surface area; 5, net filter material; 6, fibriform filter material. Data represent means and standard deviations. Open column, oxidation rate after loading ammonia for 3 months; solid column, oxidation rate after loading ammonia and Ehrlich meat extract for 3 months

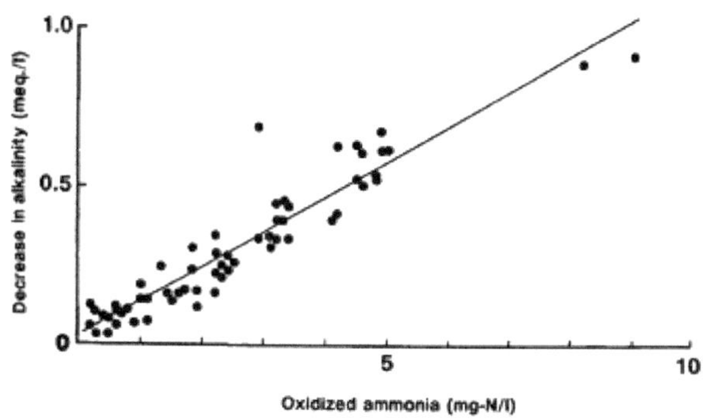

Fig. 5.8 Relationship between amount of biologically oxidized ammonia and alkalinity decrease in recirculating seawater in closed system. Reproduced from Kikuchi et al. 1994a, b, c

surface area is considered to be enough to produce 100 kg flounder in recirculation system, corresponding to 260 L volume of the net filter material.

We have little knowledge on the production of particulate substance in the closed recirculating aquaculture of marine finfish. However, several publications are recommended to remove suspended solids and leftover feed quickly from culture water to keep condition good for the growth of aquatic organisms. Based on the available information (Chen et al. 1994), we chose a combination of precipitation (settling) tank and drum screen filter as mechanical filtration (Fig. 5.9) for the closed production of Japanese flounder.

5.2.2 Management of Culture Water

Quality Changes of Recycling Culture Water Decrease in pH and alkalinity of recycling culture water due to nitrification happen at the early stage of closed

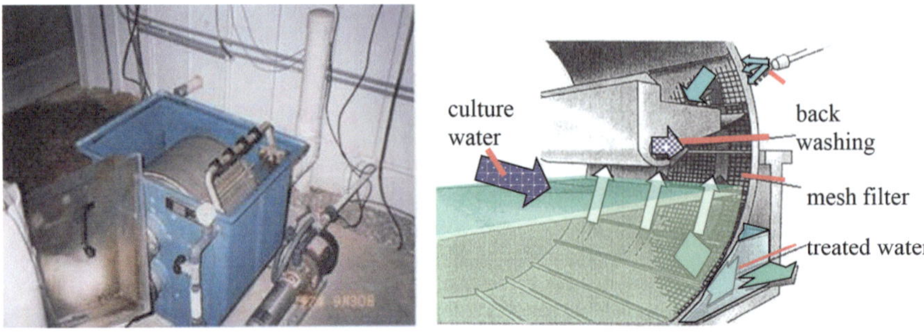

Fig. 5.9 Drum screen filter for suspended solids treatment

aquaculture depending on feeding rate and stocking density, and low pH of less than 6 disturbs nitrification as mentioned above.

Quality changes of seawater with the progress of the culture period were examined with two closed systems of 4.5 m^3 and 3.4 m^3 in total water volume each. Flounders were fed commercial pellet for 11 and 8.5 months without replacing culture water, and total biomass in each system reached to 70 and 40 kg at the end, respectively. Typical characteristics of culture water at the end of rearing were high concentrations of nitrate, phosphate, and dissolved organic carbon (DOC) in both systems (Table 5.2). Almost all excreted ammonia and urea from fish turn to nitrate, and nitrate accumulates linearly in culture water when the culture system is kept fully aerobic condition. There are little information on the effect of nitrate on the survival and growth of aquatic organisms, and 800–1000 mg-N/L is known to affect the feeding of Japanese flounder in our preliminary study. On the other hand, toxicity of phosphate is considered to be negligible, and 25 mg-P/L did not show any adverse effects on the feeding of the flounder in both systems. Furthermore, phosphate makes precipitation with calcium and magnesium ions in seawater when the concentration is 40–50 mg-P/L depending on pH and does not increase anymore (Suzuki et al. 2000). Therefore, phosphate is not considered to be an important accumulates for the long-term use of recycling culture water.

The other characteristic of long-term used culture water is an accumulation of "yellow substance" which colors culture water yellowish to brownish with the progress of culture period (Fig. 5.10). Yellow substance in the culture water strongly related to DOC concentration (Table 5.2). Little knowledge has been obtained for the toxicity of yellow substance, and 200–300 mg/L (corresponding to 40–60 mg DOC/L) was reported to disturb cleavage of oysters (Takeda and Kiyono 1990). However, 50 mg DOC/L did not show any adverse effects on the feeding of fish as well as nitrification activity of biofilter in both systems.

Among major elements in seawater, sodium ion increased with the culture period mainly by an addition of sodium bicarbonate to keep pH of recycling water (Table 5.3). Calcium and magnesium in seawater made precipitation with phosphate; however, they did not decrease because of stable supplies from feed or feces. Changes in major elements were rather smaller and are not considered to

Table 5.2 Changes in carbon, nitrogen, and phosphorus compounds in culture water

Substances	Start	At the end of rearing	
		Aquarium 1	Aquarium 2
TOC (mg/L)	1.3	56.8	30
DOC (mg/L)	1.3	56.0	29.6
Polysaccharide (uM)	0.1	29.4	20.5
Monosaccharide (uM)	0.4	16.4	13.6
NH4-N (ug/L)	0	203	73
NO2-N (ug/L)	2.8	16.7	8.3
NO3-N (mg/L)	0	305	334
Urea-N (ug/L)	7.7	909	3.7
Organic-N (uM)	0	13	15
PO4-P (mg/L)	0	24.5	25.0
Organic-P (mg/L)	0	< 1	< 1

Fish were fed commercial pellet for 11 and 8.5 months for Aquarium 1 and 2, respectively
Total biomass at the end was 70 kg for Aquarium 1 of 4500 L water volume and 40 kg for Aquarium 2 of 3400 L water volume

Fig. 5.10 Culture water at the start (left) and after 20 weeks (right) of rearing

affect the growth of fish. Changes were also shown for trace elements, and nickel, copper, and zinc at the end were much higher than those of fresh seawater (Table 5.3). Copper and zinc level in aquarium 1 exceeded the recommended values shown in Quality Standards for Fishery Water (Japan Fisheries Resource Conservation Association 2013). However, no adverse effects were observed for the feeding of the flounder in our studies.

Management of Recycling Culture Water Prevent pH (alkalinity) from deceasing is essential to keep good condition for nitrification in closed aquaculture. Calcareous filter materials such as crushed oyster shell, gravel, and limestone can supply alkalinity to recycling seawater; however, acid-insoluble organic substances growing on the surface of filter materials might reduce the buffering activity with the

Table 5.3 Changes in major and trace elements in culture water

Substances	Start	At the end of rearing	
		Aquarium 1	Aquarium 2
Na+	84.28	90.08	86.70
Mg2+	9.32	9.31	9.17
Ca2+	1.99	2.09	2.11
K+	1.77	1.87	1.82
Cl−	100	100	100
SO4−	5.18	5.40	5.35
Mn (ug/L)	0.3	0.5	0.3
Fe (ug/L)	8	11	< 0.08
Co (ug/L)	< 0.2	< 0.2	< 0.2
Ni (ug/L)	0.3	2	2
Cu (ug/L)	0.6	7.3	2.2
Zn (ug/L)	2	13	3

Major ions represent relative molar value to chlorine
Fish were fed commercial pellet for 11 and 8.5 months for Aquarium 1 and 2, respectively
Total biomass at the end was 70 kg for Aquarium 1 of 4500 L water volume and 40 kg for Aquarium 2 of 3400 L water volume

progress of culture period (Siddall 1974). Nitrification activity is generally high in closed aquaculture because fish are stocked intensively, and enough amount of alkaline substance should be supplied continuously. Therefore, natural calcareous materials are not effective as alkaline sources. Fine calcium carbonate or bicarbonate is considered to be more appropriate and be supplied based on the biologically oxidized ammonia (Fig. 5.8). Sodium bicarbonate of 0.2 g/L is required to keep pH at around 8.0 for 20 mg-N/L of ammonia oxidation. Biological denitrification also supplies alkalinity.

As mentioned above, high level of nitrate is considered to be the most concern for the long-term use of recycling culture water. Bacterial denitrification is one of the possible ways to reduce nitrate which requires anaerobic condition and organic matter. We conducted closed aquaculture of Japanese flounder with 2.25 m^3 total water volume system equipped with denitrification unit of 0.3 m^3 volume (Honda et al. 1993). Fibriform filter material of 240 L volume was filled, and denitrifying bacteria were inoculated before operation. Culture water was bypassed from the culture tank to the unit at 5.4 L/h (water in the unit fully exchanged by 2 days), and 2 M of glucose was supplied as an energy source at 4.4 mL/h. Nitrate reduced successfully after the operation of denitrifying unit and tended to be negligible at the end of rearing, although fish were fed actively (Fig. 5.11). However, operation of denitrification unit is not easy work, and nitrate sometimes turns to ammonia or nitrite depending on reduction condition. Absorption of nitrate (and phosphate) by plant is also possible. In fresh water recirculation system, closed aquaculture of tilapia or catfish and hydroponics of tomato and lettuce have been tried to combine

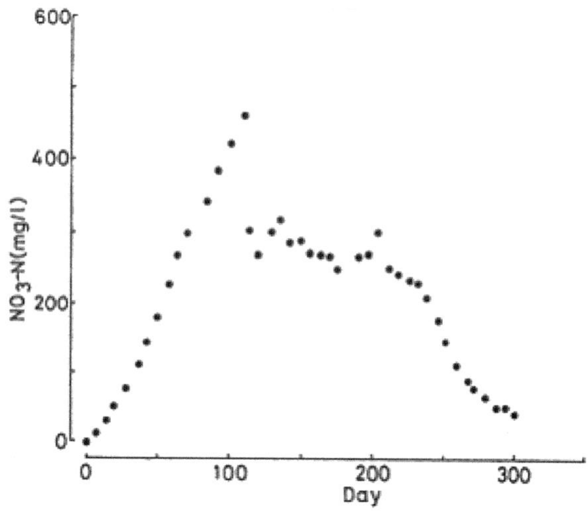

Fig. 5.11 Nitrate concentration in closed recirculating culture water of Japanese flounder. Denitrification unit started to operate at 130th day after the commencement of the rearing experiment. Reproduced from Honda et al. 1993

to utilize nitrogen and phosphate from fish culture (Fig. 5.12; Naegel 1977, Lewis et al. 1979). On the other hand, there were little trials for seawater system. We examined nitrate absorption rate of several seaweeds and found that *Ulva* sp. was one of the most appropriate due to the absorption rate and easy operation, although their market value was negligible. Absorption rate was estimated to be 0.3 mg-N/g *Ulva*/day at 20 °C and 3000 lx fluorescent light. When we utilize all nitrate from 100 kg flounder aquaculture with the seaweed, *Ulva* production is calculated to be 300 kg, 3 times higher than fish production. Exchange of culture water with fresh seawater (or artificial seawater) is considered to be more convenient if the seawater is easily available. Production of 1000 kg flounder (500 g body weight, 2000 fish) accompanies 45 kg-N of nitrate production when fish are fed commercial pallet, and 75 m^3 of seawater is required to keep nitrate level of less than 600 mg-N/L (non-harmful level). This means that we need to exchange culture water 2–3 times up to the end of production with a system of 25 m^3 in total water volume.

Effect of yellow substance on the fish growth is negligible; however, it is nuisance substance for aquariums and it sometimes produces odd smell in fish fillet. Yellow substance is generally treated with activated carbon and ozone in aquarium, although the latter must be used carefully due to its toxicity especially in seawater. UV sterilizer reduces yellowish color of culture water with a little extent as well as pathogens. Furthermore, UV can reduce suspended solids of 10–20 um in culture water which are difficult to remove by mechanical filters, although experimental data were very limited. Quick treatment of fine suspended solids is effective for easy operation by making visible condition in culture tank and less oxygen consumption. Therefore, UV is strongly recommended to equip in closed recirculating aquaculture system for marine finfish.

Discharge of Culture Water Replaced culture water is discharged to natural water body (river, inlet, and coast) or sewers if the farmer does not have specific treatment

Fig. 5.12 Aquaponics at the Department of Aquaculture, Pukyong National University. Left, hydroponics of common water hyacinth; right, culture tanks for tilapia

basin. We do not have clear regulations (standards) for discharged culture water from aquaculture facilities in Japan yet, and highest allowed concentrations for discharged water are 100 mg/L for total nitrogen, 16 mg/L for phosphorus, 160 mg/L for BOD or COD, and 200 mg/L for suspended solids if the Water Quality Standard for Waste Water is adapted. Higher values are generally allowed for sewage treatment. Regulations for discharged water differ by local authorities in Japan, and we must check specifics before the start of aquaculture production.

5.2.3 *Factors Affecting Productivity*

Temperature and Salinity Water temperature is one of the most influential environmental factors for the growth of fish. Although a few papers have attempted to elucidate the effect of temperature on the growth of Japanese flounder (Koshiishi 1981; Morizane 1984; Seikai et al. 1986), useful information for grow-out was limited. Iwata et al. (1994) conducted feeding experiments in which Japanese flounder of 4, 16, 88, and 176 g initial body weight were fed a commercial pellet diet twice daily to satiation at 10, 15, 20, 25, and 30 °C for 20 days. Based on the results of daily growth rate and feed efficiency of the cultured fish, the study suggested that the optimum temperature for the production of Japanese flounder is between 20 and 25 °C (Fig. 5.13). They also showed that high temperature (30 °C) negatively affects the growth of fish specifically for larger fish. However,

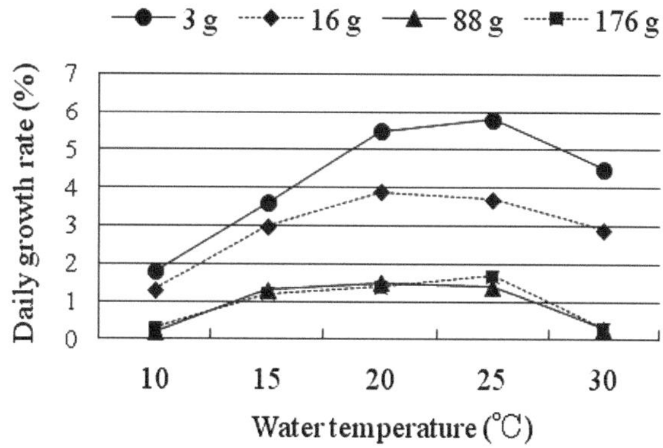

Fig. 5.13 Effect of water temperature on the growth of Japanese flounder with 4 different body weights. Fish were fed commercial pellet diet for 20 days

the culture period of 20 days was too short; more research especially with larger fish (200–500 g initial weight) is required to estimate the optimum temperature for the commercial production of Japanese flounder.

There is only a preliminary study on the effect of salinity. Salinity of more than 3.3‰ did not affect 5 days survival of 44 g fish; however, all fish died at 2.7‰. One month's culture experiment with fish of 8 g initial body weight revealed that the salinity of 4.4–34.0‰ had no adverse effects on the growth of fish nor did major blood constituents.

Stocking Density Socking density directly affects the productivity of aquaculture system, and fish are tried to produce intensively in closed recirculating aquaculture. As mentioned at the beginning, 15 kg/m^2 is the highest stocking density for 1 kg size fish in commercial production with running seawater. We conducted 5 weeks rearing experiment with 320–450 g body weight fish at initial stocking density of 5–60 kg/m^2 bottom area of the culture tank to elucidate upper limit density for normal and healthy growth. Feed intake, weight gain, and feed efficiency were statistically similar among experimental groups except 40 kg/m^2 group (Fig. 5.14). Although data obtained were fluctuated, it is indicated that 2–3 times higher density of the commercial production can be allowed if the water quality is sustained at the optimum level.

Availability of Dietary Carbohydrate and Lipid The protein content in commercial fish diets, while differing for various species, is much higher than that of the protein content for domesticated animals and ranges from 30% to more than 50% protein by dry weight. Because most fish have only a limited ability to utilize dietary carbohydrate as an energy source, they require a much higher percentage of protein in the diet. Utilization of dietary carbohydrate or lipid as an energy source has been widely examined among various kinds of fish and resulted in increasing protein efficiency ratio (PER) and decreasing feed cost (as well as nitrogen excretion) in some fish species.

We examined the availability of dietary carbohydrate as an energy source in the diet of Japanese flounder (Kikuchi et al. 1998). Juvenile fish of 3.5 g initial weight

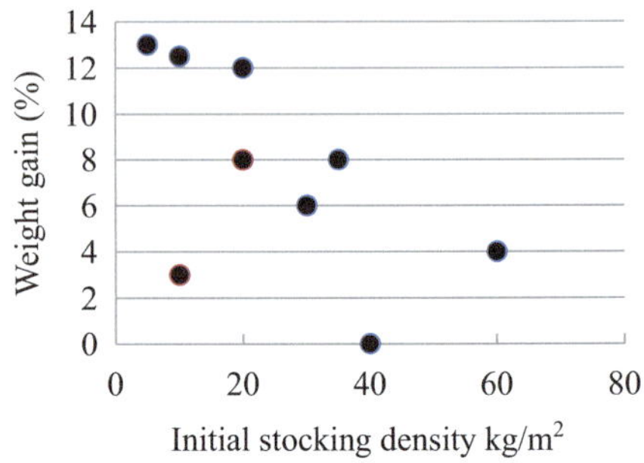

Fig. 5.14 Effect of stocking density on the weight gain of Japanese flounder. Fish of 320 to 450 g initial body weight were fed commercial pellet diet for 5 weeks at 20 °C

were fed 5 experimental diets containing different ratios of fish meal/potato starch (protein/carbohydrate; crude protein, 33–53%) twice daily to satiation for 6 weeks at 20 °C. There were no significant differences in PER among dietary treatments; however, feed efficiency decreased with the increase of potato starch in the diet. Furthermore, the final body weight and weight gain of fish fed diets with less than 40% crude protein were significantly lower than those of the other dietary groups. Adverse effects of increasing potato starch in the diet on growth and feed utilization are reported in other feeding experiments with fish of 50 and 310–360 g initial body weight, and the effects were more serious in larger fish (Kikuchi et al. 1992a). When 5 g flounder were fed diets containing glucose, maltose, dextrin, and potato starch, growth and feed efficiency of the cultured fish increased in dietary groups with dextrin and potato starch but tended to decrease with decreasing molecular weights of carbohydrate. Dietary inclusion of glucose and maltose also resulted in a marked increase of blood sugar level after feeding (Kikuchi et al. 1998), as in other fish species (Furuichi and Yone 1982).

Sato (1999) examined the utilization of dietary lipid of Japanese flounder. Four to 5 g juvenile Japanese flounder were fed 18 experimental diets with 6 protein levels (40, 45, 50, 55, 60, and 65%) and 3 lipid levels each (10–28%) twice daily to satiation for 8 weeks. Growth of fish generally depended on dietary protein level, and increasing level of dietary lipid did not produce positive effects at all protein levels. PER was statistically identical among dietary treatments, regardless of dietary protein and lipid levels. Similar results were obtained with fish of 55 and 245 g initial body weight; however, PER of fish fed the diet with the highest lipid level (20.3%) was significantly higher than that of fish fed the diet with the lowest lipid content (9.8%) in both feeding trials (Kikuchi et al. 2000). Utilization of dietary lipid of the flounder is considered to increase with the growth of fish; however, a large quantity of dietary lipid may result in adverse effects on the health of the cultured fish by increasing blood triglyceride level, liver weight, and crude lipid content of the liver and muscle (Sato 1999; Kikuchi et al. 2000). Although more information is required to clarify the optimum inclusion level of dietary lipid as an energy source, the potential utilization of dietary lipid of Japanese flounder is considered to be much lower than that of Atlantic salmon, rainbow trout, or yellowtail.

Alternative Protein Sources for Fish Meal Japanese flounder requires a high percentage of protein in the diet as mentioned above. Because of the shortage of sardines that had been the main ingredient (protein source) in the formulated diet of fish in Japan, finding an alternative protein sources is required to produce a stable supply of commercial diets at a reduced price. We examined the potential of feather meal, meat and bone meal, meat meal, and corn gluten meal as alternative protein sources for the diet of juvenile Japanese flounder (less than 10 g initial body weight). Previous results indicate that 20–40% fish meal protein can be replaced by feather meal, 20% by meat and bone meal, 60% by meat meal, and 40% by corn gluten meal (Kikuchi et al. 1994b, 1997; Sato and Kikuchi 1997; Kikuchi 1999a). However, most of these protein sources require supplementation of crystalline essential amino acids that are lacking in each alternative to achieve comparable growth to the control diet (fish meal is a sole protein source). Ten amino acids – arginine, histidine, isoleucine, leucine, lysine, methionine, phenylalanine, threonine, tryptophan, and valine – are considered to be essential for Japanese flounder (Kanazawa 1990). A preliminary study on commercial defatted soybean meal showed that substituting 20 to 50% fish meal protein for the soybean meal and supplemental amino acids in the diet did not lead to serious adverse effects on growth of juvenile flounder (Kikuchi et al. 1994c). It is also clear that 47% fish meal protein in the diet can be successfully replaced with defatted soybean meal (30% in the diet) in combination with blood meal (10%) without amino acid supplements (Kikuchi 1999b). Furthermore, inclusion of blue mussel meat, *Mytilus galloprovincialis*, (5% of diet) for an equal amount of soybean to this diet improves the growth of the flounder markedly, mostly due to increasing feed consumption (Kikuchi 1999b). Stimulation of feeding with the mussel meat is demonstrated with the diet in which 3% fish meal protein was replaced with the mussel protein (Kikuchi 1998). Blue mussel meat is an effective protein source that can replace 60% fish meal protein in the diet, with incremental increases in growth and feed utilization, without supplemental amino acids (Kikuchi and Sakaguchi 1997), although using blue mussel meat as the main ingredient of fish diet is far from practical.

These studies revealed that a considerable amount of fish meal protein can be replaced by several alternatives in the diet of Japanese flounder. However, the results were obtained from a 6–8 weeks feeding trial with fish of less than 10 g initial body weight. Therefore, a long-term culture is needed to clarify the practical use of these ingredients.

5.2.4 A Trial to Produce Japanese Flounder with a Closed Recirculating System

Based on the above mentioned experimental results and additional available information on the design and equipment for oxygen supply, solid removal, and disinfection that have been developed in western countries, we designed a pilot scale closed recirculating culture system for Japanese flounder. The system consisted of a culture tank of round shape, settling tank, drum screen filter, submerged biological filter, heating-cooling unit, oxygen generator and supplier (contactor), blower, foam fractionators, and UV sterilizer (Fig. 5.15).

Rearing experiments were conducted with the closed system of 10 m^3 in total water volume installed in a greenhouse (Fig. 5.16, Table 5.4). At the start, 1015 fish of 2 g in initial body weight were introduced to the culture tank of 4 m in diameter with 60 cm in water depth and fed commercial flounder pellet diet twice daily to apparent satiation each by hand. Water temperature was maintained at 23 ± 1 °C. Foam fractionators were not operated in the rearing experiment.

Fish fed actively for the duration of the study and grew to 456 g in mean body weight during a 259 day period (Fig. 5.17, Phase 1) with good survival rate (85%) and feed efficiency (97%) as shown in Table 5.5. Culture density at the end was 31 kg/m^2 bottom area of the culture tank, corresponding to 39 kg/m^3 total culture water. These results were relatively higher than those of previous studies for the flounder without drum screen filter and oxygen supplier (Honda et al. 1993; Furuta et al. 1998). About half of the culture water was exchanged at 148, 225 and 255 days of the rearing to keep the nitrate level at less than 600 mg-N/L. After 259 days, fish were randomly divided into two groups (430 fish each) and placed in duplicate culture systems and fed the commercial pellet. Water temperature was controlled at 19 ± 1 °C. Fish fed actively and grew to 692 g and 720 g during 78-day rearing period with a greater than 103% feed efficiency (Phase 2; Fig. 5.17, Table 5.5). Dissolved oxygen ranged from 90 to 130% of saturation level through the 337-day rearing period. Ammonia and nitrite were maintained at less than 4 mg-N/L, and no apparent adverse effects on the feeding and growth were observed. High stocking density did not result in any visible adverse effects such as injuring dorsal fins for flounders. We produced about 400 kg quality flounders with 25 m^3 of

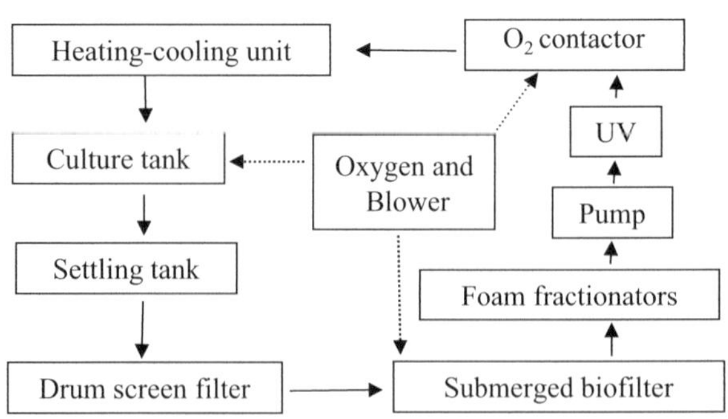

Fig. 5.15 Schematic diagram of an experimental closed system

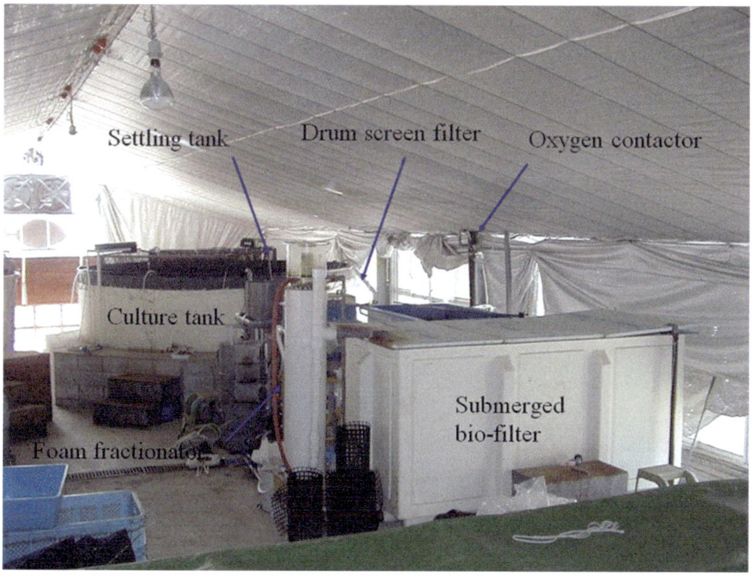

Fig. 5.16 Closed recirculating aquaculture system of 10 m^3 water volume for Japanese flounder

Table 5.4 Electric appliances equipped for the closed recirculating aquaculture system of 10 m^3 water volume for Japanese flounder

	Electric appliances	Rated power consumption	Specification
Circulation pump	Magnet pump MDH-400, Iwaki Co., Ltd	380	Maximum flow rate, 280 L/min
Water temperature controlling unit	Heat pump for seawater MR-3750H-TR, Marine River Co., Ltd	4630	Cooling, 10,700 kcal/h; heating, 11,600 kcal/h
Drum screen filter	HDF501, Hydrotech	180	Filter mesh, 60 μm; filtration, 350 L/min
Oxygen generator	AS12, AirSep	500	Oxygen supply, 5 L/min
Blower	Diaphragm blower DF-300, Taiko Kikai Industries Co., Ltd	390	Air supply 270 L/min
UV sterilizer	Sanitron SS-90SMR, Sen Light Co., Ltd	110	UV output, 90 W

total seawater in Phase 1, similar to that we designed (1000 kg with 50–75 m^3 seawater), and was extremely high efficiency (fish/water) compared with conventional flounder farm with running seawater. Additionally, growth rate, feed efficiency, and survival rate for closed aquaculture were superior to those for running seawater, although there were few information available for comparison.

Based on the results of entire culture period (337 days), the costs for electricity, seedling, and feed to produce 1 kg Japanese flounder were estimated to be 618, 203,

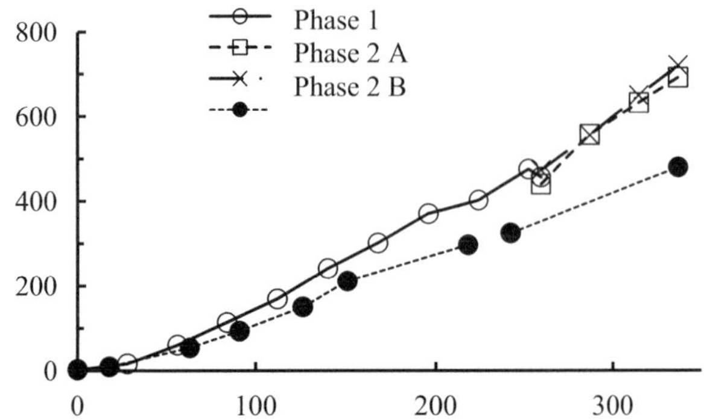

Fig. 5.17 Growth curve of Japanese flounder reared with closed recirculating aquaculture system

and 352 JPY, respectively (total, 1173 JPY). If it assumes that we construct the closed system mostly with commercial products including overseas and duration of the depreciation is 10 years, a construction cost for facility is estimated to be 420 JPY per year. Therefore, total cost except for labor is about 1600 JPY/kg flounder production, and unfortunately is not considered to be economically feasible at this moment. However, this value is based on a rearing experiment with a pilot scale facility, and construction and production costs are able to reduce by an expansion of production scale with equipments of reasonable prices and cheaper contracts of electricity.

5.2.5 *Quality of Japanese Flounder Produced with Closed Recirculating System*

Because Japanese flounder is one of the most quality fish species having high market value and mostly is eaten as fresh *sashimi* or *sushi* in Japan, the quality of raw fillet such as taste, texture, and appearance is very important. We conducted chemical and sensory analyses of the fillet of the flounder reared with closed recirculation system fed commercial pellet, wild catch, and fish produced with flow through culture tank fed raw sand lance. The flounder of 3 g initial body weight were fed the pellet for 20 months in a closed system installed in the building at nearly 20 °C and less than the stocking density of 10 kg/m^2 bottom area of fish tank. The maximum nitrate level was 200 mg-N/L due to partial exchange of culture water, and ammonia and nitrite were kept under 1 mg-N/L throughout the culture period. The flounder produced with flow through system was purchased from a commercial farm in Chiba Prefecture, Japan. The body weights of fish used for the examination ranged from 637 to 1088 g for closed fish, 841 to 1014 g for flow through, and 587 to 1193 g for wild catch. Sensory analysis of fresh fillet was done by panel test with selected panelist working for Japan Food Research Laboratories.

Moisture, crude protein, and ash contents of fillet did not differ to each other. Crude lipid of the flounder from closed aquaculture was slightly higher than that of

Table 5.5 Results of rearing experiment for Japanese flounder with closed recirculating aquaculture system

	Average body weight (g)							Culture density	
	Initial	Final	Survival (%)	Final biomass (kg)	Additional water (m^3)	Growth rate (%)	Feed efficiency (%)	kg/m^2	kg/m^3
Phase 1	2	456	84.7	393	15	2.1	96.8	31	39
Phase 2 A	440	692	98.1	292	0	0.6	104.4	23	29
B	473	720	99.5	308	0	0.5	102.7	25	31

Culture water was partially replaced with fresh seawater at 148, 225, and 255 days of the rearing in Phase 1

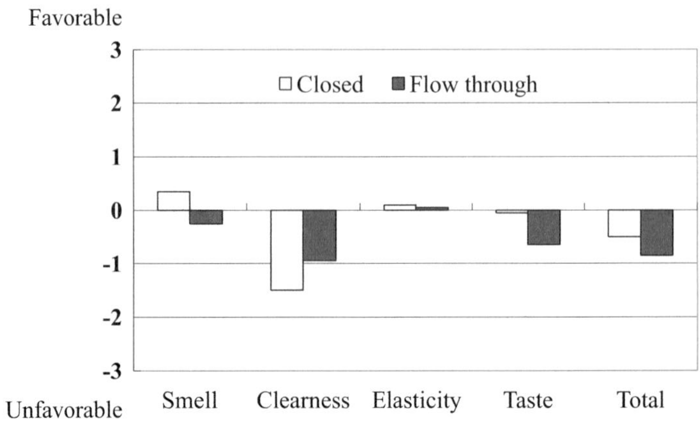

Fig. 5.18 Sensory analyses of fresh fillet (*sashimi*) of Japanese flounder reared in closed recirculation and flow through culture tank. Scores are relative value to those of wild catch flounders

the wild catch but was almost equal to the flow through culture. A similar trend to the crude lipid was shown for extractive nitrogen content and that of the cultured fish was slightly lower than the wild catch. There were no differences in free amino acids and nucleotides among three flounders tested such as glutamic acid, proline, glycine, inosine monophosphate, adenosine monophosphate, and adenosine triphosphate which are considered to be important substances to affect the taste. However, sensory analysis on smell, clearness, elasticity, and taste showed that fillet of the culture flounders were slightly inferior to that of wild catch. Scores were similar between two culture fish groups regardless of difference in feed. Fillet of closed fish were more yellowish than that of the wild catch (Fig. 5.18).

5.3 Conclusions

Research on the closed recirculating fish culture was extensively conducted during the 1950s to early 1970s especially for water treatment system of aquaria in Japan (Saeki 1958; Kawai et al. 1964, 1965; Hirayama 1970, 1974). Concurrently, aquaculture production was tried with carp and eel with basic closed system that consisted of culture tanks and biological filters (Japan Aquicultural Research group 1962). However, few research trials had been conducted on the closed aquaculture before we stared the project in 1986 (Fisheries Agency 1981; Hirayama et al. 1988). Based on the basic information on the flounder aquaculture, development of closed aquaculture techniques for flounder had been tried by private sectors as new business in the early 1990s in Japan. However, almost all the efforts were finished with failure, and we have no commercial aquaculture farms for marine fish using closed recirculation techniques up to now. Although it cannot be overemphasized that serious recession after 1992 in Japan is considered to affect seriously for the stagnation of developing technologies. The other major possible reason is that closed aquaculture was not proved to be promising business mostly due to the following:

1. Production trials with pilot scale facilities to elucidate the economical feasibility has not been tried: there were little information on the construction cost of average closed aquaculture facility nor the operation cost to produce flounder, basic information for farmers who wish to start closed aquaculture.
2. Environmentally friendly does not earn money: we have little environmental regulations for net cage aquaculture along the coast in Japan; therefore, less discharge of organic matter to natural environment, a major characteristic of closed aquaculture, cannot be an advantage from economical point of view.
3. Less availability of devices for closed aquaculture: we did not have the vendors specific for aquaculture gears and should obtain some devices from other industries with high prices; for instance, the market price of a heat pump for seawater in Japan was 4–5 times higher than that in the United States on power basis in near around 2000; there is lots of available information on aquaculture devices in the home page of "Pentair Aquatic Eco-Systems," and we can get preferable supplies with reasonable prices in the Unites States.
4. Higher cost of electricity: simple comparison of the electricity fee between Japan and the United States is not realistic; however, electricity occupied more than half of the total (electricity, feed, and seedling) operation cost in our study and might be serious issues to increase profitability.
5. Fluctuation in market price of fish: the market price of Japanese flounder was nearly 3000 JPY/kg in 1990 and decreased to less than half in 2000; such a drastic change in market price makes aquaculture to be unstable (risky) business.

These situations unfortunately continue nowadays, and closed recirculating aquaculture is not recognized as a promising business in Japan yet. First priority for the commercial success of closed aquaculture is that doing a feasibility study with a pilot scale facility (1) and clarify the major issues in reducing construction and operation costs (3, 4) required for commercial production. Minimum market price of the product should be discussed with the data of growth rate and proper stocking density of target species in relation with construction and operation costs to stabilize the business (5). Recent scientific and quantitative information on the closed aquaculture should be provided continuously to the farmers or entrepreneur, and the role of academic sectors is considered to be very important for the future development of closed aquaculture.

References

Chen S, Stechey D, Malone R (1994) Suspended solids control in recirculating aquaculture systems. In: Timmons M, Losordo T (eds) Aquaculture water reuse systems: engineering design and management. Elsevier, Amsterdam, pp 61–100

Fisheries Agency (1981) Studies on the water reuse system for eel production. Fisheries Agency, Tokyo, p 265. (In Japanese)

Furuichi M, Yone Y (1982) Availability of carbohydrate in nutrition of carp and red sea bream. Bull Jpn Soc Sci Fish 48:945–948

Furuta T, Iida T, Trongvanichnam K, Sakaguchi I, Wakabayashi H (1995) Indirect enzyme-linked immunosorbent assay (ELISA) for the detection of antibody in serum of Japanese flounder. Fish Sci 61:663–667

Furuta T, Kikuchi K, Iwata N, Honda H (1998) Culture of Japanese flounder, *Paralichthys olivaceus*, with a closed seawater recirculating system. Suisanzoshoku 46:557–562

Hirayama K (1970) Studies on water control by filtration through sand bed in a marine aquarium with closed circulating system-IV. Acidification of aquarium water. Nippon Suisan Gakkaishi 36:26–34

Hirayama K (1974) Water control by filtration in closed systems. Aquaculture 4:369–385

Hirayama K, Mizuma H, Mizue Y (1988) The accumulation of dissolved organic substances in closed recirculation systems. Aquac Eng 7:73–87

Honda H, Watanabe Y, Kikuchi K, Iwata N, Takeda S, Uemoto H, Furuta T, Kiyono M (1993) High density rearing of Japanese flounder, *Paralichthys olivaceus* with a closed seawater recirculation system equipped with a denitrification unit. Suisanzoshoku 41:19–26

Iida T, Wakabayashi H (1992) Preliminary trials of vaccination against *edwardsiellosis* of hirame (Japanese flounder), *Paralichthys olivaceus*. In: Shariff M, Subasinghe RP, Arthur JR (eds) Diseases in Asian aquaculture I. Fish Health Section, Asian Fish Soc, Manila, pp 407–412

Iwata N, Kikuchi K, Honda H, Kiyono M, Kurokura H (1994) Effects of temperature on the growth of Japanese flounder. Fish Sci 60:527–531

Japan Aquicultural Research group (1962) Symposium on closed recirculation system. Suisanzoshoku Special Issue, Hiroshima. p 76. (In Japanese)

Japan Fisheries Resource Conservation Association (2013) Seventh quality standards for fishery water 2012 edition, p 108

Jobling M (1981) Some effects of temperature, feeding and body weight on nitrogenous excretion in young plaice *Pleuronectes platessa* L. J Fish Biol 18:87–96

Kanai K, Tawaki S, Uchida Y (1988) A ecological study of *Edwardsiella tarda* in flounder farm. Fish Pathology 23:41–47

Kanazawa A (1990) Nutrition and feed in Hirame culture. Suisanzoshoku 38:398–399

Kawai K, Yoshida Y, Kimata M (1964) Biochemical studies on the bacteria in aquarium with circulating system-I. Changes in the qualities of breeding water and bacterial population of the aquarium during fish cultivation. Nippon Suisan Gakkaishi 30:55–62

Kawai K, Yoshida Y, Kimata M (1965) Biochemical studies on the bacteria in the aquarium with a circulating system-II. Nitrifying activity of the filter-sand. Nippon Suisan Gakkaishi 31:65–71

Kikuchi K (1998) Blue mussels in the diet of juvenile Japanese flounder. UJNR Tech Rep 26:269–274

Kikuchi K (1999a) Partial replacement of fish meal with corn gluten meal in diets for Japanese flounder *Paralichthys olivaceus*. J World Aquacult Soc 30:357–363

Kikuchi K (1999b) Use of defatted soybean meal as a substitute for fish meal in diets of Japanese flounder (*Paralichthys olivaceus*). Aquaculture 179:3–11

Kikuchi K, Sakaguchi I (1997) Blue mussel as an ingredient in the diet of juvenile Japanese flounder. Fish Sci 63:837–838

Kikuchi K, Takeda S, Honda H, Kiyono M (1990) Oxygen consumption and nitrogenous excretion of starved Japanese flounder *Paralichthys olivaceus*. Nippon Suisan Gakkaishi 56:1891
Kikuchi K, Takeda S, Honda H, Kiyono M (1991) Effect of feeding on nitrogen excretion of Japanese flounder *Paralichthys olivaceus*. Nippon Suisan Gakkaishi 57:2059–2064
Kikuchi K, Honda H, Kiyono M (1992a) Effect of dietary protein level on growth and body composition of Japanese flounder, *Paralichthys olivaceus*. Suisanzoshoku 40:335–340
Kikuchi K, Takeda S, Honda H, Kiyono M (1992b) Nitrogenous excretion of juvenile and young Japanese flounder. Nippon Suisan Gakkaishi 58:2329–2333
Kikuchi K, Honda H, Kiyono M (1994a) Ammonia oxidation in marine biological filters with plastic filter media. Fish Sci 60:133–136
Kikuchi K, Furuta T, Honda H (1994b) Utilization of feather meal as a protein source in the diet of juvenile Japanese flounder. Fish Sci 60:203–306
Kikuchi K, Furuta T, Honda H (1994c) Utilization of soybean meal as a protein source in the diet of juvenile Japanese flounder, *Paralichthys olivaceus*. Suisanzoshoku 42:601–604
Kikuchi K, Sato T, Iwata N, Sakaguchi I, Deguchi Y (1995) Effects of temperature on nitrogenous excretion of Japanese flounder. Fish Sci 61:604–607
Kikuchi K, Sato T, Furuta T, Sakaguchi I, Deguchi Y (1997) Utilization of meat and bone meal as a protein source in the diet of juvenile Japanese flounder. Fish Sci 63:29–32
Kikuchi K, Sato T, Deguchi Y (1998) Effect of dietary carbohydrates on the growth of juvenile Japanese flounder. Suisanzoshoku 46:541–546
Kikuchi K, Sugita H, Watanabe T (2000) Effect of dietary protein and lipid levels on growth and body composition of Japanese flounder. Suisanzoshoku 48:537–543
Koshiishi Y (1981) On the feed conversion efficiency of juvenile plaice, *Paralichthys olivaceus*. Bull Jpn Sea Nat Fish Res Inst 32:73–75
Lewis WM, Yopp JH, Schramm HL Jr, Brandenburg M (1979) Use of hydroponics to maintain quality of recirculated water in a fish culture system. Trans Am Fish Soc 107:92–99
Morizane T (1984) Fundamental study on the culture of plaice *Paralichthys olivaceus*-II. Effect of water temperature condition on the growth of young fish. Suisanzoshoku 32:127–131
Muroga K, Nakai T (1990) Diseases of the Japanese flounder, *Paralichthys olivaceus* in seedling production process. Suisanzoshoku 38:396–397
Naegel LCA (1977) Combined production of fish and plants in recirculating water. Aquaculture 10:17–24
Nakastugawa T (1983) Edwardsiella Tarda isolated from cultured young flounder. Fish Pathol 18:99–101
Nijhof M, Bovendeur J (1990) Fixed film nitrification characteristics in seawater recirculation fish culture system. Aquaculture 87:133–143
Ramnarine IW, Pirie JM, Johnstone ADF, Smith GW (1987) The influence of ration size and feeding frequency on ammonia excretion by juvenile Atlantic cod, *Gadus morhua* L. J Fish Biol 31:545–559
Rijn J, Rivera G (1990) Aerobic and anaerobic biofiltration in an aquaculture unit-nitrite accumulation as a result of nitrification and denitrification. Aquac Eng 9:217–234
Saeki A (1958) Studies on fish culture in the aquarium of closed-circulating system. Its fundamental theory and standard plan. Nippon Suisan Gakkaishi 23:684–695
Sato T (1999) Development of formulated feed for juvenile Japanese flounder. Doctoral dissertation, Tokyo University of Fisheries, Tokyo, p 134
Sato T, Kikuchi K (1997) Meat meal as a protein source in the diet of juvenile Japanese flounder. Fish Sci 63:877–880
Seikai T, Tanangonan JB, Tanaka M (1986) Temperature influence on larval growth and metamorphosis of the Japanese flounder *Paralichthys olivaceus* in the laboratory. Nippon Suisan Gakkaishi 52:977–982
Siddall SE (1974) Studies of closed marine culture systems. Progress Fish Cult 36:8–15

Suzuki Y, Maruyama T, Sato D, Kanda T, Michishita T (2000) Water quality and mass balances in a culture of Japanese flounder using a closed recirculating system with foam-separation and nitrification units. Nippon Suisan Gakkaishi 66:1–9

Takeda S, Kiyono M (1990) The characterization of yellow substances accumulated in a closed recirculation system for fish culture. In: Hirano R, Hanyu (eds) Proceedings of the Second Asian Fisheries Forum, Tokyo, April 1989, pp 129–132

Chapter 6
Kuruma Shrimp *Marsupenaeus japonicus*

Yoshihiro Suzuki

Abstract Kuruma shrimps *Marsupenaeus japonicus* were reared in a closed recirculating system, which consisted of a culture tank, a foam separation unit, a nitrification unit, and a denitrification unit. The shrimps used in the test were judged in advance to be free of the white spot disease virus (WSDV) by the method of loop-mediated isothermal amplification (LAMP). In the growth test of juvenile shrimp using a bench-scale system, the growth of shrimps, which were fed a commercial diet, was satisfactory, with the average weight increases of up to 11 times in 4.5 months. The individual density at the end of the culture period was 51 individuals/m^2. The foam separation unit maintained oxygen saturation in the water used for rearing at 101%. Furthermore, contaminants such as suspended solids, chromaticity substances, and bacteria absorbed on the stable foam were removed from the culture water by foam separation. The turbidity in culture water was kept at less than two units. Total ammonia nitrogen and nitrite oxidation were accomplished rapidly and simultaneously in the nitrification unit. When the denitrification process was operated, nitrate that accumulated in the culture water (20 mg-N/L) was reduced to 4 mg-N/L. Mating and spawning stages could be attained in the closed recirculating system less than 1m^3 of total water volume. In addition, the large-sized system (10 m^3) was built, and the possibility of utilization for the actual shrimp production was also proved. Based on these results, the intensive aquaculture of kuruma shrimp can be achieved using a closed recirculating system under virus-free conditions without emission.

Keywords A closed recirculating system • Juvenile growth • Mating • Spawning • Production • Virus-free

Y. Suzuki (✉)
Department of Civil and Environmental Engineering, Faculty of Engineering, University of Miyazaki, Miyazaki 889-2192, Japan
e-mail: suzuki@civil.miyazaki-u.ac.jp

T. Takeuchi (ed.), *Application of Recirculating Aquaculture Systems in Japan*, Fisheries Science Series, https://doi.org/10.1007/978-4-431-56585-7_6

6.1 Introduction

Recently, the aquaculture industry is likewise urged to convert to a new system that introduces the concept of zero emission, and this requires the development of a closed recirculating system using innovative technology for fish production. Especially, production of kuruma shrimp *Marsupenaeus japonicus* is one of the most important fish species for aquaculture in Japan (Hamasaki and Kitada 2006). However, zero-emission aquaculture system in an actual kuruma shrimp farm has not yet been developed. If intensive shrimp culture and production in a perfectly closed recirculating system become technically possible, the development of zero-emission systems can be realized.

With the combined aim of increased fish production and reduced nutrient load in aquatic environments, we have advanced the development of a zero-emission system composed of foam separation, nitrification, and denitrification units. Culture trials of Japanese flounder in saltwater (Maruyama et al. 1999; Suzuki et al. 2000) and eel in freshwater (Suzuki et al. 2003) were carried out using this almost perfectly closed system. The survival rate was very high, that is, more than 3 months despite the high fish density. The advantage of this system is that it is equipped with an effective foam separation unit as part of its main purification process. Oxygen supply, removal of suspended substances, and deaeration can be achieved simultaneously by the foam separation process (Maruyama et al. 1991, 1996). By applying the principles of the fish culture system, we tried the development of a closed recirculating system for culturing kuruma shrimp. An ideal aquaculture system is one which purifies the rearing water under virus-free conditions while obtaining high biomass productivity and production of virus-free seeds and seedlings.

6.2 System for Bench-Scale Study

A closed recirculating system with foam separation, nitrification, and denitrification units consisted of a shrimp rearing tank (water volume, 0.54 m^3; water surface area, 1.1 m^2), a foam separation tank (0.17 m^3) equipped with an inhalation-type aerator (200 V, 0.2 kw), a nitrification tank (0.13 m^3), and a denitrification tank (0.06 m^3). The total amount of water in this system was 0.9 m^3. The construction of the system is identical shown in Sect. 3.2. A heater for adjusting the conditions of the rearing water (28 °C) was set in the rearing tank. First, sand-filtered saltwater was introduced to the system, and one cycle was carried out for 22 min. The salinity in rearing water was adjusted to 28 psu using seawater and tap water. The system was located indoors and the rearing tank was shielded from light.

The difference between fish and kuruma shrimp rearing is that kuruma shrimp lives in the sand during daytime. Therefore, the sand must be laid at the bottom of the rearing tank. The sand used in the rearing tank accumulated evacuated matters and residual feed, and large amount of labor is required to clean the sand. To reduce the accumulation of polluted substances in the sand, the rearing tank was made with a double bottom, and the rearing water was made to flow upward under the sand layer. Coral sand (grain size, 2 mm) was used for this system.

6.3 Growth of Juvenile Shrimp and System Performance (Suzuki et al. 2010a)

6.3.1 Rearing

Juvenile kuruma shrimps (total gross weight 100 g, 125 individuals, about 0.8 g/individual) were placed in the rearing tank. The shrimps used in the test were judged in advance to be free of the white spot disease virus (WSDV) by the method of loop-mediated isothermal amplification (LAMP) (Kono et al. 2004). Throughout the rearing experiment, the shrimps were fed a commercial diet (crude protein, above 51.0%; crude lipid, above 5.0%; fiber, less than 5.0%; crude ash, less than 22.0%; calcium, above 0.5%; phosphorus, above 0.080%; Kyowahakko Co., Japan) daily. In the initial stage, 5 g of the feed was given once daily every evening. Shrimps were cultured for 135 days.

6.3.2 Analytical Methods

To determine the quality of rearing water, a sample was collected every 2 or 3 days from the rearing tank before feeding. Dissolved oxygen (DO), turbidity as kaolin standard (Mitsubishi Kagaku Co., SEP-PT-706D), total organic carbon (TOC, Shimadzu Co., TOC-5000), color as cobalt-platinum standard, absorbance at 260 nm (E260, Shimadzu Co., UV-2200), total ammonia nitrogen (TAN) (HACH Co., DR-2000), NO_3-N (HACH Co., DR-2000), NO_2-N (HACH, DR-2000), total nitrogen (T-N), phosphate (PO_4-P), and total phosphorus (T-P) were analyzed. The standard platinum-cobalt method of measuring color was used, in which the unit of color is that produced by 1 mg-Pt/L in the form of chloroplatinate ion. The collapsed-foam water samples were also obtained, and TOC, color, E260, suspended solids (SS), T-N, and T-P were analyzed. The analytical methods followed that of the Japanese Industrial Standard (JIS K 0102) or HACH Co. analytical manual.

Nitrogen in the solid samples, such as feed, shrimp tissue, ecdysis shell, and dried sludge, were analyzed using an elemental analyzer (CHNSO Analyzer 2400, Perkin Elmer Co.). Phosphorus in solid matter was decomposed in a mixture of perchloric acid and nitric acid and analyzed in the same way as T-P in rearing water.

6.3.3 Growth of Juvenile Shrimp

The shrimp growth during the experimental period (135 days) is shown in Fig. 6.1. Throughout the rearing period, the shrimp fed actively, and their total weight increased over time. The individual weight increased from an average of 0.80 g to 8.72 g during the experimental period. The growth rate was 1.98 g/month, and the rearing density was 51 individuals/m^2 by the end of the study. Although there was no dead individual throughout the rearing period, total population decreased from 125 to 61 individuals. The survival rate through the trial was 49% because of cannibalism.

6.3.4 Quality of Rearing Water

The changes in DO of the rearing water are shown in Fig. 6.2. The oxygen saturation percentage was kept above approximately 101% throughout the experimental period. The maximum and minimum DO concentrations were 7.0 mg/L and 6.0 mg/L, respectively, with a mean of 6.55 mg/L. While this system does not provide an extraneous source of oxygen except for the aerator in the foam separation tank, a high DO concentration was properly maintained in the rearing water.

The changes in the concentrations of TAN, NO_2-N, and NO_3-N in the rearing water are shown in Fig. 6.3. The TAN concentration was kept low throughout the study period at less than 0.4 mg-N/L. The NO_2-N concentration was maintained at a very low level less than 0.02 mg-N/L for 135 days. In contrast, NO_3-N was formed via TAN oxidation in the absence of denitrification, and NO_3-N steadily accumulated in the rearing water. However, when the denitrification process was initiated on the 78th day, NO_3-N concentration began to decrease after about 1 week and was reduced to 4 mg-N/L by the end of the study. In the period without denitrification,

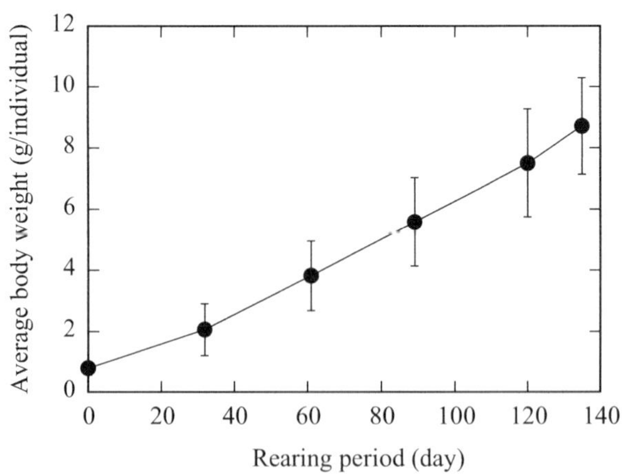

Fig. 6.1 Shrimp growth during the rearing period (Modified from Suzuki et al. 2010a)

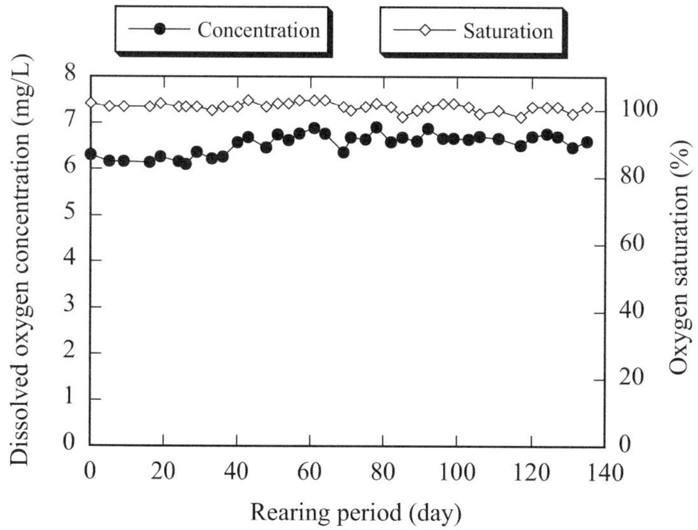

Fig. 6.2 Dissolved oxygen level in the rearing water (Modified from Suzuki et al. 2010a)

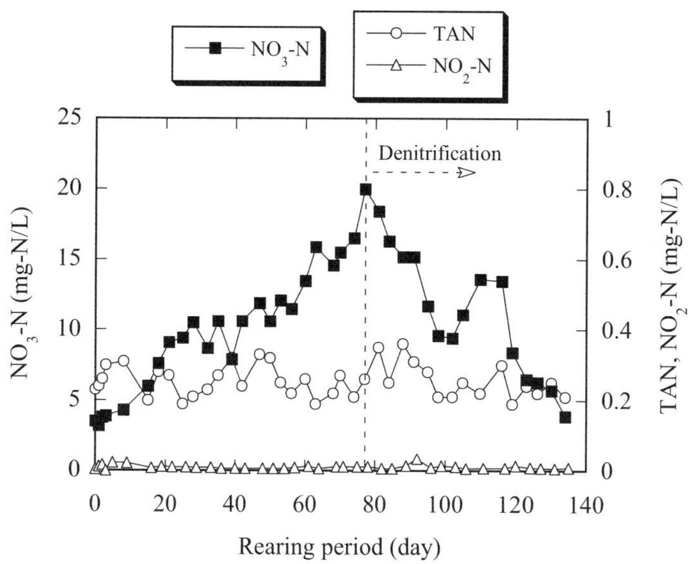

Fig. 6.3 Concentration of nitrogen compounds in the rearing water during the rearing period (Modified from Suzuki et al. 2010a). *TAN* total ammonia nitrogen

the cumulative amount of feed intake (x) and the amount of NO_3-N (y) in the system showed a good correlation ($y = 0.031x$, $r = 0.934$). For example, when 100 g of feed was given to the shrimp, 3.1 g of NO_3-N accumulated continuously in the rearing water without denitrification process. The fundamental parameters such as pH, water temperature, and salinity were kept at 8.0, 28 °C and 28 psu, respectively, throughout the trial. By the neutralizing by coral sand, a mechanical control of pH was unnecessary for this shrimp system. Though the concentration of calcium ion in rearing water was not determined, it is assumed that calcium was supplied with dissolution of calcium carbonate from the coral sand.

6.3.5 Characteristics of Foam Separation Process

The foam which concentrated the polluted substances was generated from the duct of the foam separation unit during the rearing period. The average quantity of water discharged per day was 293 mL ($n = 36$, 0.02% per day, 0.293 L/1300 L = 0.0002). Total volume of foam water was less than 40 L during the rearing period. It has been reported that foam generation of fish mucus is dependent on the concentrations of mucus and coexisting solvent ions (Suzuki et al. 2003). It was proven that the mucus substance acts on the foam separation process for not only fish rearing but also shrimp rearing.

The SS were significantly concentrated in the separated foam water, and the turbidity of the separated foam water was two orders of magnitude higher than that of the rearing water (Fig. 6.4). The turbidity of the rearing water was maintained in the range of one to two units, whereby almost no suspended substances could be observed. The turbidity in the foam water changed irregularly and varied from 50 to 600 mg/L, making it necessary to remove suspended substances from the system by a foam separation process. SS was not analyzed in the rearing water since only small amounts were observed, and the turbidity of the rearing water was retained at less than 1.0 unit. Moreover, a brown material was significantly concentrated in the foam water. The color unit of the foam water (average, 369 units; $n = 36$) was 100 to 1000 times higher than that of the rearing water (Fig. 6.5). The foam separation process was able to remove the color components, which are difficult to remove by biological treatment or physical filtration. Furthermore, an analysis of the effect of bacterial removal was undertaken (Fig. 6.5). The bacteria were concentrated markedly and suspended in the foam water.

6.3.6 Mass Balances

About 25% of the total weight of feed remained in the system as SS. The total of the residual amount of SS was considered as 100%. The total N and P contents in the feed were considered as 100%. The mass balances of this system for SS, N, and P are shown in Fig. 6.6. The N and P contents in the dried feed were 9.4% and 1.5%, respectively. The N and P contents in the shrimp body were 11.7% and 1.0%, respectively.

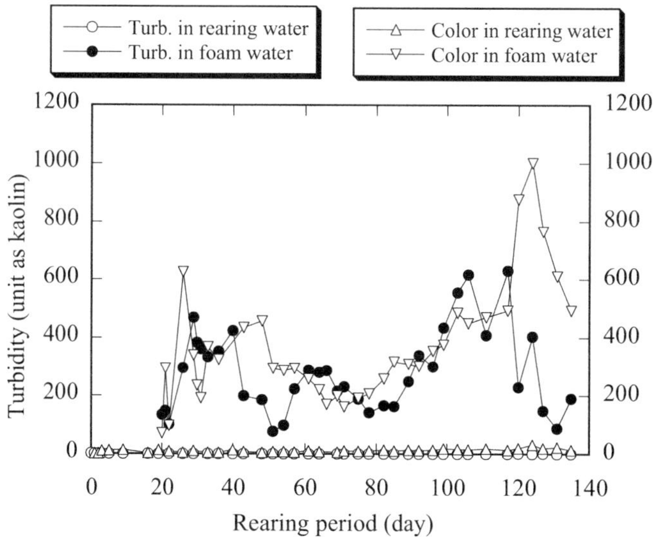

Fig. 6.4 Changes in turbidity and color in the rearing water and the foam water during the rearing period (Modified from Suzuki et al. 2010a). *Turb.* turbility

Fig. 6.5 Changes in bacteria counts in the rearing water and the foam water during the rearing period (Modified from Suzuki et al. 2010a)

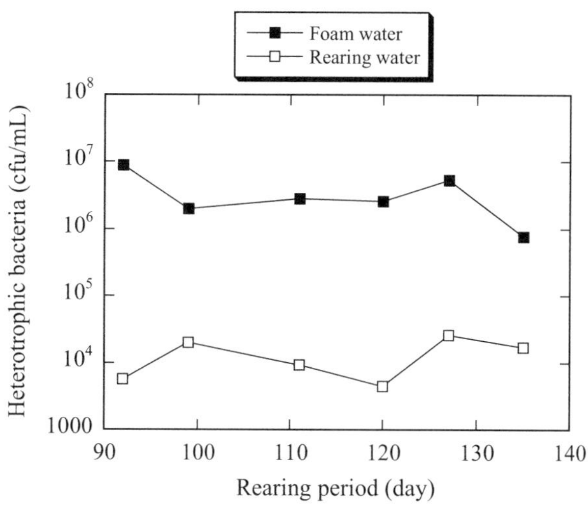

In the case of the SS, about 60% was accumulated in the nitrification tank, and 5.9% was removed by foam separation unit (Fig. 6.7a). The nitrification tank functioned as both nitrification and sedimentation units. The point to be noticed is that the accumulation of SS in the sand was small, less than 3%. In general, a huge amount of labor is needed for cleaning and maintenance of the sand in the shrimp culture tank. The upflow type of rearing tank is possible to drastically ease the maintenance of rearing tank or pond for shrimp culture.

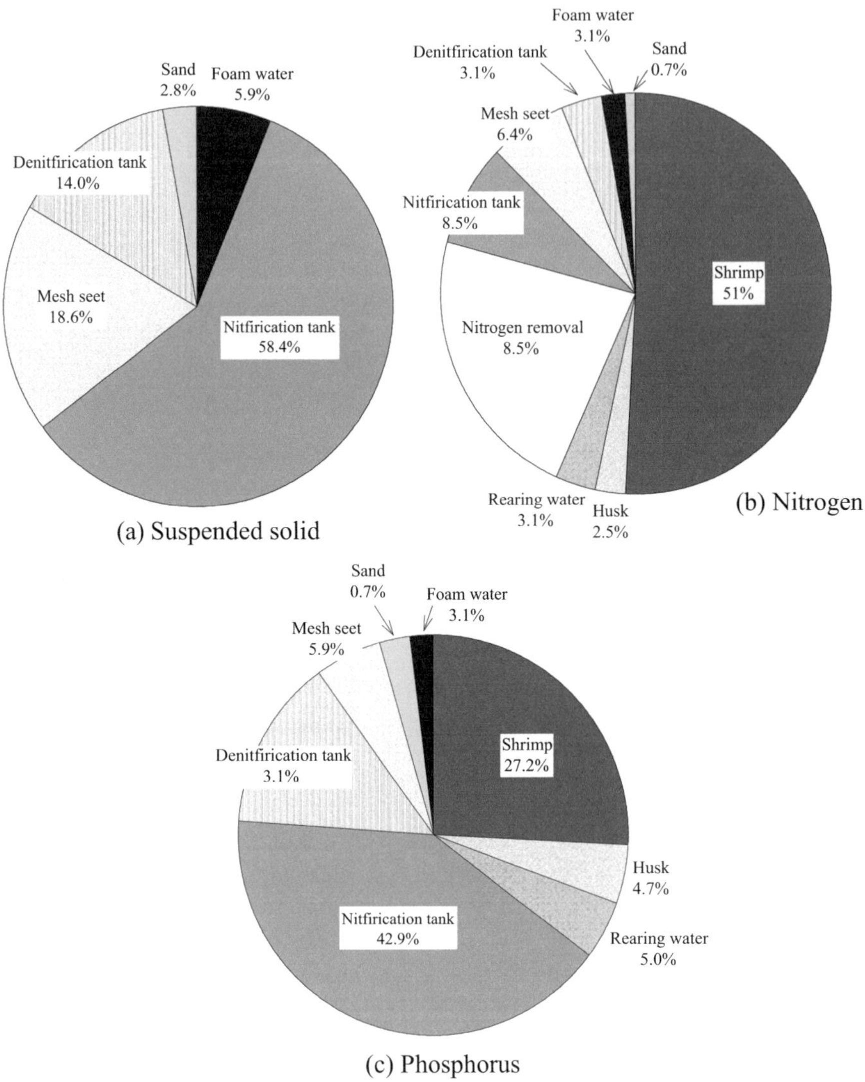

Fig. 6.6 Mass balances of suspended solid (**a**), nitrogen (**b**), and phosphorus (**c**) in the closed recirculating system (Modified from Suzuki et al. 2010a)

In the case of total nitrogen, 51% was utilized for shrimp growth, 3.1% was accumulated in the rearing water as NO_3-N and organic nitrogen, 2.2% was removed by foam separation, and 9.5% was accumulated in the nitrification and denitrification tanks as sediment (Fig. 6.7b). Regarding mass balances in the culture, the assimilation of nitrogen in the fish body varied from 25 to 35% of the

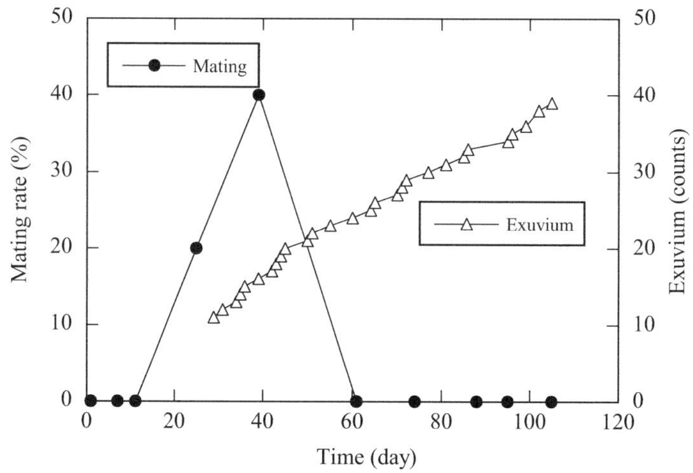

Fig. 6.7 The relationship between mating rate and cumulative number of shrimp exuviae (Modified from Suzuki et al. 2008)

total nitrogen input without regarding the difference in fish species (Folke and Kautsky 1989; Hall et al. 1992; Maruyama and Suzuki 1998; Skjølstrup et al. 1998; Suzuki et al. 1999). The nitrogen assimilation in shrimp was higher than that in fish. Almost all the nitrogen that must be treated in this system was present as a dissolved fraction. In this study, the remaining 34.2% of nitrogen in the system was removed as nitrogen gas by denitrification. Denitrification could have removed the residual NO_3-N in the rearing water if the operation was continued for a few days after the shrimp was harvested.

In the case of total phosphorus, 27% was utilized for shrimp growth, 5% was accumulated in the rearing water, 2.1% was removed by foam separation, and 60% was accumulated in the nitrification and denitrification tanks as sediment (Fig. 6.7c). Because of analytical error, the total percentage exceeded 100%. In case of a recirculating aquaculture system for saltwater fish, it has been reported that the phosphorus accumulates in high concentration in the sludge, because the phosphorus reacted with calcium and magnesium in the saltwater and formed insoluble compounds (Suzuki et al. 2000).

6.4 Mating Test (Suzuki et al. 2008)

6.4.1 Rearing for Mating

Ten individuals of adult kuruma shrimp (five male individuals, five female individuals, about 40 g/individual) were placed in the rearing tank. The shrimps used in the test were judged in advance to be free of the white spot disease virus (WSDV) by the LAMP method in the same as rearing of juvenile shrimp. For 95 days in the rearing experiment, the shrimps were fed a commercial diet once daily every evening (4.5 g/day). In the final period of mating test remaining 10 days, the shrimps were fed both frozen lugworm daily (4.5 g/day). To judge the mating evidence, formation of a stopper at the thelycum on the female shrimp was checked every 1 to 2 weeks. Adult shrimps were cultured for 105 days.

6.4.2 *Mating*

The foam separation unit maintained oxygen saturation in the water used for rearing at 99% (6.45 ± 0.16 mg/L, mean ± SD, $n = 37$). The turbidity of the rearing water was maintained in 0.038 ± 0.133 units. There were few suspended substances (SS) in rearing water throughout the mating test. Ammonia and nitrite oxidation were accomplished rapidly and simultaneously in the nitrification unit. TAN and nitrite concentrations were kept at less than 0.2 mg-N/L. Furthermore, contaminants such as SS, chromaticity substances, organic substances, and proteins absorbed on the stable foam were removed from the culture water by foam separation. This system was able to maintain good water quality for culturing shrimps under a closed condition without water exchange when shrimps were fed both a commercial diet and frozen lugworm.

The survival rate in the mating test for 105 days was 100%. Throughout the rearing period, the shrimp fed actively, even when which of the commercial diet and frozen lugworm was fed. The individual weight increased from an average of 42 g to 54 g during the experimental period. The relationship between mating rate and cumulative number of ecdysis shrimp shell is shown in Fig. 6.7. The mating rate was defined as the rate of female shrimps which formed a stopper to the total number of female shrimps. In the closed recirculating system less than 1m^3 of total water volume, 40% of the population of female shrimps formed a stopper at the thelycum within 1–2 months after the mating test was started. The increase in a mating rate was not observed by feeding frozen lugworm. After the mating test, the shrimps cultured in the system were judged WSDV-free by the LAMP method. It was verified that female shrimps possessing a stopper could be produced under virus-free conditions.

6.5 Spawning Test (Suzuki et al. 2010b)

6.5.1 *Rearing for Spawning*

The inner side of a plastic cage (39 cm × 56 cm × 28 cm) was covered with the mesh sheet (opening size 1 mm), and the cage installed in the rearing tank of a system (Fig. 6.8) for individual rearing of shrimp. A cylinder type of spawning cage consisted of 1 mm mesh sheet in the inner side, and 100 μm mesh sheet in the outside was also prepared.

The female shrimps (nine individuals, average body weight 65 g) used in the test were unqualified for seedling production in the actual shrimp farm because they had a spermatophore but immaturity of their ovary. Before the spawning test, blood was taken from each shrimp, and the shrimps were judged in advance to be free of WSDV by the LAMP method. The nine females were treated with unilateral eyestalk ablation using a rubber tag and cultured in the individual cages set in the rearing tank (Fig. 6.9). Unilateral eyestalk ablation is the advanced treatment for promoting maturity of the ovary (Sano et al. 2002). The females were fed live lugworm (5 g/individual). Ovarian development was visualized by employing a flashlight to distinguish the development ratio of the ovary in the dorsal exoskeleton. Ovary maturation was checked every day throughout the spawning test. When the ovary matured, the female was transferred to a spawning cage. After spawning, the eggs were collected from the mesh sheet of the cage, and then the number of

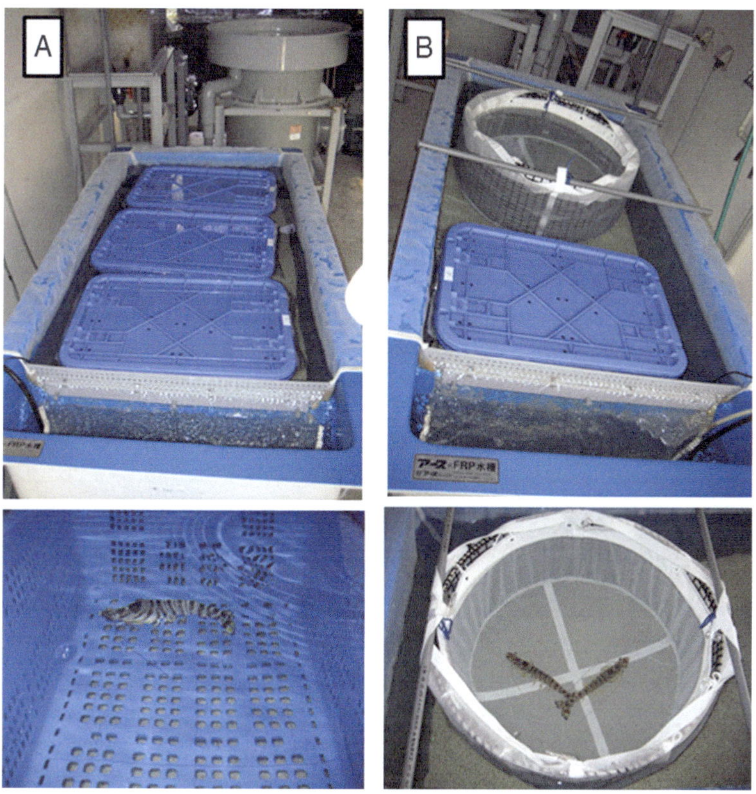

Fig. 6.8 Exterior of individual rearing cage (**a**) and spawning cage (**b**) (Modified from Suzuki et al. 2010b)

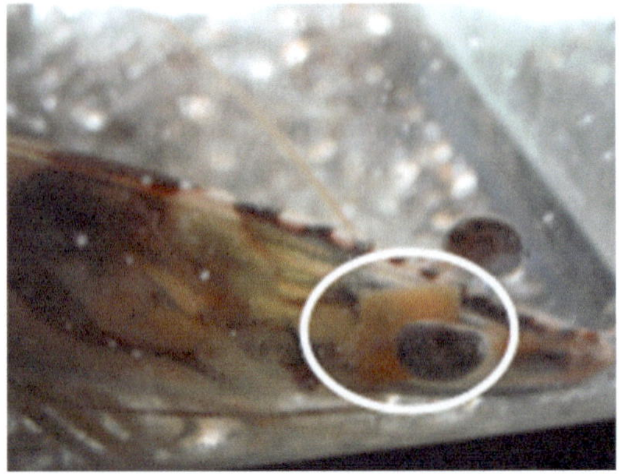

Fig. 6.9 Female shrimp treated with unilateral eyestalk ablation using a rubber tag

eggs was counted under the microscope. The spawning test was carried out for 15 days. All individuals in the system survived throughout the test period (15 days). After the spawning test, the females cultured in the system were judged WSDV-free again by the LAMP method.

6.5.2 *Spawning*

This system was able to maintain good water quality for culturing shrimps under a closed condition without water exchange through the spawning test.

All the female shrimp treated with unilateral eyestalk ablation fed actively, and then the ovary matured even in the small individual cages. The observation data of the development ratio of the ovary (ovary width/body width) and spawning for each female are shown in Table 6.1. Five individuals within nine females spawned within 12 days. There were two individuals which spawned two to three times until the period. The average number of eggs in spawning was 5.1×10^4/tail. The development ratio of the ovary just before spawning was over 0.5, and then the ovary shape became clear and changed to black. The photographs of both individuals which spawned and did not spawn are shown in Fig. 6.10. It is useful to judge the timing spawn by the development ratio of the ovary and the ovary color. After the spawning test, the females cultured in the system were judged WSDV-free by the LAMP method. In the small scale of closed recirculating system, it was verified that female shrimps could be spawned under virus-free conditions.

Table 6.1 Changes in the ratio of the ovary in dorsal exoskeleton and spawning for individual female shrimp in the spawning test

Identification	A	B	C	D	E	F	G	H	I
Body weight (g-wet)	41.10	84.19	61.66	92.18	46.90	64.47	68.65	57.46	71.02
Body length (cm)	18.5	23.2	20.2	20.1	19.4	18.8	20.0	19.5	20.5
Virus check	Negative	Negative	Negative	Negative	Negative	Negative	Negative	Negative	Negative
Rearing period (d)	Ratio (−)	Ratio (−)	Ratio (−)	Ratio (−)	Ratio (−)	Ratio (−)	Ratio (−)	Ratio (−)	Ratio (−)
1	0.31	0.66	0.31	0.63	0.65	0.65	0.40	0.58	0.50
2	0.31	0.67	0.40	0.70	0.78	0.70	0.44	0.63	0.61
3	0.29	0.74	0.37	0.75	0.82	0.67	0.44	0.65	0.58
4	Ecdysis	Spawning	0.32	Spawning	0.78	0.66	0.29	0.59	0.71
5		0.38	0.32	0.40	0.85	0.66	0.33	0.64	0.72
6		0.38	0.36	0.47	0.88	0.67	0.37	0.62	0.54
7		Ecdysis	0.36	0.46	Spawning	0.61	0.36	0.71	0.52
8			0.38	0.28	0.31	0.65	0.39	0.71	0.53
9			0.35	0.37	0.24	0.60	0.36	0.74	Spawning
10			0.19	0.47	0.30	0.57	0.38	Spawning	0.50
11			0.21	0.28	Ecdysis	0.42	0.25	0.44	0.50
12			0.17	0.34		0.37	0.28	0.63	Spawning
13			Ecdysis	0.32		0.26	0.26	Spawning	0.44
14				0.31		0.27	0.31	0.19	0.55
15				0.33		Ecdysis	Ecdysis	0.21	Spawning

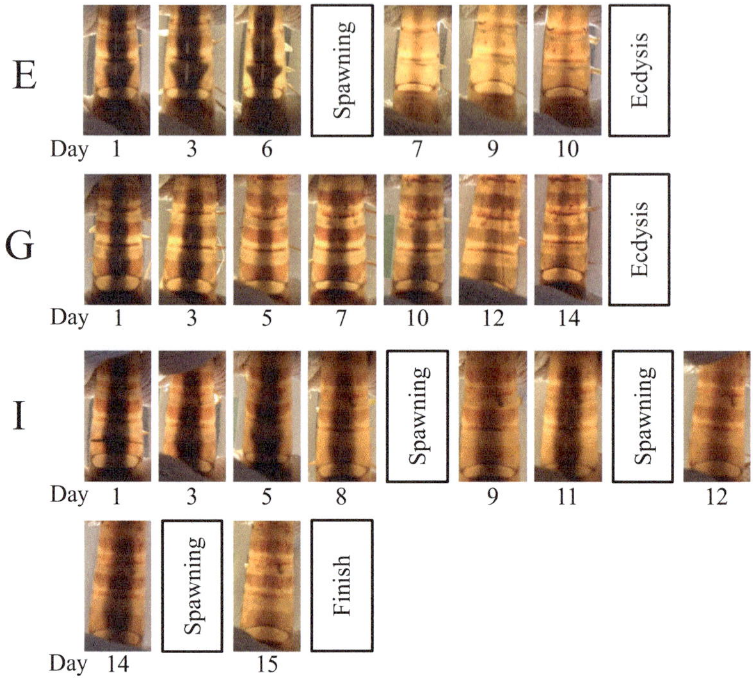

Fig. 6.10 Development of the ovary for female shrimp until the rearing period. Shrimp E, ecdysis after spawning; shrimp G, ecdysis without spawning; shrimp I, spawning three times (Modified from Suzuki et al. 2010b)

6.6 Actual Scale System

When the large-sized (50–100 g/individual) female shrimp, which mated and matured, can be reared and maintained under virus-free conditions throughout the year, it will be possible to produce virus-free shrimp as a year-round product. Using the existing aquaculture pond in an actual farm, a large-sized system (rearing pond, 10 m^2, 0.6 m depth) equipped with all the functions such as foam separation, nitrification, and denitrification was designed and constructed (Fig. 6.11). The amount of the maximum rearing individuals was designed to fully breed the total of 200 shrimps as over 100 g/individual size. All processing units in the system were attachment types. Since repairs were not constructed in the existing pond, it is easy to restore to the conventional aquaculture pond.

Adult shrimps (male 100 individuals, female 100 individuals, about 30 g/individual) were placed in the rearing pond. The shrimps used in the test were judged in advance to be free of the white spot disease virus (WSDV) by the LAMP method in

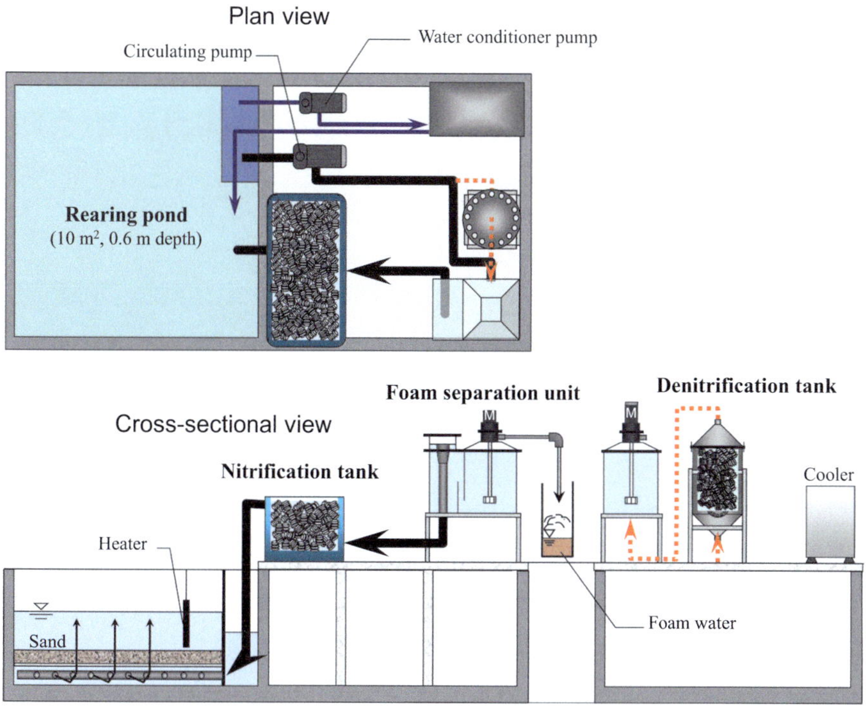

Fig. 6.11 Actual scale system with foam separation, nitrification, and denitrification units

the same as the bench-scale study. Throughout the rearing experiment, the shrimps were fed a commercial diet (Goldprawn, Higashimaru Co., Japan) daily. About 30–70 g of the feed was given once daily every evening. Shrimps were cultured for 4 months.

As the same in the bench-scale study, the rearing water qualities were maintained satisfactorily throughout the test of the actual scale system. The foam separation unit maintained oxygen saturation in the water used for rearing at 100%. The turbidity of the rearing water was maintained in less than 1 unit. The ammonium ion and nitrite concentrations were kept at less than 0.2 mg-N/L. The shrimp grew up 30 g to 50 g at the end of actual scale experiment. More than 20 individuals of female, which mated and matured, were observed within this experiment period (Fig. 6.12). The survival rate in the mating test for 105 days was 70%. The cause of the decline in the survival rate was cannibalization of individuals. In the actual scale system, it is possible to maintain the female shrimp used for the base of seeds-and-saplings production under virus-free conditions throughout the year.

Fig. 6.12 Female shrimp formed a stopper at the thelycum and matured ovary in the actual scale system

6.7 Conclusions

Our proposed system achieved kuruma shrimp culture in a perfectly closed cycle for more than 4 months. Under good water quality and virus-free condition, shrimp growth and survival rate were satisfactory during the several study periods such as growth, mating, and spawning tests. Oxygen was efficiently supplied to the rearing water by a foam separation unit, and oxygen saturation was maintained at 100% throughout the experiment. Simultaneously, the foam separation process removed the brown colloidal substances generated by shrimp mucus. The nitrification tank removed SS and likewise rapidly nitrified ammonia. While NO_3-N accumulated in the rearing water in the absence of denitrification, after initiation of denitrification, NO_3-N was effectively removed and reduced to less than 10 mg-N/L. In addition, the large-sized system was built, and the possibility of utilization for the actual shrimp production was also proved.

In this study, the rearing trial serves as a starting point. Further development to minimize the capital and operating costs of this system will be necessary prior to potential commercial viability. This system has an application potential for the production of kuruma shrimp under a perfectly closed rearing condition free of WSDV. Furthermore, the closed recirculating system can be utilized for maintaining pathogen-free brood stocks since this system makes it easy to control the condition of a specific pathogen-free environment.

Acknowledgment This work was supported in part by grants from the Research and Development Program for New Bio-industry Initiatives, Japan.

References

Folke C, Kautsky N (1989) The role of ecosystems for a sustainable development of aquaculture. Ambio 18:234–243

Hall POF, Holby O, Kollberg S, Samuelsson M (1992) Chemical fluxes and mass balances in a marine fish cage farm. IV. Nitrogen. Mar Ecol Prog Ser 89:81–91

Hamasaki K, Kitada S (2006) A review of kuruma prawn *Penaeus japonicus* stock enhancement in Japan. Fish Res 80:80–90

Kono T, Savan R, Sakai M, Itami T (2004) Detection of white spot syndrome virus in shrimp by loop-mediated isothermal amplification. J Virol Methods 115:59–65

Maruyama T, Suzuki Y (1998) The present stage of effluent control in Japan and pollutant load from fish culture to environment—possibility of intensive recirculating fish culture systems. Nippon Suisan Gakkaishi 64:216–226. (In Japanese with English abstract)

Maruyama T, Okuzumi M, Saheki A, Shimamura S (1991) The purification effect of the foam separating system in living fish transportation and preservation. Nippon Suisan Gakkaishi 57:219–225. (In Japanese with English abstract)

Maruyama T, Okuzumi M, Satoh Y (1996) The purification of rearing saltwater of Japanese flounder with the closed foam separation-filtration system. Nippon Suisan Gakkaishi 62:578–585. (In Japanese with English abstract)

Maruyama T, Suzuki Y, Sato D, Kanda T, Michisita T (1999) Performance of a closed recirculating system with foam-separation and nitrification units for intensive culture of Japanese flounder. Nippon Suisan Gakkaishi 65:818–825. (In Japanese with English abstract)

Sano M, Minagawa M, Tamaki M, Hayashibara T, Shimizu H (2002) Effects of water temperature on the spawning interval of kuruma prawn *Penaeus japonicus* after unilateral eyestalk ablation. Suisanzoshoku 50:433–434. (In Japanese with English abstract)

Skjølstrup J, Nielsen PH, Frier FO, McLean E (1998) Performance characteristics of fluidized bed biofilters in a novel laboratory scale recirculation system for rainbow trout: nitrification rates, oxygen consumption and sludge collection. Aquac Eng 18:265–276

Suzuki Y, Maruyama T, Takemoto S, Oda R (1999) Performance of a closed recirculating system with foam separation, nitrification and denitrification units for the intensive culture of eel. J Jpn Soc Water Environ 22:896–903. (In Japanese with English abstract)

Suzuki Y, Maruyama T, Sato D, Kanda T, Michisita T (2000) Water quality and mass balances in a culture of Japanese flounder using a closed recirculation system with foam-separation and nitrification units. Nippon Suisan Gakkaishi 66:1–9. (In Japanese with English abstract)

Suzuki Y, Maruyama T, Numata H, Sato H, Asakawa M (2003) Performance of a closed recirculating system with foam separation, nitrification and denitrification units for the intensive culture of eel: towards zero emission. Aquac Eng 29:165–182

Suzuki Y, Suzuki T, Sagisu Y, Mekata T, Kono T, Koshio S, Yokoyama S, Sakai M, Itami T (2008) Mating test of kuruma shrimp in a closed recirculating system including foam separation, nitrification and denitrification units. Suisanzoshoku 56:479–485. (In Japanese with English abstract)

Suzuki Y, Takeshima T, Mekata T, Kono T, Sakai M, Itami T, Maruyama T (2010a) Performance of a closed recirculating system with foam separation, nitrification and denitrification units for intensive culture of kuruma shrimp, *Marsupenaeus japonicus*: a bench scale study. Asian Fish Sci 23:483–496

Suzuki Y, Sagisu Y, Suzuki T, Mekata T, Kono T, Koshio S, Yokoyama S, Sakai M, Itami T (2010b) Spawning test of Kuruma shrimp in a closed recirculating system. Suisanzoshoku 58:161–166. (In Japanese with English abstract)

Chapter 7
White Shrimp *Litopenaeus vannamei*

Marcy N. Wilder and Setsuo Nohara

Abstract The importation of shrimp into Japan at more than 250,000 tons/year is thought to be related in certain aspects to the serious environmental problems that have been caused by shrimp farming in Southeast Asian countries. This includes mangrove deforestation and marine pollution due to the discharge of wastes and leftover feed from intensive and semi-intensive farms. Moreover, because of the advent of a new disease, early mortality syndrome (EMS), which first appeared in China and Vietnam in 2009, shrimp production volumes have decreased, and the industry is becoming rapidly unstable. In order to minimize the impacts of this industry on the environment, an industry-government consortium was formed in Japan to promote the sustainable and safe technical development of shrimp farming on a practical level. Major development themes are as follows: (1) establishment of *Litopenaeus vannamei* freshwater aquaculture technology based on physiological studies, including the engineering of a high-density recirculating shrimp-production plant, (2) development of techniques for evaluating and reducing shrimp stress, and (3) development of a low-cost feed that does not degrade water quality. In addition, we have conducted preliminary work on seed production technology using closed systems for purposes of serving domestic supply. A commercial plant was set up in Niigata Prefecture on the basis of the results of this research, and production commenced in 2007. Here, we describe our research efforts, the challenges we faced, and the progress made so far.

Keywords Automatic cleaner • Biological filtration • Freshwater rearing • *Litopenaeus vannamei* • Oxygen cone • Recirculating aquaculture systems • Wave-generating system

---

M.N. Wilder (✉)
Japan International Research Center for Agricultural Sciences, Tsukuba, Japan
e-mail: marwil@jircas.affrc.go.jp

S. Nohara
IMT Engineering Inc., Tokyo, Japan

T. Takeuchi (ed.), *Application of Recirculating Aquaculture Systems in Japan*, Fisheries Science Series, https://doi.org/10.1007/978-4-431-56585-7_7

7.1 Introduction

7.1.1 Growth of the World's Aquaculture Industry

In the 19 years from 1990 to 2009, the global aquaculture industry grew at 8% annually in terms of both production volume and production yield (Table 7.1). By 2009 the value of the global aquaculture industry had grown to $110 billion (FAO 2011). This is against a projected growth in the world's population from 6.1 billion in 2000 to 9 billion in 2050. The need for food production is thus increasing because of population growth, and global expansion of grain production is approaching its limit. In addition, with the effects of diseases such as bird flu and mad cow disease and the current trend toward replacing red meats with other proteins, the demand for fish has risen and is being boosted worldwide by a boom in health-conscious fish eating (FAO 2011).

As income levels improve, and lifestyles become richer in developing areas, it is seen that there is an increased trend toward eating more animal protein, including fish, rather than grains (Fisheries Agency 2007: Fig. 7.1). However, to produce 1 kg of beef, 11 kg of feed (corn equivalents) is needed, 3.5 kg is required to produce 1 kg of pork, and 2.3 kg of feed is required to produce 1 kg of chicken (Kumamoto Pref. 2014). Because of the above-described population growth, if we take into account the increase in meat production, it can be calculated that if twice as much grain is not produced in 2050 than was produced in 2000, production will not cover the demand for grain or meat. However, grain production has reached its technological limits in terms of yield per unit area, and the availability of arable land has also reached its limit; the only other alternatives are to produce grain by using special technologies or genetically modified strains (Kumamoto Pref. 2014).

7.1.2 Selection of Target Species Suitable for Land-Based Aquaculture in Japan

The costs associated with conventional aquaculture are generally lower than those incurred by land-based aquaculture; this includes not only the costs of electricity and other utilities, and feed, but also construction and design costs of the aquaculture facility. We considered that if a target species of sufficiently high added value is not selected, then the business will not be viable. In our project, selection criteria included the following: (1) species capable of fast growth (reaches market size in less than 1 year) as this reduces production costs and various associated risks; (2) species exhibits good feed-utilization efficiency (feed conversion ratio of 2.0 or less); (3) species has stable supply of seed throughout the year (especially if specific pathogen-free (SPF) seed are available, disease risk can be minimized), and species is of high value and is widely marketable. Furthermore, if the unit cost of production exceeds 1000 yen/kg, then due to the high cost of energy in Japan, there will be difficulties in making the enterprise viable as a whole.

Table 7.1 Transitions in global aquaculture production and value (FAO 2011)

	1990	1995	2000	2005	2009	2009/1990	Growth rate (/year)
Production volume (tons)	16,840,078	31,232,447	41,723,758	57,825,241	73,044,604	4.34	8%
Production value (US $1000)	27,167,197	44,126,958	52,899,513	72,995,975	110,149,041	4.05	8%
Unit price ($/kg)	1.61	1.41	1.27	1.26	1.51		

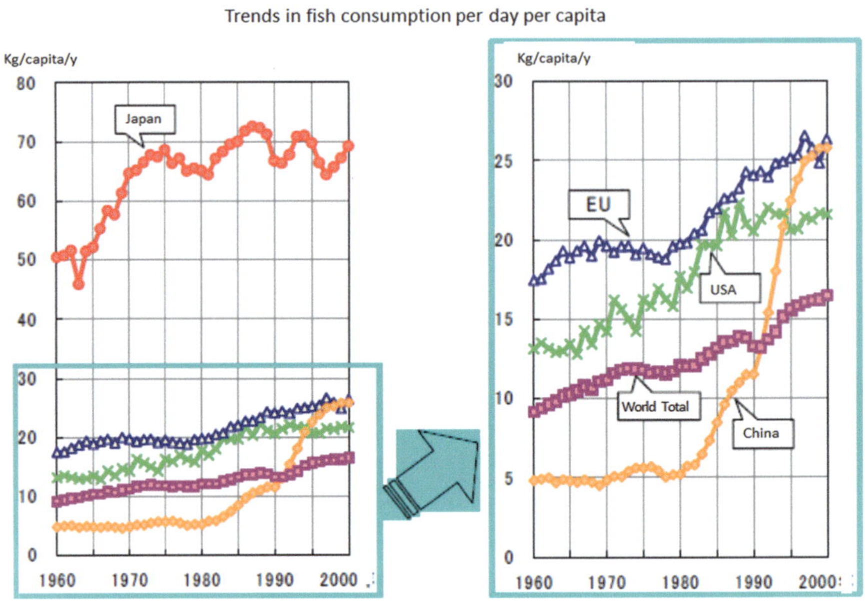

Fig. 7.1 Seafood supply and demand globally and in Japan (Fisheries Agency Japan 2007)

7.1.3 Why Were Shrimp Selected?

We selected shrimp as the target species in our recirculating aquaculture system based on the criteria described above. Nearly 280,000 tons of shrimp are consumed per year in Japan, of which nearly 90% is imported (including processed goods) and the remainder is supplied by domestic conventional fishing activity of species that cannot be cultured (Fisheries Agency 2012). Moreover, shrimp is of high value (especially processed goods have added value) and sells well irrespective of season. In addition, because Japan relies on imports for such a large portion of its total consumption, if shrimp culture were to be more actively promoted, competition with existing fishermen would be minimal. Therefore, shrimp production can help to improve Japan's current level of food self-sufficiency. Worldwide, shrimp farming accounted for 13% of the total value of all aquaculture production in 2009 and has now become a major industry (Table 7.2) (FAO 2011).

The system that we have developed for shrimp aquaculture is unique throughout the world, although the species we are targeting is *Litopenaeus vannamei* (Pacific white shrimp) (Fig. 7.2), which is native to Ecuador and other countries in middle and South America. Currently, SPF shrimp seed for *L. vannamei* are available for five types of viral pathogen that have been recognized to have spread worldwide. *L. vannamei* is the only species of shrimp for which seed are available free of all of the major pathogens throughout the year (Moss and Arce 2003). This is one of the reasons for which we have chosen this species for our technology development.

Table 7.2 Trends in shrimp farming (FAO 2011)

	1990	1995	2000	2005	2009	2009/1990	
Production volume	680,255	928,281	1,136,953	2,667,614	3,004,802	4.42	8%
Production value	4,224,209	6,055,871	7,161,168	10,430,824	14,647,123	3.47	7%
Unit price	6.21	6.52	6.30	3.91	4.87		

Fig. 7.2 White shrimp (*Litopenaeus vannamei*) ready for the market

Furthermore, we have also confirmed that *L. vannamei* can be raised on low-protein diets through consortium-based research on optimal protein and fatty acid levels (Wilder et al. 2009). In brief, we established that lowering protein content to a level of 38% in the feed does not sacrifice growth rates at various life stages, and at this level of protein, fatty acid levels can be set to 7%. Indeed, it is common knowledge that *L. vannamei* does well on low-protein diets which utilize higher proportions of vegetable-based protein compared to other economically important shrimp species. In this way, we have also been able to lower costs associated with feed in commercial production.

Because *L. vannamei* usually lives in estuaries (low-salinity water), it can be cultured in either saltwater or freshwater; culture is therefore possible even in inland areas where saltwater is not available. The growth of this species is faster than that of other shrimp; in our system, seed shrimp (0.002 g) reach market size of about 15–18 g in about 4 months. Unlike *Marsupenaeus japonicus* and *Penaeus monodon*, *L. vannamei* does not bury itself in the sand and can grow well while swimming, as do fish. Therefore, in the case of *L. vannamei* culture, it is not necessary to provide sand at the bottom of the tank, and therefore we do not encounter the problem of feces and uneaten feed becoming mixed in with sand. This simplifies the process of removing solid materials, and water quality management becomes relatively easy, making the species suitable for high-density intensive aquaculture. *Litopenaeus vannamei* now accounts for 80% of all shrimp being produced by aquaculture globally (FAO 2014), but most of it is done by conventional aquaculture, not by recirculating systems.

In theory, closed-cycle systems used in land-based aquaculture can be used for food production with minimal water use. Animals and plants consume certain quantities of water during their rearing period, but although shrimp and fish need to be in a tank with water, they do not actually consume the water. It takes 20,700 L of water to produce 1 kg of beef and 4500 L of water to produce 1 kg of chicken. For wheat production, 2000 L/kg is needed, and there has been a general increase in the amounts of water used in agriculture in general (Oki et al. 2003). Conventional outdoor pond systems used for shrimp aquaculture in Southeast Asia generally require 10,000 L/kg. In contrast, the indoor-type shrimp production system that we have developed requires only 315 L/kg, thus enabling food production with a minimum amount of water (Nohara 2009).

7.1.4 Development Strategy (Industry-Government Collaboration)

We developed a unique recirculating aquaculture system for shrimp culture through industry-government collaboration (sponsored by a research grant from Japan's Bio-oriented Technology Research Advancement Institution: BRAIN) and considering the pros and cons of other types of recirculating systems in the United States and Europe. We initially set a target to meet the need for mass consumption of shrimp in Japan, where the rate of shrimp self-sufficiency is less than 10%, by developing a domestic production technology that will also help to establish a safe food supply. The system is also designed for adaptation to Japan's energy situation. By realizing high-density, intensive production of *L. vannamei* with the added benefit of short production cycles, it is possible to maximize the use of energy. In addition, the system uses a simplified manual for feeding methodology, with the use of a dedicated feed that is high in vegetable protein and low in environmental impact, thereby being suitable for high-density intensive aquaculture. Finally, we have demonstrated that this system reduces the stress of animals under high-density rearing as described in the research results section below. Our ultimate goal is to achieve production of *L. vannamei* at the world's highest density (10 kg/m^3) for a full year and to thus to promote further stability of the business model.

The overall research was coordinated by the Japan International Research Center for Agricultural Sciences (JIRCAS). IMT Engineering Inc. was responsible for the design and engineering of the system, to reflect the physiological parameters required by *L. vannamei* in a closed system, including dissolved oxygen, and salinity and calcium levels as elucidated by basic research by all of the project partners. The National Research Institute of Aquaculture developed methods for evaluating and reducing stress in shrimp in the system. Higashimaru Co. Ltd. developed an optimized feed for closed systems (Fig. 7.3). In a second research grant, JIRCAS, IMT Engineering Inc., and Marinetech Co. Ltd. explored methodology for producing *L. vannamei* seed domestically, including the development of preliminary maturation technology.

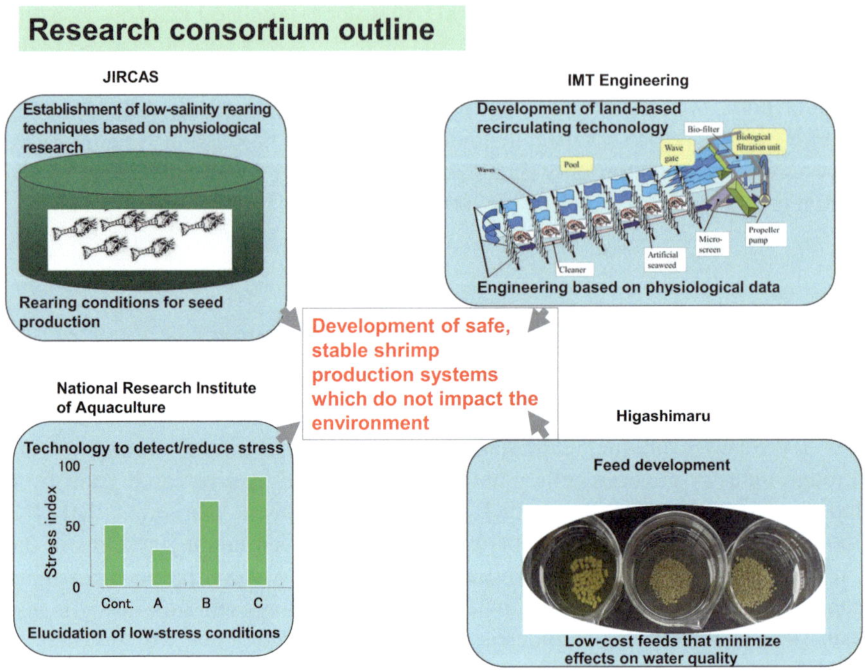

Fig. 7.3 Development strategy for shrimp culture system based on industry-government collaboration showing the roles of the participating organizations

7.2 Main Achievements

7.2.1 *Osmoregulatory Mechanisms in* L. vannamei

Examination of the osmoregulatory mechanisms in *L. vannamei* revealed that rearing in low-salinity water (salinity 5 ppt, hardness 1400 ppm) constitutes the most suitable and economic conditions for shrimp culture starting from an early stage (Fig. 7.4). For acclimation to this level of low salinity, experiments revealed that more than 1 day is required (Jayasankar et al. 2009).

7.2.2 *Reproductive Mechanisms in* L. vannamei

As part of our examination of reproductive mechanisms in *L. vannamei*, a detailed analysis of peptides derived from the sinus gland in the eyestalk revealed that six out of seven peptides had vitellogenesis-inhibitory activity (Fig. 7.5). From this point on, we succeeded in identifying vitellogenesis-inhibiting hormone (VIH) (Tsutsui et al. 2007), and JIRCAS has been working on developing a technology

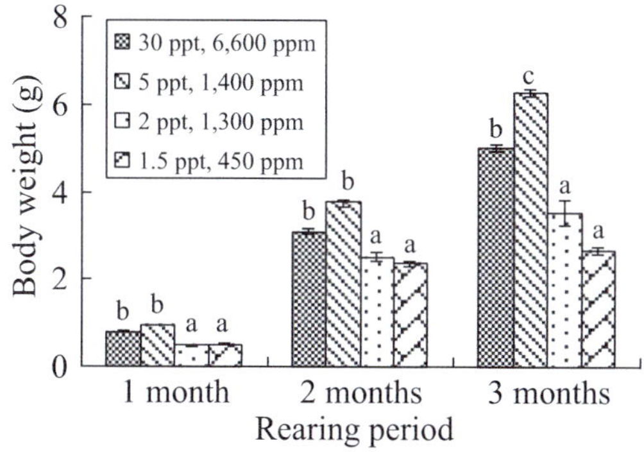

Fig. 7.4 Increase in body weight of *Litopenaeus vannamei* (starting at 0.014 g) reared in different salinities and hardness levels for 3 months (Jayasankar et al. 2009)

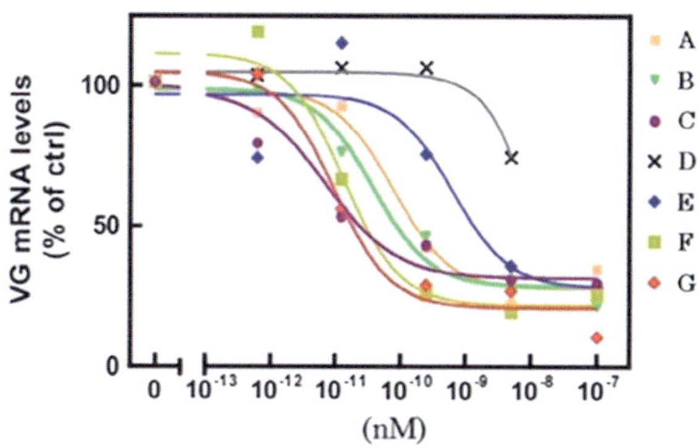

Fig. 7.5 Vitellogenesis-inhibitory activity of peptides (A–G) derived from the sinus glands of *L. vannamei* eyestalks (Tsutsui et al. 2007)

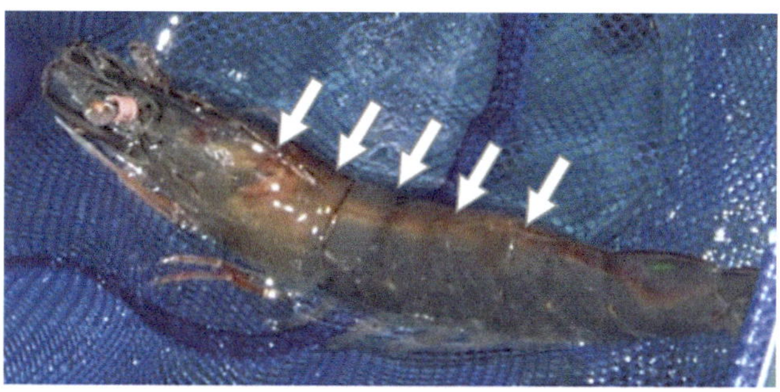

Fig. 7.6 Parent shrimp used for artificial insemination (arrows indicate mature ovaries)

for using this hormone as part of a means of increasing the efficiency of female reproduction in captivity. In addition, in order to help stabilize domestic shrimp production in the long term, we have been attempting to develop seed production technology for the domestic market and have succeeded in inducing the maturation of parent shrimp (Fig. 7.6).

7.2.3 Optimum Water Temperature and Oxygen Consumption Levels

For purposes of optimizing costs in our shrimp-production plant, we determined the optimum water temperature at each growth stage of *L. vannamei*, as well as oxygen consumption levels under high-density intensive water recirculating (Fig. 7.7: oxygen requirement is three times that of other prawn species). We also optimized flow rate and established methodology to control water quality. As a result, we were able to set up the "Indoor Shrimp Production System (ISPS)" and obtain patents for the relevant technology for operating this system. A schematic diagram of the ISPS is shown in Fig. 7.8.

7.2.4 Unique Equipment

The ISPS consists of unique equipment (wave-generating gate, micro-screen, sediment-removal equipment, oxygen mixer, artificial seaweed, low-head and high-flow cycle pump, and four-armed spoon net) that we developed under this project and is included in our commercial-scale production plant that was established in Niigata Prefecture. The plant achieves a 75% final survival rate, thereby realizing a density of 9.43 kg/m^3. We have also compiled operational manuals that can be easily employed by plant personnel with a minimum of training.

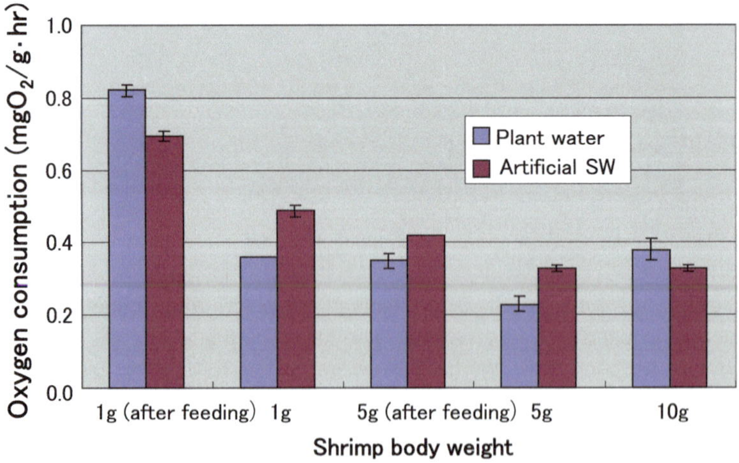

Fig. 7.7 Oxygen consumption in *L. vannamei* (0.4–0.5 mg/g·h)

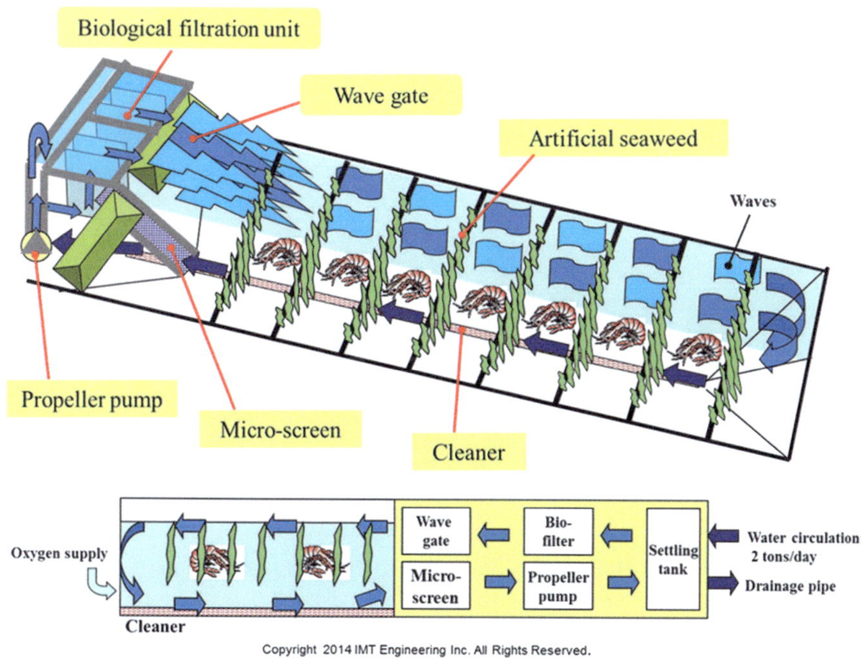

Fig. 7.8 Schematic diagram of the patented "Indoor Shrimp Production System (ISPS)"

7.2.5 *Evaluation of Stress*

We have also evaluated whether shrimp experience stress under this system using markers that indicate ability to cope with disease. In more detail, by applying stress such as reduced levels of dissolved oxygen levels, increased concentration of ammonia concentration, and subjection to fasting or handling, it is possible to ascertain the stress response of the shrimp based on the expression levels of immune defense-related genes (Wilder et al. 2009). We therefore investigated the relationship between stress index and stocking density in laboratory tests or actual rearing in the commercial plant. We found that even if shrimp were cultured at the highest target densities, if water quality was maintained adequately, then the stress indices could be kept within an appropriate range (Fig. 7.9).

7.2.6 *Basic Nutritional Requirements of* L. vannamei

We elucidated the basic nutritional requirements of *L. vannamei*, and taking into consideration the rearing environment in low-salinity water, we determined the basic feed composition for culture of this species including calcium and phosphorus

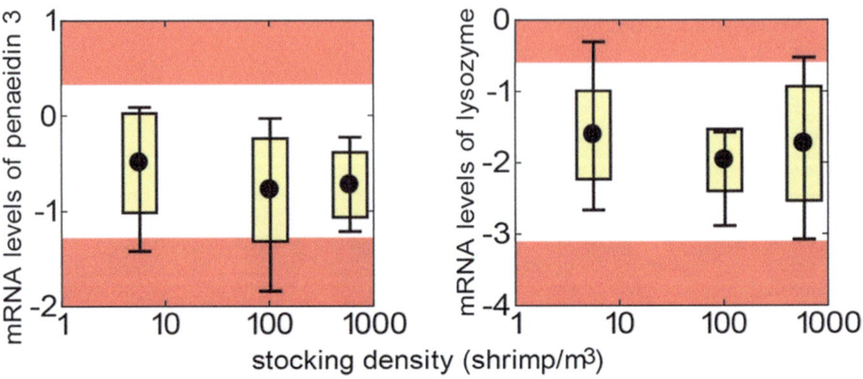

Fig. 7.9 Stress indices and rearing densities of shrimp grown in the commercial plant (black circles, average values; yellow squares, standard deviation; bars, range). Red bands indicate areas that fall outside the appropriate range of stress to maintain shrimp in a healthy condition under commercial production (Logarithmic scale displays values relative to beta-actin expression levels)

requirements (Wilder et al. 2009). In addition, we examined protein combinations in the basic feed and were able to decrease feed costs by increasing the amount of plant-based protein over that of animal-based protein. We therefore established an economical composition that yielded a thick-textured feed. We were also able to improve the underwater shape and retention of the feed by using appropriate binders, and thus the stability of the feed, allowing us to prevent degradation of the feed and maintain good water quality.

7.2.7 *Summary of Development Strategy*

In conclusion, having studied the examples of others engaging in recirculating shrimp culture, we amalgamated our results to produce a complete system unique throughout the world. The machinery component of the ISPS technology was patented in 2007, and in 2010 we acquired a patent for the "soft technology" operation of the system pertaining to water quality and health management. This research was made possible by research funding supplied for 8 years by BRAIN.

By integrating the knowledge of each partner in the research consortium, we were able to create a full technology package for producing shrimp on a commercial scale. Our first commercial plant was built in Myoko City, Niigata Prefecture, and commercial operations commenced in September 2007 (Figs. 7.10 and 7.11). Since December 2009, the product has been sold mainly in the Kanto region and locally under the trade name Myoko Snow Shrimp®.

Fig. 7.10 Exterior view of the commercial ISPS plant in Myoko City, Niigata Prefecture

Fig. 7.11 View of the inside of the ISPS plant showing the two 600-ton grow-out pools

7.3 Features of the Development System

The ISPS plant currently in operation at Niigata is comprised of four 20-ton nursery tanks that are used for the first month of rearing and two 600-ton grow-out pools used to bring shrimp to market size. By using these facilities in rotation, it is possible to achieve six to seven production cycles per year, yielding a total production volume of 24–40 tons of shrimp. A detailed technical description of the various components of the plant is given below.

7.3.1 ISPS Technology Components

Water Circulation Technology

Up until the present, horizontal water circulation systems have been mostly used in conventional land-based aquaculture, and the associated costs of electricity in operating pumps found in this type of system is generally very high. This has made it difficult to use such systems in countries like Japan where energy costs are high. We therefore developed a vertical pumping system that agitates the water by using wave force and exhibits only 1/10 the energy usage of conventional land aquaculture (Fig. 7.12). Moreover, the agitation and water cycling control the environment in the pools to be uniform (Fig. 7.13).

Water circulation is achieved by raising the water to be fed into the system to a height of 1 m by the vertical pump, where it is briefly stored in the tank of the wave-generating apparatus (Fig. 7.14). The water gate that opens into the grow-out pool is opened once per minute, generating a wave. The wave travels across the surface of the pool, hits opposite side of the pool, and travels back along the bottom of the

Fig. 7.12 Vertical pump with capacity of 250 tons water/hour

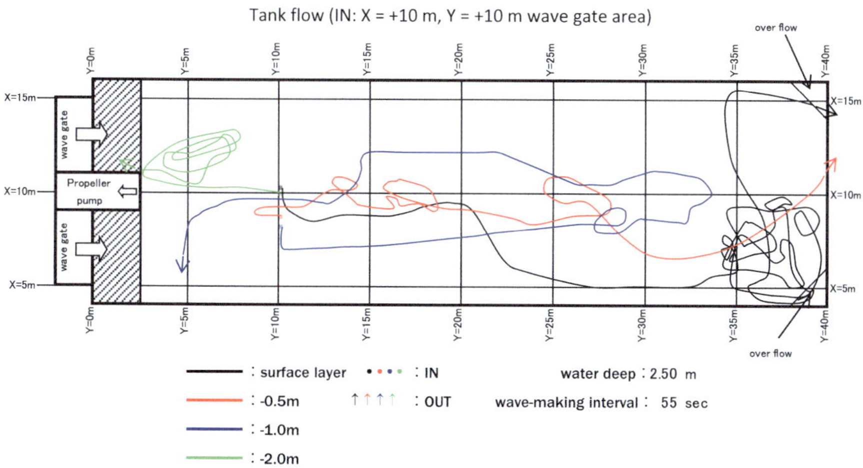

Fig. 7.13 Flow of water in the grow-out pool

Fig. 7.14 Wave-generating apparatus

pool, causing the water to be recirculated. This action also makes the artificial seaweed sway and stirs the rearing water, thus causing the quality of the water to be homogeneous throughout the pool. Moreover, through the use of a timer, the opening and closing of the wave-generating apparatus can be adjusted, thus providing the shrimp with optimal wave strength. When the water is released from the wave-generating tank, a waterfall-like effect is brought about, such that excess carbon dioxide in the water is dissipated into the air.

The actual production pools used in the ISPS plant in Myoko are 40 m in length and 12 m width and are built to scale according to the schematic diagram shown in Fig. 7.15. Both pools (A and B) are equipped with a wave-generating apparatus and

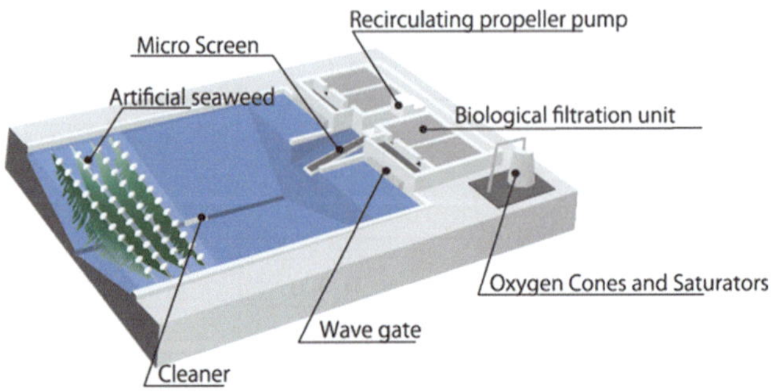

Fig. 7.15 Plan model of the Indoor Shrimp Production System

a vertical pump. The water is deepest in the center of the pool (2.5 m). As shown in Fig. 7.13, there were four observation levels for water flow analysis: the surface layer, 0.5 m, 1.0 m, and 2.0 m depth ("—" symbol is used to indicate depth, e.g., below water surface).

In further detail, the following observations were made regarding water flow in the grow-out pools. First of all, parameters such as wave-generating intervals, water level, and quantity of water outflow could be controlled via the wave-generating apparatus. However, while pools A and B had the same basic structure, it was noted that there were subtle differences in slope. Thus, slight differences in water level in the pool appeared to have arisen from differences in the flow in each aquarium. Furthermore, (1) a rise in water level occurred with the return wave; when the next wave was generated, because the water level was higher, the wave was smaller and flow was halved; (2) the flow rate tended to accelerate when the water was shallow on both sides of the pool; (3) the optimum water level was 2.50 m and the optimum wave-making interval was 55 s; (4) if the wave-generating interval in the pool was 55 s, the first wave returned in about 25 s and the second in 35 s; (5) under these conditions, active flow was observed not only in the surface layer but also (depending on the direction) in the waves in the 0.5 m layer; (6) from the differences in observations between pools A and B, we also found that flow changes depended on the water level at that time; (7) large flows did not occur when the water level was excessively high; and (8) the motion became unvarying when the water level was too low; this phenomenon was governed by the left and right movements of the layers.

Finally, by changing the water level and adjusting the extent to which the wave-generating gate is opened, the flow rate within the pools could be modified, and this enabled us to adjust the flow rate to the shrimps' growth as well as exercise capacity. This created an optimum feeding environment for the animals being reared. In this way, if water flow is present in the rearing tanks, shrimp attempt to swim against the current; thus, the rearing tanks can be utilized in a complete three-dimensional capacity. This leads to greater rearing density and increased production capacity.

Prevention of Cannibalism and Removal of Suspended Materials by Using Artificial Seaweed and Screen Filters

Materials suspended in the water are removed by passage through a screen filter with an 800-μm flow-through mesh (Fig. 7.16). In addition, the placement of artificial seaweed in the grow-out pool provides shelter for shrimp that show reduced exercise capacity after molting. This helps to prevent cannibalism and thus improves survival rates. Artificial seaweed absorbs suspended materials and cleans up the water as a type of biological filtration (Fig. 7.17).

Fig. 7.16 Screen filter

Fig. 7.17 Artificial seaweed

Use of Probiotics to Stabilize Water Quality

Floating objects (800 m^2/m^3) made from polyethylene (PE) are floated inside the filtration tank, thereby increasing the contact time with the water being recirculated and achieving effective biofiltration. We used commercial probiotic kits and regularly inserted nitrifying bacteria into the system. This assisted in keeping the water quality stable.

Water Pollution Load Reduction by Use of a Sediment Recovery Unit and Calculation of Leftover Feed Quantities

Because the pool has an inverted triangle shape (Fig. 7.18), sediment (leftover feed, feces, dead shrimp, and molting exuvia) naturally collects in an elongated pit at the bottom of the pool. On every occasion before feeding the shrimp, any sediment present in the pit is scraped off using the automatic cleaner (Fig. 7.19) and is discharged as solid material by air lift. This eliminates the occurrence of sludge,

Fig. 7.18 Cross section of the production pool showing the suspension of artificial saltwater

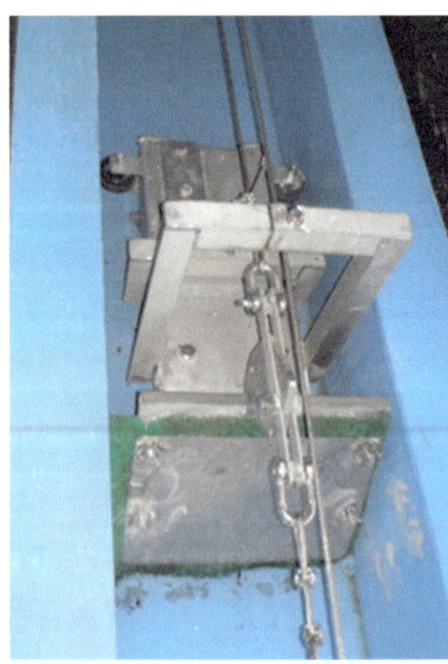

Fig. 7.19 Automatic cleaner being used to remove sediment

which adversely affects water quality in the pool. By using this device, we can always determine the amount of leftover feed and can adjust feeding quantities accordingly. We can also obtain raw materials for making fertilizer which can be recycled for used in agriculture and animal husbandry.

Supply of Oxygen to Pools

In order to supply oxygen to the pools, pressurized air is separated into oxygen and nitrogen and the nitrogen is adsorbed onto zeolite. Remaining pure oxygen is extracted, using a special apparatus, an "oxygen cone" (Fig. 7.20). We have also installed an oxygen mixer that efficiently produces supersaturated oxygenated water containing oxygen at 24 ppm by applying pressure inside a vessel (water normally contains oxygen at a level of 7–8 ppm.) By injecting the super-oxygenated water from four locations into the pool, we can achieve a uniform oxygen environment within the pool. *L. vannamei* is one of the most actively swimming shrimp species and requires about three times the oxygen of other prawns. Therefore, blowers and water mills used in conventional shrimp aquaculture generate insufficient oxygen for high-density *L. vannamei* culture.

Fig. 7.20 Oxygen cone

Operational Manuals for Training and Health Management
Conventional aquaculture conducted in coastal areas is often affected by various uncontrollable factors such as typhoons, occurrence of red tide, and decreases in water temperature. Because such parameters affect the growth and survival, rearing is quite difficult for operators with no previous training in aquaculture techniques. We have compiled a series of manuals for operating the ISPS technology, based on demonstration of operational performance and basic research for a total of more than 10 years. Therefore, even a person without prior experience can operate an ISPS plant and produce shrimp through the use of these manuals.

The following manuals constitute the total operational package for the ISPS technology:

1. Introduction of seed into the facility and first-stage acclimation (methods of adaptation to freshwater from saltwater)
2. Basic training procedures (day-to-day operations, feeding, water quality standards)
3. Harvesting, farming, and shipping procedures
4. Measurement of water quality, mineral content, and hardness of the pool water
5. Bacterial counting procedures for rearing water
6. Feeding methodologies and regimens
7. Production schedule
8. User's manual for plant equipment
9. Maintenance procedures for plant equipment
10. Troubleshooting manual for plant equipment
11. Recording sheets for all pertinent data
12. Hygiene management
13. Daily maintenance check sheets

7.3.2 Learning Curve: Problems Encountered during the Research Development Phase

Management of Water Quality
There was often the increased risk of ammonia and nitrite levels to become elevated in the grow-out pools in situations where the breakdown of nitrogenous compounds would be compromised. In initial endeavors, we employed a fixed honeycomb filter medium in the biofilter component of the system, but the material often clogged through overuse, halving the nitrification capacity of the system. We also had to clean the filter once every production cycle, which was very time-consuming, requiring a large number of personnel and increasing operating costs (Nohara 2012). Therefore, for 2 years we ran an experiment to compare various filter media, and we have now changed to a floating system that uses a

floating carrier within the biofilter tank. This greatly increases nitrification capacity, and the filter medium does not require frequent cleaning, thus reducing operating costs.

Light Conditions
Because we initially operated the system disregarding light conditions, as would be the case in conventional outdoor pond aquaculture, we had not considered the optimization of the light environment to be critical. However, lighting experiments in which we changed the illuminance and the light source revealed significant effects of light conditions on survival and growth rates. Multi-halogen lamps were found to be the best type of light source, and to obtain optimal results, illuminance should be less than 150 lx, and light should be received for 12 h daily (Nohara et al. 2012a, b).

Mineral Balance
Regarding the importance of mineral balance in near-freshwater rearing, *L. vannamei* has attracted attention as a target species of aquaculture because it can be acclimated to either saltwater or near-freshwater conditions. In a manual published by the Harbor Branch Oceanographic Institute in Florida (Wyk et al. 1999), it is stated that the shrimp can be reared in freshwater. However, in many countries, freshwater is often hard, unlike the soft water of Japan. This difference has a marked impact on shrimp rearing, as minerals such as calcium that are present in hard water are indispensable to shrimp rearing.

The question arises, why do shrimp need minerals during rearing? Adult or near market size *L. vannamei* molt about once every 2 weeks, and smaller animals molt much more frequently. Calcium is deposited in the carapace of crustaceans including shrimp in the form of calcium carbonate and allows the shell to become hardened after molting. Calcium is normally obtained from the rearing water and feed. If calcium, and to a certain extent, magnesium, are lacking, the animal may fail to develop a healthy carapace after molting. This can especially lead to "black spot disease," where the animal develops melanized lesions all of its body, for example, after injury such as scraping against the sides of the rearing facilities (Wyk et al. 1999). This leads to decreased market value and lower growth rates during culture. We experienced such problems firsthand in the early days of our research development. Thus, we adapted the rearing water in the pools to be of low salinity (5 ppt) and high hardness (1400 ppm). However, solving this problem involved more than just adjusting the water hardness: the balance of all principle elements, e.g., calcium, magnesium, potassium, and sodium was highly important. If this balance was not optimal, even if the hardness was maintained at high levels, molting failure could occur. Experimental work relating to water constitution is described in Jayasankar et al. (2009).

7.4 Challenges Relating to Business Promotion

7.4.1 High Cost of Production

In the case of production in cold climates such as that of Niigata Prefecture, heating costs have a direct effect on production costs. Therefore, we initially planned to use excess heat released from an adjacent waste incineration plant that was scheduled to be built. However, these plans did not materialize, so we are currently using our own gas boiler. This accounts for 23% of operating costs and has thus pushed up the unit cost of production. In addition, because this is a land-based aquaculture facility and the scale of production is limited, personnel costs are relatively high. A rearing water temperature of 28 °C is necessary for this species of shrimp, so in future business endeavors in Japan, cost-effective production is expected to require the use of natural heating sources, such as from hot springs, excess heat from waste incineration plants, biomass power generation, and the like. If a low-cost heat source can be tied to shrimp production, business activity will improve markedly.

7.4.2 Finding Steady Customers

We originally thought that a production volume of about 30 tons/year could easily be consumed locally. However, this point was a much greater hurdle than we had expected, owing to our lack of experience in the food industry. The price of our shrimp was very expensive, being double that of imported frozen shrimp. We were not able to make the consumer aware that the ISPS technology constitutes a safe and biosecure production system that does not use any chemicals or additives during the rearing process. Therefore, at first, we were therefore not able to sell sufficient quantities of the product locally. However, through the use of frequent advertising in the mass media, including on television shopping, we are now selling 60% of the product in Niigata Prefecture. The remaining 40% is being sold mainly to restaurants in the Tokyo metropolitan area.

7.4.3 Current Efforts

L. vannamei as a species is becoming well known in Japan and is advertised frequently in the mass media, so we consider this to be a timely opportunity to expand sales further. We are collaborating with local producers of food goods in Niigata Prefecture to produce not only conventional products, e.g., fresh or frozen whole shrimp, but also value added products such as "tsukudani" and ready-to-eat shrimp curry. It is hoped that such products will become famous as local specialties. The outbreak of EMS that was first observed in 2009 in China and Vietnam has spread to other major shrimp-producing countries throughout Southeast Asia (Hirono 2014). In Thailand, it severely impacted the country's shrimp production volume, especially in 2013. Because of this situation, prices of imported frozen shrimp in Japan have soared, and the price difference between our product and frozen imported products is becoming smaller. In addition, there have been a number of food-related scandals if late in Japan, most famously the incident in October 2013, in which several major hotel chains and restaurants featured the high-priced, naturally fished "shiba-ebi" (*Metapenaeus joyneri*) on their menus, but actually used inexpensive frozen imported *L. vannamei* (perhaps up to 90% of the shiba-ebi served in Japan was actually *L. vannamei*). This scandal did not present any problems with food safety, but it was a form of cheating the consumer and making excess profits. However, the flip side is that the Japanese public has learned that *L. vannamei* can actually be quite tasty.

7.4.4 Ideal Business Model

Currently, the commercial ISPS plant in Myoko consists of two 600-ton rearing pools, but in order to supply full shipments of shrimp year-round, it would be preferable to have at least six rearing pools of 750 tons as an ideal business unit. This would make it possible to ship shrimp all the time and would make operations more efficient, including costs associated with personnel. In such a situation, the unit cost of production would be lowered from 2300 yen/kg to 2153 yen/kg, becoming more competitive with the retail cost of 1700 yen for frozen imported shrimp. However, even at such a level, if the retail price does not exceed 3500 yen/kg, no profit will be possible, and our market research indicates it would be difficult to sell all of the product at this level. We have done several exercises to calculate the cost reduction achieved by expanding the production scale (Table 7.3). Increasing the capacity of the theoretical 3-pool plant to 30 × 750-ton grow-out pools would further reduce the unit cost of production to 1300 yen/kg. In this case, the selling price could be set at 2500 yen and yield a sufficiently high profit, such that the ISPS technology would be attractive to potential investors. Therefore, along the lines of the above, in land-based aquaculture, if production capacity for the targeted aquatic species does not realize economy of scale, then the enterprise will not be profitable and is not likely to continue. However, the benefits of land-based aquaculture, not only for shrimp but also for various fish species, in terms of environmental preservation and stimulating the local economy, are indeed significant. It is our hope that such production systems can be made more viable by recognizing the important of not only having good technology but also a good sense of business management.

Table 7.3 Economic analysis of ISPS business employing six grow-out pools

System scale and breakdown of costs	Grow-out pools (750 tons)	6 units
	Nursery tanks (35 tons)	4 units
Annual production volume	20 production cycles using 750-ton pools in 1 year	80 tons
Item	**Cost yen/per year**	**% of total cost**
Shrimp seed	31,250,000	18.0%
Feed	34,020,000	20.0%
Other materials	15,250,000	9.0%
Personnel	21,600,000	12.5%
Electricity	26,500,000	15.5%
Heating	27,600,000	16.0%
Other (packing, business matters)	8,400,000	5.0%
Maintenance	7,500,000	4.0%
Total operating expenses	172,120,000	
Unit production cost	2152 yen/kg	

7.5 Global Developments

Shrimp culture has been a major worldwide industry for over 20 years, but land-based recirculating systems are very recent, and most of them have been experimental up until now. At present, the world's closed recirculating systems for shrimp aquaculture can be categorized into three main categories: 1) raceway system developed by the Harbor Branch Oceanographic Institution in Florida (Wyk et al. 1999; Fig. 7.21) and also being developed by a number of other workers, such as Dr. Addison Lawrence (Lawrence 2010); 2) biofloc systems such as described by Dr. Yoram Avnimelech of Israel (Avnimelech 2009), and our ISPS technology (Kang and Nohara 2013). The raceway system was tested dating from nearly 20 years ago in the United States but had not been used commercially. However, since around 2011, large-scale systems, such as that belonging to the company Blue Oasis near Las Vegas which aims to supply shrimp to the hotel industry, are being implemented; however, the details of the technology and actual business are difficult to obtain.

Regarding the biofloc systems, in particular, they are being developed with the support of the South Korean government. Commercial-scale plants began operating at two locations last year. However, these plants are difficult to operate without personnel who have advanced knowledge and previous experience (Kang and Nohara 2013). Representative examples are shown in the text (Figs. 7.22 and 7.23).

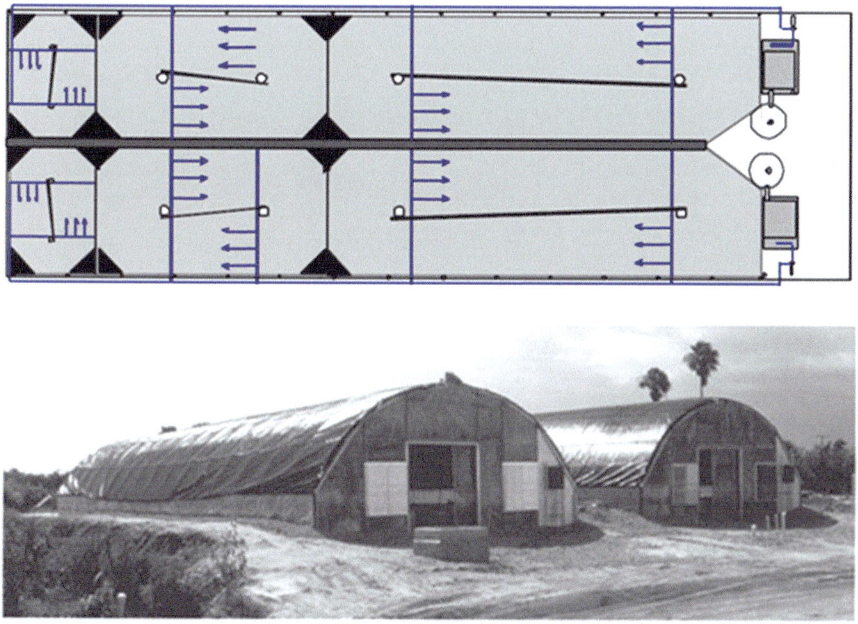

Fig. 7.21 Shrimp production using a raceway system (Wyk et al. 1999) and schematic diagram of water flow in the system

Fig. 7.22 Biofloc system in South Korea (Reproduced from Kang and Nohara 2013)

Fig. 7.23 Experimental shrimp-rearing facilities at the South Korea National Research Institute of Fisheries Science (Reproduced from Kang and Nohara 2013)

7.6 Integrated Approaches to Developing Land-Based Aquaculture

In order to further promote land-based aquaculture in Japan, it is often considered that amalgamating production facilities with restaurants and exhibition halls, as a way of also drawing in tourists, would be highly effective. We are also considering such schemes, but it is also worthwhile to couple shrimp production with that of other salable agricultural goods. Here, we introduce aquaponics as an example. A detailed explanation of aquaponics is given in Chap. 11.

7.6.1 Approaches to Aquaponics in Japan

Between 2003 and 2005, IMT Engineering Inc. attempted aquaponics trials using the wastewater from shrimp culture at an experimental facility located in Tsukuba City (this facility was the predecessor to the current commercial plant in Myoko). The crops that were attempted were water spinach and watercress (Nohara 2014). In the experimental facility, a 1200-ton grow-out pool was in operation, and we set up a very simple hydroponic trough (50 m^2) having a moderate gradient and measuring 2.5 m × 20 m with a depth of 10 cm (Figs. 7.24, 7.25, and 7.26). In the first trial, watercress was grown with no major problems, although the nitrogen removal rate was not as high as expected. Thus, in the second trial, a water recirculation system

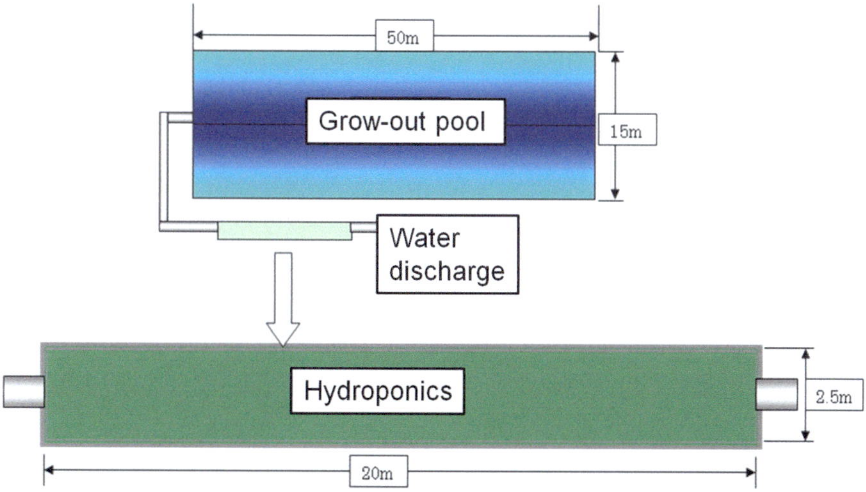

Fig. 7.24 Schematic diagram of IMT Engineering's experimental aquaponics system in Tsukuba

Fig. 7.25 Outside view of the hydroponic trough in Tsukuba

Fig. 7.26 Water spinach in the hydroponic trough

was added to the hydroponic trough. This increased the nitrogen removal rate from 30% to a maximum of 53%.

7.7 Summary

Research and development on land-based shrimp culture of shrimp is advancing rapidly throughout the world, and the past couple of years have seen a great deal of investment in other countries. However, there have not been many workers involved in the field in Japan for the following reasons: (1) a support structure for promoting land-based aquaculture has not been established in this country; (2) energy costs (electricity, heating) are high; (3) the market does not yet consistently value safe and secure aquaculture products; and (4) equipment and materials needed for advanced land-based aquaculture are expensive and difficult to obtain.

However, since the Great East Japan Earthquake of 2011, the domestic situation has changed greatly. In particular, the public's faith in the safety of seafoods obtained from natural sources is declining. Nevertheless, while aquatic foods produced by land-based aquaculture are safe and tasty, they are still more expensive than fish from the sea, but people willing to purchase them are beginning to increase in number. If government support could be obtained as a component of regional development, then land-based aquaculture of shrimp should become a promising business, especially if costs can be reduced through, for example, the effective use of underutilized energy sources.

Acknowledgments We thank the Bio-oriented Technology Research Advancement Institution (BRAIN) of Japan's National Agriculture and Food Research Organization for supporting this research from FY 2004–2011. We thank Dr. Takuji Okumura of the National Research Institute of Aquaculture of the Japan Fisheries Research and Education Agency for collaboration under the BRAIN project and assistance with Fig. 7.9 in this text.

References

Avnimelech Y (2009) Biofloc technology: a practical guide book. The World Aquaculture Society, Baton Rouge. 182 pp

FAO (2011) Global aquaculture production. http://www.fao.org/fishery/statistics/global-aquaculture-production/enFishery Statistics

FAO (2014) Yearbooks of Fishery Statistics Summary Tables. ftp://ftp.fao.org/fi/stat/summary/default.htm

Fisheries Agency (Japan) (2007) White paper on the fisheries industry (FY 2007), published by the Government of Japan, 123 pp. (In Japanese)

Fisheries Agency (Japan) (2012) Fisheries production/importation statistics for FY 2011, http://www.jfa.maff.go.jp/j/kikaku/toukei/index.html. (In Japanese)

Hirono I (2014) The latest on EMS/AHPND in whiteleg shrimp. Aqua Culture Business, Feb. 2014 issue, pp 3–5. (In Japanese)

Jayasankar V, Jasmani S, Nomura T, Nohara S, Wilder MN (2009) Low salinity rearing of the Pacific white shrimp *Litopenaeus vannamei*: acclimation, survival and growth of postlarvae and juveniles. Japan Agricultural Research Quarterly (JARQ) 43:345–350

Kang BJ, Nohara S (2013) Land-based recirculating aquaculture in South Korea: The latest trends for eel, Pacific white shrimp, and puffer fish. Youshoku (Aquaculture) Business, July 2013: 34–38. (In Japanese)

Kumamoto Pref (2014) Chikusan-hiroba (Kumamoto Prefecture's Animal Husbandry Forum). http://kumamoto.lin.gr.jp/i-menu.html. (In Japanese)

Lawrence A (2010) System and method for super-intensive shrimp production system. US Patent Application, submitted May 7, 2010; US20100294202 A1

Moss SM, Arce SM (2003) SPF defined: Pathogen-free status of shrimp limited. Global Aquaculture Advocate, Dec. 2003 issue: p. 86

Nohara S (2009) "The Blue Revolution" and a role for land-based recirculating aquaculture: Japan's first Indoor Shrimp Production System. Suisankai (Fisheries World), Oct. 2009: 17–20. (In Japanese)

Nohara S (2012) The latest trends in land aquaculture technology of vannamei. Agric For Fish Technol Res J 35:29–34. (In Japanese)

Nohara S (2014) Aquaponics: The latest advances overseas, and status in Japan. Youshoku (Aquaculture) Business, February, 2014: 48–52. (In Japanese)

Nohara S, Iguma K, Yoshino K, Okayasu K (2012a) Light conditions for rearing Litopenaeus vannamei, part 1. The Japanese Society of Fisheries Science Spring Meeting. Tokyo, Japan, March 26–30. Abstract no. 935. (In Japanese)

Nohara S, Iguma K, Yoshino K, Okayasu K (2012b) Light conditions for rearing Litopenaeus vannamei, part 2. The Japanese Society of Fisheries Science Spring Meeting. Tokyo, Japan, March 26–30. Abstract no. 936. (In Japanese)

Oki T, Sato M, Kawamura A, Miyake M, Kanae S, Musiake K (2003)Virtual water trade to Japan and in the world, Virtual Water Trade, Hoekstra, A.Y.. Ed. Proceedings of the International Expert Meeting on Virtual Water Trade, Delft, The Netherlands, 12–13 December 2002, Value of Water Research Report Series No.12: 221–235, February 2003

Tsutsui N, Ohira T, Kawazoe I, Takahashi A, Wilder MN (2007) Purification of sinus gland peptides having vitellogenesis-inhibiting activity from the whiteleg shrimp *Litopenaeus vannamei*. Mar Biotechnol 9:360–369

Wilder MN, Nohara S, Okumura T, Fukuzaki R (2009) Development of safe, stable production systems for whiteleg shrimp, final project report for interdisciplinary research supported by the bio-oriented research advancement institution (BRAIN). Research Results Collection for Projects Completed in FY 2008:17–20. (In Japanese)

Wyk PV, Davis-Hodgkins M, Laramore R, Main KL, Mountain J, Scarpa J (1999) *Farming Marine Shrimp in Recirculating Freshwater Systems*, prepared by the Harbor Branch Oceanographic Institution. Contract No. 4520, Florida Department of Agriculture and Consumer Services

Chapter 8
Abalone *Haliotis* spp.

Yoshikazu Koizumi and Youichi Tsuji

Abstract Abalone is considered as one of the most valuable fishery products elsewhere in the world. However, landings from legal fisheries are declining year by year because of overexploitation, illegal harvesting, and habitat degradation. Thus, abalone farming plays increasingly an important role in maintaining abalone supply. Methods of abalone production vary country by country and are still developing. Although most developed land-based abalone aquaculture by using closed recirculating systems is more advantageous than conventional abalone production, water treatment systems are essentially required to maintain water quality without saltwater replacement. Water quality standards and water treatment technologies for abalone rearing have been acquired through various experiments. The key to succeeding with abalone culture by closed recirculating aquaculture system is avoidance of accumulation of solids, nitrogen compounds, and recalcitrant dissolved organic matters. Moreover, calcium supplement and adjustment of pH and alkalinity by the addition of some chemicals are required as considering chemical interreaction. Reduction of capital expenditure and operating expense, increase of abalone productivity, and sales and marketing management must be intensively promoted in order to disseminate the technology of closed recirculating system.

Keywords Abalone culture • Business strategy • Denitrification • Nitrification • Organic matters • Water quality • Water treatment system

---

Y. Koizumi (✉)
Applied Life Science Research Center, Tamagawa University Research Institute, Machida, Japan

Food Research and Developments Japan Corporation, Saitama, Japan
e-mail: koyoshikazu@agr.tamagawa.ac.jp

Y. Tsuji
Food Research and Developments Japan Corporation, Saitama, Japan

T. Takeuchi (ed.), *Application of Recirculating Aquaculture Systems in Japan*, Fisheries Science Series, https://doi.org/10.1007/978-4-431-56585-7_8

8.1 General Information on Abalone

8.1.1 Species and Geographical Distribution

Abalone is phylogenetically classified into Phylum Mollusca, Class Gastropoda, Order Archaeogastropoda, Family Haliotidae, and Genus *Haliotis*. There are approximately 75 species of abalone in the world oceans, but 20 species are commercially valuable (Uki 1989). Four species are intended as fishery in Japan, *Haliotis discus discus*, *H. discus hannai*, *H. gigantea*, and *H. madaka*.

H. d. hannai and the others are distributed in the northern and southern part of Japan, respectively. In Korea, the measurement of isothermal line is conducted in which water temperature in 25 m depth reaches 12 °C in February (FAO 1990). With the isothermal line as a basis for identification, abalone species distribution can be roughly divided into two parts: northern and southern species. Hara and Fujio (1992) investigated phylogenetic relationship among four types of Japanese abalone by isozyme analysis. As a result, the relationship between *H. gigantea* and the others was subspecies level. The relationship among the other types of abalone was subspecies or local variety level. Furthermore, Hara and Sekino (2005) revealed that *H. d. discus* and *H. d. hannai* were obviously differentiated by using microsatellite DNA markers.

Abalones are found at depth up to 30 m in the intertidal zone, and their distributions are segregated depending on geographic and water depth. The distribution of three southern species of Japanese abalone tends to be separated by water depth. Generally, *H. d. discus* and *H. madaka* dominated shallower than 10 m and deeper than 10 m, respectively. *H. gigantea* evenly spreads between 0 and 20 m (Ino 1952).

They eat a wide range of natural grown seaweeds in their distribution. They prefer to eat mainly the Class Phaeophyceae (brown algae) such as the Order Laminariales including *Laminaria*, *Macrocystis*, and *Eisenia* (also known as kombu, giant kelp, and arame, respectively), the Family Alariaceae including *Undaria pinnatifida* (also known as brown seaweed or wakame), and the Phylum Rhodophyta (red algae). Growth efficiency of the abalone of the above species varies among seaweed type (Uki et al. 1986).

In the world, commercially captured abalone includes *H. rufescens* (red abalone) in west coast of North America, *H. ruber* (black lip) and *H. laevigata* (green lip) in Australia, and *H. midae* in South Africa.

8.1.2 History and Culture

Archaeological excavation of shell mounds have evidenced that abalone has been eaten since initial Jomon period (about 10,000 years ago) in Japan. Since then, abalone is deeply rooted in Japanese culture, trade, cuisine, and history. There are many records regarding digging abalone shells from mounds, exporting dried abalone to China in the Edo era (more than 300 years ago), discovery of abundant abalone habitats, and progresses in development of the instruments of capture and in biological researches.

Abalone as ancient artifacts, fishery, food, and expressions of someone's feelings were dealt with in anthologies of WAKA poems (Nonaka 2011). In Edo era, dried seafood including abalone, sea cucumber, and shark fin was most important for the trade with China. Huge amount of dried abalone between 100 and 200 t (1000–2000 t of equivalent live abalone with yield ratio 10%) was exported at that era (Nonaka 2011). It takes a plenty of time and labor intensive to make Japanese dried abalones. Japanese dried abalones are branded and mostly exported to China (25,849 kg), Taiwan (266 kg), Macau (650 kg), and Singapore (446 kg) even now (Trade Statistics of Japan 2013). In particular, 92% of them are exported to Hong Kong and are dealt with high prices. The cooked dried abalone in Hong Kong is recognized the most expensive and traditional dish as it also takes a few days and special techniques to cook dried abalone. Cooked dried abalones are served with prices 85–1770 US dollars per kilogram (1 US dollar equals to 7.8 HK dollar, as of June 2019) depending on size ranging 20–120 g (dried weight) in Hong Kong (personal communications). After the Meiji Restoration (1968), new plentiful habitats of abalone in deeper depth than before have been found with developing diving equipment and hydroscope, resulting in overexploitation. From the Taisho to the early Showa era (1912–1940), abalone resource protection started with capture regulation of size (shell length), setting of close season, and fisheries conservation.

In 1952, Takashi Ino (1915–1984) reported about early development of *H. d. discus* and *H. d. hannai* and taxonomic position of *H. d. hannai*. This pioneering academic research has strongly promoted artificial reproduction of abalone. Kikuchi and Uki (1974) reported spawning induction by using UV-irradiated saltwater and contributed to the establishment of a seed production system in many countries including the USA, Mexico, Taiwan, China, Korea, Chile, and South Africa.

Abalone can be considered as one of the most expensive seafood in the world. People especially in Japan, China, and Korea savor taste of abalone in many ways such as sashimi, steamed, grilled, congee, and so on. In these countries, the technology for breeding and culturing abalone is also well promoted and actively conducted.

8.1.3 Current Trends of Abalone Production in Japan, China, and Korea

The total supply of abalone to the world market from all sources including fisheries and farming increased drastically from 23,541 t in 1970 to 149,017 t in 2015 (FAO Fishery and Aquaculture 2017). But landings from legal fisheries are declining from

23,532 t in 1970 to 7,227 t in 2015 (Fig. 8.1a) because of overexploitation, illegal harvesting, and habitat degradation. Thus, abalone farming increasingly plays an important role in filling in a gap between supply and demand.

Commercialized abalone farming started with a production of 9 t in Korea at the beginning of 1970s when farm production was almost negligible. Farming production in Taiwan gradually increased from 130 t in 1980 to 2,496 t in 2001 but then decreased until 79 t in 2011.

The huge increase in abalone farming occurred in China. Figure 8.1b shows that the farm production in China exponentially increased from 9,810 t in 2003 to

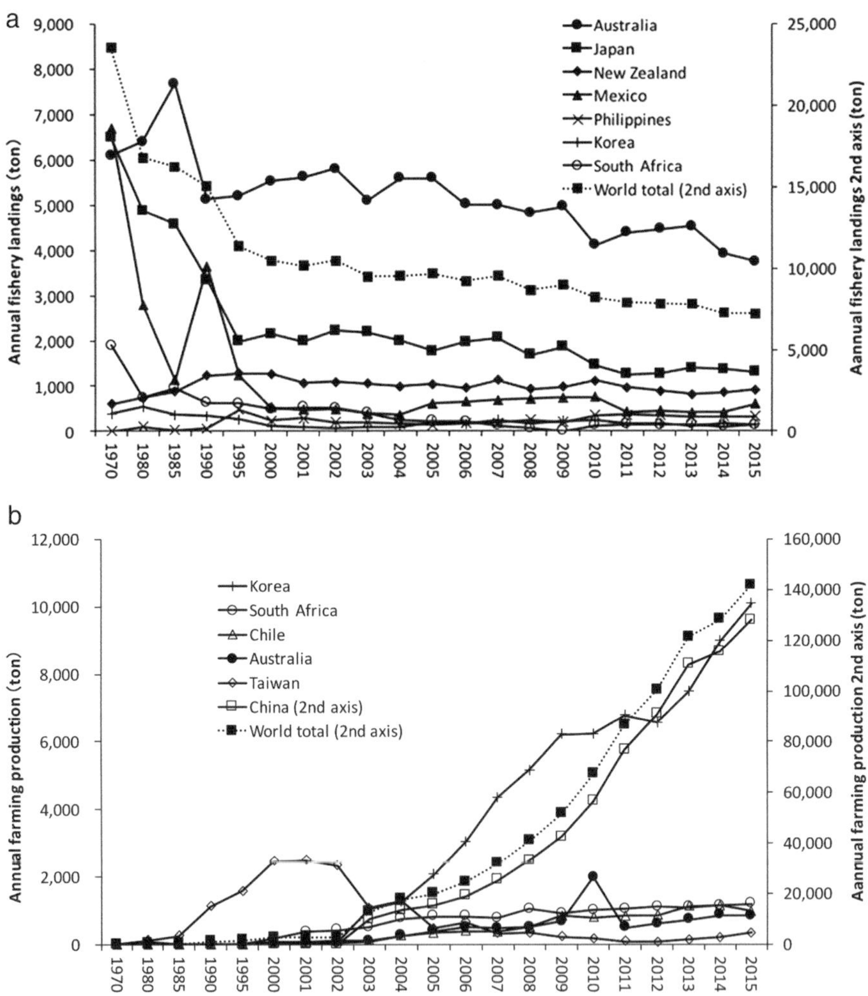

Fig. 8.1 Change in fishery landings (**a**) and farming production (**b**) of abalone in main countries (http://www.fao.org/fishery/en)

127,967 t in 2015. Chinese farm production in 2015 amazingly accounts for 90.2% of total around the world. There are more than 300 abalone farms operated in China. The majority of facilities are land-based flow-through aquaculture systems with free-flowing natural saltwater. The bulk of all production is in the southern province of Fujian and Guangdong. Much of the seed production is near Dalian in the northern province of Liaoning. Fujian produces about half of all abalone that are produced and sold in China. And their significant proportion is *Haliotis diversicolor supertexta* which is smaller and cheaper than *H. discus* (Cook and Gordon 2010).

Farm production in Korea is proportionally growing from 20 t in 2000 to 10,090 t in 2015 (Fig. 8.1b). Abalone fisheries have also been conducted at the level of a few hundred tons production. Most Korean abalones come from Wando County in South Jeolla Province because seaweeds are abundantly harvested in this area. The total fishery landings in this area account for about 70% of all taken in Korea. About 85% of that farm production is exported to Japan (Cook and Gordon 2010). The majority is marine culture using floating cages because initial cost for facilities is low. There are about 200 farm producers and 3,700 related workers. Feeding of seaweeds and uplifting of culturing cage are efficiently conducted by crane operations.

In Japan, artificial propagation and releasing seeds are intensively conducted for recovery of natural abalone resources. Japanese abalone culture is mainly for releasing hatchery-bred juveniles into potential on-growing sites of natural sea and allows them to grow mainly on natural feed organisms. This method considers the fact that it takes 3–4 years for abalones to attain the desired marketable size. However, the efforts of releasing seeds do not show an effective recovery of abalone resources because of rocky-shore denudation, weak resource management, and illegal catch.

8.1.4 Significance of Abalone Culture by Using a Land-Based Recirculating System

As described above, methods of abalone production vary country by country. There are hatchery-bred juvenile releasing to sea, marine culture using floating cages, land-based culture using the flow-through system, and semi-closed and closed recirculating system. The properties of those ways for abalone production were shown in Table 8.1.

Economical efficiencies of juvenile releasing were estimated from number of seeds, cost for seeds, cost for recapture, and earnings from recaptured abalone, showing the result that annual percentage of economical yield ranged from 4.2 to 30.9% with regional varieties in Japan (Nonaka 2011). Seed-releasing business could be profitable between the middle of the 1980s and the beginning of the 1990s. Economical efficiencies depend on retrieval rates. Mie Prefecture Research Institute (2013) reported that retrieval rates largely depended upon year of releasing,

Table 8.1 Comparison of properties among various abalone production systems

	Fishery	Culture			
	Sea based	Sea based	Land based		
	Juvenile releasing	Floating cage	Flow through[a]	Semi-closed recirculating[b]	Closed recirculating[c]
Initial cost for facilities	negligible	a little	middle	high	highest
Fishery right	necessary	necessary	when use natural seawater	as same as flow through	none
Geographical constraint	many	many	depend on water quality	as same as flow through	negligible
Water	natural saltwater (SW)	natural SW	pumped SW	pumped SW or artificial SW	pumped SW or artificial SW
Energy[d]	none	none	a lot	much less than flow through	less than semi-closed
Operating expense	negligible	depreciation and labor cost	higher than sea based	higher than flow through	higher than semi-closed
Feed	natural diet	usually seaweed	seaweed or artificial diet	seaweed or artificial diet	seaweed or artificial diet
Labor intensive for growing	little	tough	less than sea based	as same as flow through	comfortable working environment
Harvesting	tough	easier than catching in ocean	easier than sea based	as same as flow through	as same as flow through
Amount of water emission	–	–	a lot	much less than flow through	extremely little
Pollution level of emitted water	–	–	low	high	low
Environmental pollution	none	middle	high when artificial food is used	as same as flow through	extremely low
Risk for natural disaster	fishery landings are decreasing	high	middle	low	extremely low
Yield ratio	very low	predation and disease	much higher than sea based	high when water temperature is controlled	as same as semi-closed
Poaching	inevitable	inevitable	none	none	none

[a]Flow through indicates daily exchanging of more than 100% of holding seawater quantity

[b,c]Semi-closed and closed indicate daily exchanging of less than 10% and 0.1% of holding seawater quantity, respectively

[d]Energy includes electricity and fuel for water temperature control and power pump

fishing grounds, and species. Retrieval rates of *H. gigantea* in developed fisheries tend to be higher than that of *H. d. discus* in natural fisheries. Retrieval rates of *H. gigantea* and *H. d. discus* ranged from 5.6% to 20.1% and from 0.9% to 3.7%, respectively. The institute estimated that targeted profitable retrieval rates were 10% and 7% in *H. gigantea* and *H. d. discus*, respectively. It was presumed that abalones over 12 cm in length were harvested and sold at a price of 54 US dollars and 72 US dollars per kilogram, respectively. However, fishery landings are gradually decreasing year by year (Fig. 8.1a) meaning that releasing has not yet led to recover abalone resources. Fishery regulation is primarily desired in order to recover abalone resources.

The poaching (illegal exploitation of abalone) is a serious problem contributing to population disruption around the world. Annual illegal abalone capture is estimated from 120 to 155 t accounting for 32.2–52.7% of total landing in Miyagi Prefecture of Japan (Yamakawa 2006). Another estimation of all amount of poaching in Japan was 903 t, accounting for approximately 45% of total landings, which is calculated from the difference between amount of total supply (3,799 t) and amount of fishery landings (1,996 t) and imports (900 t) in 2004 (Taya 2007). Fishtech Inc. reported that illegal take is also a big problem in other countries. Metric ton of 1500, 1500, 1000, 400, and 200 are estimated as amounts of illegal take in Australia, South Africa, New Zealand, Mexico, and the USA, respectively (Fishtech Inc. 2012).

Risk of natural disaster such as Tsunami and Typhoon could not be negligible in the case of sea-dependent farming. For example, all kinds of fish farming damage by tsunami of the Great East Japan Earthquake in 2011 were estimated as much as 980 million US dollar accounting for 25% of annual national farming production value (Anonymous 2011). Korean abalone farming in Wando County is annually exposed to typhoon damage. In Fujian province of China, mass mortality due to red tide derived from eutrophication occurred in 2012, and total financial damage was 26,500,000 US dollar (Anonymous 2012). In Hokkaido prefecture of Japan, mass mortality in abalone farming using cages under the sea was caused by abnormal low sea temperature of less than 5 °C, resulting in total financial damage of 334,000 US dollar in 2014 (Anonymous 2014). Climate change has recently affected fisheries.

Sea-independent aquaculture using a land-based recirculating aquaculture system (RAS) could contribute to overcoming the weakness of natural sea-dependent fisheries. Abalone culture by RAS is especially advantageous as described below:

- The methods for abalone hatchery and culture have been almost established.
- Abalone can be cultured with relatively high density because movement of abalone is less than that of swimming fish.
- Risk for natural disaster and poaching can be evitable.
- Abalone could grow faster than natural environments by controlling water temperature, amount of food, and light intensity.
- Harvesting from RAS is much easier than that by diving into sea.
- There is a possibility to provide steady supply with scheduled production and intensive traceability.

However, RAS have some potential disadvantages. There is a risk of disease breakout once pathogens enter RAS because recirculating system is kind of an incubator of microorganisms. Thus, it is necessary to definitely prevent RAS from invasion of pathogen. There is the risk of mechanical failure and the associated water quality issues. Entrainment spawning in the RAS leads to water quality decrease and heavy gamete loss. If these accidents happen, serious problems are inevitable because individuals are reared in the high density within the RAS. Nevertheless, research and development for successful abalone farming by using RAS should be pursued in order to increase diversity in fishery production styles and contribute to food security.

It takes 3–4 years for abalones to attain the desired marketable size based on the culture method of releasing hatchery-bred juveniles into natural waters and allowing them to grow mainly on natural feed organisms. Land-based intensive farming techniques under the control of water temperature, water qualities, and feeding are largely expected to promote the efficiencies of abalone production. Additionally, water treatment systems to maintain water quality without exchanging fresh saltwater would contribute to achieving veritable environmentally friendly closed RAS (CRAS) and inland fish farming. Environmental load from rearing water emission is estimated by multiplying emitted water quantity by concentrations of nutrients such as nitrogen and phosphorous compounds. Thus, water treatment systems for elimination of accumulated substances such as nitrate and persistent organic matters are required for RAS.

8.2 Abalone Culture

8.2.1 Seed Production

Artificial abalone breeding plays an important role in maintaining supply quantity of abalone. Abalone hatchery has been well studied and manualized in some publications (Uki 1989; FAO 1990; Heasman and Savva 2007; Leighton 2008). In Japan, the three species of abalone other than *H. madaka* are mainly bred. And annual total production of abalone seeds for release was approximately 30 million individuals from 1984 to 2006. *H. d. hannai*, *H. d. discus*, *H. gigantea*, and *H. madaka* account for 57.8%, 26.3%, 15.8%, and 0.1% of total seed production in 2006, respectively (Nonaka 2011).

Abalone reproduces sexually by external fertilization. *Haliotis* spp. shows annual spawning cycle. It is thought that temperature, food availability, and photoperiod are the major factors influencing the development of the gonads. In the sea, season of peak spawning is at water temperature around 20 °C irrespective of northern and southern part in Japan. In the hatcheries, timing of spawning can be controlled by conditioning of broodstock. There is the theoretical minimum temperature at which gonad growth and development begins. For *H. d. hannai*, 7.6 °C is biological zero point (Uki 1989). Conditioning times is determined from the summation of the difference between the biological zero point and the water temperature (Effective Accumulative Temperature). It is considered that *H. d. hannai* are fully mature at 1500 °C or more (Uki 1989). After gonad fully develops, spawning is triggered by desiccation and immersion into UV-irradiated saltwater (Uki and Kikuchi 1984).

Fertilized ova are maintained in hatching tanks at 18–20 °C of water temperature then start hatching when the accumulative temperature reaches 160 °C. After metamorphosis, larvae settle on clear plates which are coated with diatoms in advance. Diatoms and mucous secreted by the foot of juvenile or adult abalone on the plates induce larvae settlement. Larvae grow by eating diatoms and mucous. When abalone juveniles grow close to or above 10 mm in shell length, they can be transferred directly from settlement plates to intermediate culture tank. Abalones are grown up to about 30 mm during the intermediate culture period. For further information on conditioning broodstock, spawning method, metamorphosis process, optimum density of rearing larvae or juvenile, culture and maintenance of diatoms, and weaning method, consult other guidelines (Uki 1989; FAO 1990; Heasman and Savva 2007; Leighton 2008).

In hatcheries by using RAS, temperature can be maintained at a constant level very accurately. It is advantageous that this permits accurate prediction of full maturity. Furthermore, temperature control makes it possible to introduce the spawning of broodstock on a year-round basis, leading to increasing production of marketable abalone even in lean season.

8.2.2 Grow-Out Culture

It is important for the sea-based culture, the land-based flow-through culture, or the semi-closed RAS culture to consider ecological properties of farming sites where clear and non-polluted saltwater is available. The growth rate of abalones strongly depends on the water temperature and availability of feed. Water temperature is preferably kept close to optimum water temperature ranging from 15 to 20 °C (Sasaki 2005). Water temperature below 6 °C or over 28 °C is not acceptable for abalone to take food. In the case of CRAS culture, maintenance of water quality by water treatment system is indispensable in addition to water temperature control. Nevertheless, comparing with flow-through system, recirculating system largely contributes to saving energy for water temperature control.

Introduction of excellent juveniles is also important to accomplish high productivity. Equal-sized juveniles in about 20–30 mm of shell length are introduced to the rearing tank when water temperature is relatively low at the level around 20 °C. Survival rate decreases with juvenile introduction in high temperature season. Optimum rearing density of big abalones depends on the culturing systems and the structure of feeding attachment plates. Further investigation is required to increase the productivity of abalone in marketable size (>60 mm).

Feeding management is also important to maximize availability of feed. Abalone fed with artificial diet significantly grows faster than that with natural seaweed. Abalone with average size of 19.8 ± 3.2 mm fed with artificial diet and *Undaria* sp. could grow up to 40.4 ± 5.5 mm and 36.6 ± 3.5 mm ($n = 30$) for 118 days (from May to September), respectively. Abalone is considered to forage for food nocturnally or under the cover of darkness. Recording of abalone movement by an infrared camera showed that active food consumption started about 1 h later after lights out and continued for 4–5 h then hardly moved (Fig. 8.2). Duration of abalones actively foraging for food corresponds to circadian rhythm in the oxygen consumption rates (Uki and Kikuchi 1975). Thus, it is efficient to carry out feeding in the evening when abalone is fed with artificial diet because uneaten diet left in the rearing tank begins deterioration and pollutes the rearing water.

Pest control is always necessary especially in high-density rearing. Amyotrophia caused by unidentified virus, bacterial disease caused by *Xenohaliotis californiensis*, *Francisella* ap., and *Viblio* sp., and fungal disease caused by *Haliophthoros* sp. are known as abalone infection. *Xenohaliotis* disease, which is specific in *H. d. discus* among Japanese indigenous abalone species, became epidemic in the USA and Mexico since 1980, and its pathogenesis was also reported in Europe and Chile. In the hatchery in Tottori, western Japan, about 13,000 juveniles of *H. d. discus* were infected and incinerated en masse in 2012 for the first time. Investigation of the wild abalone revealed that less than 1% of *H. d. discus* and *H. gigantea* were potentially infected by *Xenohaliotis* sp.. Thus, inspection of infection is important prior to introduction of juveniles into the farming to prevent from outbreak in the high-density rearing tank. Screening by genetic diagnosis is recently developed.

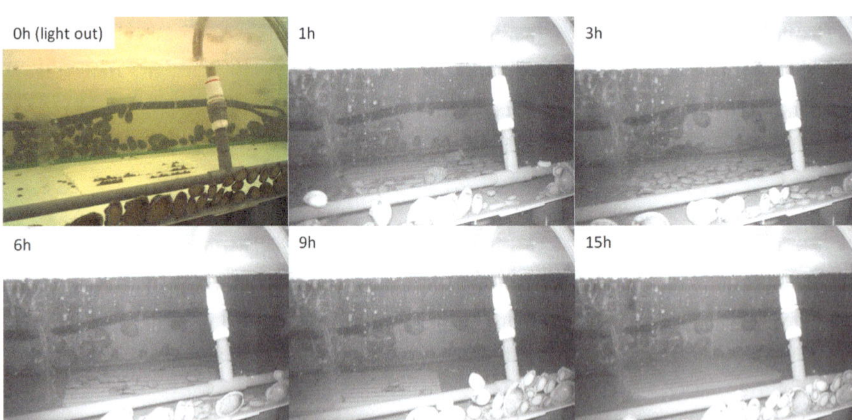

Fig. 8.2 Photographs from monitoring feeding behavior of abalone using an infrared camera: 0 h (light out), 1 h, 3 h, 6 h, 9 h, and 15 h

8.3 Water Quality Requirements and System Overview for Land-Based Abalone Farming

The breeding and culture technologies have already well developed. There are many manuals and guidelines for hatchery and culturing of abalone (Uki 1989; FAO 1990; Heasman and Savva 2007; Leighton 2008). Thus, it is possible to achieve a land-based recirculating aquaculture of abalone as long as water quality is maintained. There are internal and external factors contributing abalone growth. Internal factors are aging and genetical features. External factors are water temperature, water quality, diet, and rearing density. Water quality required for abalone culture is focused in this chapter as many published papers have regarded other than water quality.

8.3.1 Importance of Recirculating Water Treatment

In flow-through land-based abalone farming, it is sufficient to exchange rearing water with pumped saltwater of more than 100% a day for maintaining water quality. In the case of using natural saltwater, it is necessary to take measures for temporal change in water temperature, rainy change in specific gravity, and possibility of pathogen contamination.

In RAS, besides water exchanging rate decreasing until less than 10% a day, treatment waste matters derived from residual food, feces, and excretion is indispensable. In particular, accumulation of "ammonia" and "solid waste" is the main reason for fail farming with RAS. Ammonia is treated by ammonia- and nitrite-oxidizing bacteria which oxidize high-toxicity ammonia to low-toxicity nitrate (nitrification). Solid waste should be removed from the rearing tanks as soon as possible then filtrated to separate solid from water. It is very important to design facilities for consistent and steady recirculation because water treatment is mostly carried out by recirculation. Hydraulic retention time (HRT, t) is traditionally used for evaluation of recirculating:

$$t\,(\text{h}) = \text{Volume}\,\left(\text{m}^3\right)/\text{Flow rate}\,\left(\text{m}^3/\text{h}\right)$$

For example, t of 15–30 minutes has been recommended for design of effective settle basins (Liao and Mayo 1974). Efficiencies of nitrification are logarithmically raised with increase of turnover, but too many turnovers may cause plateaus (Fig. 8.3). In a nitrification biofilter, optimum turnover rate should be designed as considering cost-effectiveness of nitrification activities, specifications and price of pump, and electricity expense. Inverter control of pump power is very effective for energy and cost savings as same as those of blowers and air conditions. Water treatment efficiencies, concentration of dissolved oxygen, and water temperature

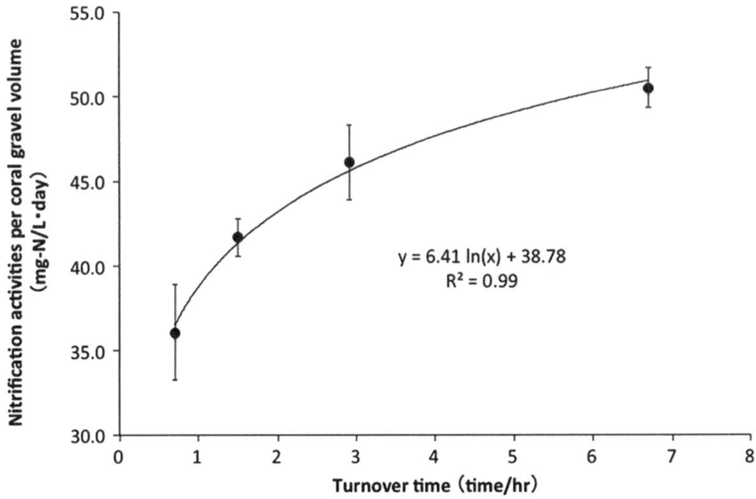

Fig. 8.3 Relationship between nitrification activities and turnover time in a closed RAS using coral biofilter

can be alterable by inverter control depending on growth stages, rearing densities, and environmental variable conditions.

Nitrification treatment by biofilter makes it possible that quantity of daily exchanging water is abundantly reduced until 5–10% of total holding water quantity. But limitations are still remained because a larger amount of fresh saltwater is necessary and costly when farming facilities become huge. For example, a 100 m^3 RAS with a daily exchanging rate of 10% saltwater is calculated to require 3650 m^3 saltwater annually. This cost is estimated to more than 200,000 US dollars on the basis of unit prices of artificial saltwater (60–80 US dollars per m^3). Thus, saltwater acquisition is too costly to commercialize land-based abalone farming because abalone is very sensitive to water quality especially for salinity and concentrations of nitrate and persistent colored organic matter as described in the next subsection of this chapter. Profitable abalone farming facilities are limited to be built near the sea unless exchanging rearing water with fresh saltwater is required. Furthermore, measures against eutrophication caused by waste discharge containing nutrient matters and solid waste should be considered because integrated amounts of nutrients from a lot of diluted waste water is not negligible. Utilizing artificial diet particularly aggravates eutrophication of rearing water because of richness in nutrients.

CRAS including denitrification and colored organic matter treatment systems has been therefore intensively studied since 1990s in order to extremely reduce environmental loads and energy (van Rijn et al. 2006). Strict water quality control and water treatment systems are required because abalone should be cultured without daily exchanging saltwater. Only loss of vapored fresh water is basically supplemented with freshwater. Transported natural saltwater or artificial saltwater is used as initial introduced saltwater in CRAS. Quality of freshwater used for artificial saltwater is recommended to be analyzed according to water quality criterion, for example, heavy metals, volatile organics, pesticides, and so on. It is possible that heavy metals and residual toxic organics are concentrated biologically and by vaporization even if toxic contaminations are very low. This is also common in the case of quality control of feed.

8.3.2 System Overview for Abalone Culture in RAS

The principal flow pattern for abalone culture in RAS is shown in Fig. 8.4. The system basically consists of rearing tanks, solids removal, nitrification, denitrification, disinfection, pumps, blower, and temperature control units. The water is cascaded from the rearing tank located in the highest elevation to the nitrification tank by gravity. As there are many varieties among water treatment units, appropriate principles and sizes of those equipments should be selected depending on fish species, rearing densities, and the scale of RAS (Timmons et al. 2002). For example, there are some options for nitrification by biological filters such as fixed bed, moving bed, and trickling filters. A fix bed biofilter has an advantage that it captures fine particles in addition to nitrification and degradation of dissolved organic matters. Thus, the fix bed biofilter is used as the base of nitrification, and the moving bed biofilters are added when further nitrification is required.

A rectangular tank is suitable and water depth of approximately 40 cm is sufficient for abalone rearing. Water level of the culture tank is regulated by overflow pipe arrangement (Fig. 8.4). Low-water rearing enables for multiple-stage tank arrangement that leads to increase abalone productivity per unit area.

Solids capture is carried out using three types of methodology depending on size classification of solids. Settleable, suspended, and dissolved solids are removed by a settling tank (gravity separation), a solid capture tank (filtration), and a foam fractionation apparatus (flotation), respectively. Abalone rearing water can be drained to the solid capture tank for tank cleaning since abalone have a desiccation tolerance of a few hours. After draining the water, the remaining solids in the rearing tank can be washed out by backwater from the solid capture tank by opening

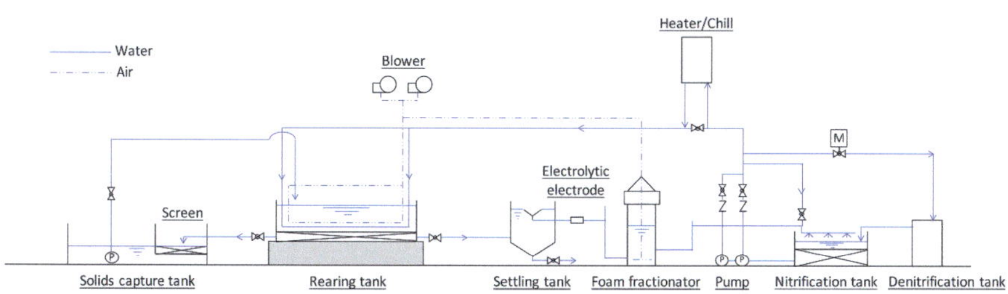

Fig. 8.4 Schematic overview of RAS for abalone culture

the drain bulb to the solids capture tank. In the case of a certain scale of RAS (approximately more than 50 m^3 of total water volume), microscreen filters such as a drum filter and a disk filter are recommended on a cost-benefit analysis.

Because water circulation shutdown, aeration system failure, and electricity blackouts can cause immediate and serious damage to abalone, backup pumps, blowers, and generators should be installed (Table 8.2). Details of nitrification, denitrification, and dissolved organic matter treatment by electrolytic electrode will be described in the following subsections of this chapter.

8.3.3 Parameters and Standards of Water Quality Control for Rearing Abalone in RAS

Water quality required for fish farming has been well described in published books and papers (Timmons et al. 2002; Poxton 2003). In this chapter, important parameters among them are focused particularly in the case of rearing abalone in RAS. Table 8.2 shows evaluation of importance of water quality parameters, reasons of failure, and measures for recovery. Importance of water quality parameters was evaluated with multiplication of altering speed by tolerability for abalone because change and significance in water quality varied depends on water quality parameters. Some parameters might change within a few minutes by mechanical failure, but others might change within a month by biological metabolisms. Measuring frequency for each parameter is also recommended in Table 8.2 based on evaluation for importance of water quality control. Standards of physicochemical water quality for rearing abalone are summarized in Table 8.3.

DO and Water Temperature The most important parameter is dissolved oxygen (DO) because DO rapidly decrease until lethal concentration once troubles of an aeration device occur. It is much safer to install a DO monitoring and alarming system and to prepare a generator and backup devices for aeration just in case of emergency.

DO has a strong relationship with water temperature. Four species of Japanese commercially valuable abalone are distributed in the northern and southern part of Japan and adapt to water temperature in their habitats. Northern abalone (*H. d. hannai*) generally grows slower than southern abalone (*H. d. discus*) in nature because of lower water temperature. However, it is possible to culture *H. d. hannai* under the natural condition of southern part of Japan because *H. d. hannai* was differentiated from *H. d. discus* (Hara and Fujio 1992; Hara and Sekino 2005). Optimum water temperature for both species ranges between 15 and 20 °C (Sasaki 2005). It is more cost-effective to control water temperature in this range according to seasonal change because temperature adjustment needs a lot of energy. Fujinaga et al. (1999) investigated the relationship between temperature, amount of food consumption, and weight gain. As a result, food consumption increased until 25 °C,

Table 8.2 Evaluation of importance of water quality parameters, reasons for failure, and measures for recovery

Parameters	Altering speed[a]	Tolerability[a]	Importance[b]	Measuring frequency	Reasons for failure	Measures for recovery
Water temperature	○	○	○	monitoring	change in temperature, trouble of thermocontrol	repair of thermocontrol, insulation of building
Salinity	△	○	△	every month	vaporization, addition of too much freshwater	addition of artificial saltwater or freshwater
pH	○	○	○	monitoring	decrease by respiration and nitrification	addition of sodium bicarbonate, installation of denitrification device
DO	◎	◎	◎	monitoring	respiration, nitrification, trouble of aeration devices	repair of devices, installation of backup devices
Alkalinity	△	△	△	every month	increase by respiration	aeration, degassing
Total ammonia (NH_3 and NH_4^+)	○	◎	◎	everyday	excretion of biological wastes and residual diet	enrichment of ammonia oxidizer, optimization of water quality for nitrification
Nitrite	○	◎	◎	everyday	incomplete nitrification	enrichment of nitrite oxidizer, optimization of water quality for nitrification
Nitrate	○	○	○	every 1 to 2 weeks	increase by nitrification	exchanging water, installation of denitrification device
Phosphate	△	△	△	every 1 to 2 months	excretion of biological wastes and residual diet	exchanging water, addition of coral (calcium)
Suspended solids (SS)	△	○	△	every 1 to 2 months	trouble of filtration and solid removal systems	maintenance or upgrade of solid removal devices
Color	○	◎	◎	everyday	accumulation of recalcitrant organic compounds from wastes	installation of dissolved organic matter treatment devices

(continued)

Table 8.2 (continued)

Parameters	Altering speed[a]	Tolerability[a]	Importance[b]	Measuring frequency	Reasons for failure	Measures for recovery
Total organic carbon	△	○	△	every 1 to 2 months	accumulation of organic compounds from wastes	installation of any size of waste removal devices
Calcium	○	○	○	every 1 to 2 weeks	decrease by shell formation	addition of calcium hydroxide, calcium bicarbonate, or calcium chloride
Ion balance[c]	△	○	△	every 1 to 2 months	increase of various ions from wastes	exchanging water, supplement of chemicals, absorption of nutrients by seaweeds
Pathogen	△	○	△	every 6 months	introduction of infected individual, resident bacteria	epidemic control. installation of sterilization devices, antibiotics
(Free/combined residual chlorine)	△	◎	○	everyday	trouble of electrolyzation device	setting change of device, neutralization by sodium thiosulfate
(Ozone)	△	◎	○	everyday	trouble of ozonizer	setting change of device

[a]Evaluation criteria (altering speed/tolerability):◎3pt very quick (a few hours)/not tolerate, ○2pt quick (a few days to weeks)/a little tolerate, △1pt slow (a few months)/tolerate

[b]Evaluation criteria (importance):◎6–9pt, ○3–4pt, △1–2pt multiplication of score of (altering speed) and (tolerability)

[c] Ion balance means relative concentrations of Na^+, K^+, Ca^{2+}, Mg^{2+}, NH^{4+}, Cl^-, $SO_4{}^{2-}$, Br^-, $NO_2{}^-$, $NO_3{}^-$, $PO_4{}^{3-}$ compared with natural seawater

Table 8.3 Physicochemical standards of water quality for rearing abalone

Parameters	Standard	Unit
Salinity	31–38	‰
Water temperature	15–22	°C
DO	6.5–8 (90–100%)	mg O_2/L
pH	8.0–8.5	
Alkalinity ($CaCO_3$)	200–300	mg $CaCO_3$/L
TAN[a]	<1.0	mg-N/L
NH_3-N	<0.025	mg-N/L
NO_2^--N	<0.5	mg-N/L
NO_3^--N	<50	mg-N/L
SS	<10	mg/L
Calcium	360–460	mg/L
Chromaticity	<10	Abs 390 nm[b]

[a]TAN means total ammonia (NH_3 + NH_4^+)
[b]Measured by using a color meter (WA-PT-4DG, Kyoritsu Chemical-Check Lab., Corp., Japan)

but maximum weight gain could not be expected over 25 °C because metabolic activities also increased.

DO increase with low temperature and decrease with high salinity. Saturated concentration of DO is 7.83–7.15 mg/L under the air pressure at the optimum temperature of 15–20 °C and salinity of 32‰ (chloride ion concentration of 19,400 mg/L). In the case of flow-through culturing system, robust water supply contributes to providing sufficient oxygen and to remove solid waste. By contrast, oxygen must be supplied by vigorous aeration using blower in RAS. Compared with flow-through system, RAS is advantageous in reducing energy and cost for temperature control. Uki and Kikuchi (1975) investigated the relationship between oxygen consumption, body weight, and temperature in *H. d. hannai* culture. There is proportional relationship between temperature and logarithm of oxygen consumption under 20 °C, and oxygen consumption did not change over 20 °C. Oxygen consumption of *H. d. hannai* with body weight of 1.5 g and 151 g were 0.184 and 7.70 mL O_2/(individual•hour) (0.245 and 10.2 mg O_2/individual •hour), respectively. Moreover, oxygen consumption tended to increase between around sunset and midnight.

Salinity Range of salinity is required between 31 and 38‰ as same as natural sea. Salinity of the sea around Japan is ~34‰. It is known that salinity and composition of animals are close to those of aquatic environments where the animals evolved. Among marine animals, invertebrates such as mollusk including abalone rarely have an experience to evolve in freshwater environments. Body fluid and osmotic pressure of mollusk are likely similar to saltwater. Thus, abalone does not tolerate low salinity of saltwater. Salinity of water should be paid attention during abalone culturing. Salinity of saltwater in inner bays is unstable because heavy rain is possible to dilute saltwater. *H. d. hannai* start to die after about 15 h when they are put in 15‰ of diluted saltwater. And *H. d. discus* is significantly inhibited to grow under the condition of less than 26‰ of saltwater (Kobayashi 2002). If salinity is measured by specific gravity, conversion is needed based on temperature. For example, 35‰ of saltwater at 24 °C is converted to 1.0240 of specific gravity. Salinity measurement using a conductivity meter is recommended because precise measurement of specific gravity is difficult by general instruments.

Nitrogen Compounds Nitrogen is an essential nutrient for all living organisms and is found in proteins, nucleic acids, adenosine phosphates, and so on. However, nitrogen is required in relatively small quantities, and physiological needs are easily satisfied. Excess quantities become nitrogenous wastes, and removal is necessary. Abalone creates various nitrogenous waste products through gill diffusion, gill cation exchange, and urine and feces excretion. Nitrogenous wastes also accumulate from the organic debris of dead and dying organisms and uneaten feed. Decomposing these nitrogenous compounds is particularly important in intensive RAS because of the toxicity of ammonia and nitrite.

Fujinaga et al. (1999) investigated the amount of nitrogenous compounds excreted from *H. gigantea* under the no feeding condition. As a result, an individual of 37 mm excreted 379, 64, and 34 ug-N/(individual•day) of ammonia, urea, and feces, respectively. By the same, an individual of 70 mm excreted 2290, 372, and 201 ug-N/(individual•day), respectively. The amount of excretion of ammonia and urea increased with water temperature.

Unionized ammonia (NH_3) is considered to be the most toxic form of ammonia, while ionized ammonia (NH_4^+) is much less toxic. Fish can generally cope with quite large concentrations of ammonia in the form of ionized ammonia. An increase in pH, temperature, or salinity increases the proportion of the unionized form of ammonia. Unionized ammonia concentrations should be held below 0.025 mg/L and total ammonia-nitrogen (TAN: unionized and ionized ammonia) concentrations below 1.0 mg/L for long exposure (Table 8.3) (Leighton 2008; Timmons et al. 2002).

Nitrite is the intermediate product in the process of nitrification of ammonia to nitrate and also extremely toxic because it affects the blood hemoglobin's ability to carry oxygen (Timmons et al. 2002). Levels in a system for abalone should not exceed 0.5 mg-N/L (Table 8.3) (Leighton 2008).

Nitrate is the end product of nitrification and is the least toxic of the nitrogen compounds. However, accumulation of nitrate at high concentration could impact on mortality and growth. Toxicity of nitrate to fish has not been much investigated because acute toxicity is not serious. However, chronic toxicity, inhibition of nitrification, and proliferation of phytoplankton are gradually promoted by abundant nitrate accumulation (Poxton 2003). In the case of abalone, about 100 mg-N/L of nitrate concentration affected their mortality (Fig. 8.5). In flow-through system, nitrate levels are usually controlled by daily water exchanges. In CRAS with low water exchange or high hydraulic retention times, denitrification has become increasingly important.

Reaching level of nitrate concentrations in RAS can be estimated from "nitrogen increasing rate (mg-N/day)" and "water exchanging rate (L/day)." For example,

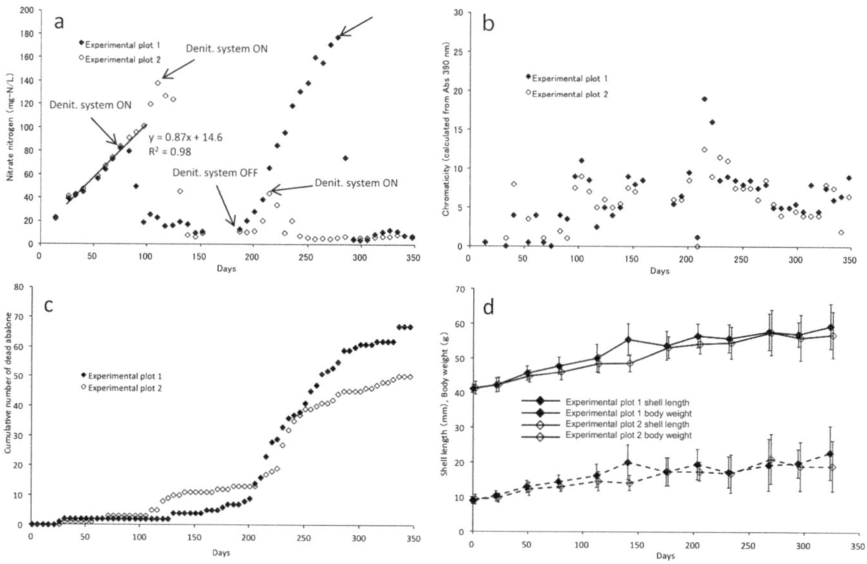

Fig. 8.5 Monitoring of nitrate concentrations (**a**), chromaticity (**b**), mortalities (**c**), and individual shell length and body weight (**d**) in a closed RAS rearing *Haliotis discus hannai*

concentration of nitrate proportionally increased at the rate of 0.87 g-N/(L•day) when 100 individuals of *H. d. hannai* (average shell length 41.2 mm and weight 9.1 g) were reared in RAS, which has 390 L of effective volume of water including 95 L of filtering media, with 10 g of daily artificial diet feeding (1.1% of total body weight) at constant water temperature of 20 °C without water exchange (Fig. 8.5a). Nitrification smoothly occurred at this time because concentrations of TAN and nitrite were under the detection limits. But nitrogen of urea, uneaten food, and feces were not measured. Excretion speed of nitrogen by individual abalone was roughly estimated at 2600 ug-N/(individual•day). The value from this estimation is much higher than that by Fujinaga et al. (1999) because of the difference of feeding condition.

Effects of nitrate concentrations on growth and mortality of abalone were evaluated in CRAS by using newly developed automated denitrification system. Experiment was carried out in two plots. Nitrate concentration rapidly decreased by running denitrification system when nitrate concentration reached 80 mg-N/L (Fig. 8.5a). Abalone started to die with increase of nitrate concentration and mortality significantly increased when nitrate concentration reached more than 100 mg-N/L (Fig. 8.5c). Mortality of abalone was repeatedly declined with decrease of nitrate concentration in both experimental plots. Significant negative impact of high nitrate concentration on abalone mortality was revealed with reproducibility. Levels of nitrate concentration ideally should not exceed 50 mg-N/L within RAS. Denitrification system is indispensable in the case of intensive CRAS. Furthermore, high concentration of nitrate significantly affects not only mortality but also growth of abalone (Fig. 8.5d). There was lag time between water quality degradation, growth stagnation, and mortality. It is well known that abalone starts to die after 5–7 days when factors affecting the mortality occur, such as deterioration in water quality, growth stagnation, and body damage. And it takes 7–10 days to recover after excluding wrong factors.

Solids Waste solids accumulating in RAS come from uneaten diet, feces, dead bodies, and microbial floc. These solids are classified by size into settleable (>100 μm), suspended (0.001 to 100 μm), and dissolved (<0.001 μm) (Timmons et al. 2002). Studies indicate that fish produce 0.3–0.4 kg total suspended solids for every 1 kg of feed fed (Timmons et al. 2002). Waste solids influence the efficiency of all unit process in RAS and can directly affect fish health by damaging fish gills and harboring pathogens. The upper limit for fish is tentatively recommended at 10 mg SS/L (Timmons et al. 2002). Suspended solids do not settle out by gravity settling basin, therefore require a treatment process as filtration. Dissolved organic matter is difficult to remove. In particular, recalcitrant dissolved organic matter such as humic substances is difficult to be degraded by microorganisms.

Effect of accumulated organic matter on abalone has not been much investigated as well as nitrate because most abalone farming are carried out by flow-through aquaculture system or marine culture. But accumulated organic matter also significantly affects molluscan mortality. The effects of accumulation organic matter and of added nitrate on bloody clam (*Anadara broughtonii*) mortality were investigated at the water temperature from 10 to 13 °C using 15 L of small RAS (Fig. 8.6). Mortality was monitored under the conditions with combination of nitrate addition, denitrification, and activated carbon treatment. Dead clams were confirmed one by one and removed at the same time. Total organic carbon (TOC) and chromaticity of

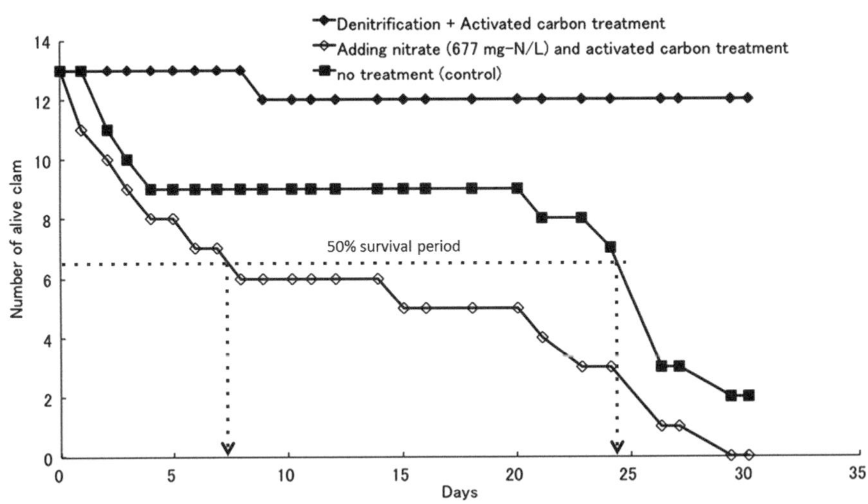

Fig. 8.6 Effects of added nitrate and of dissolved organic matter removal on mortality of bloody clam

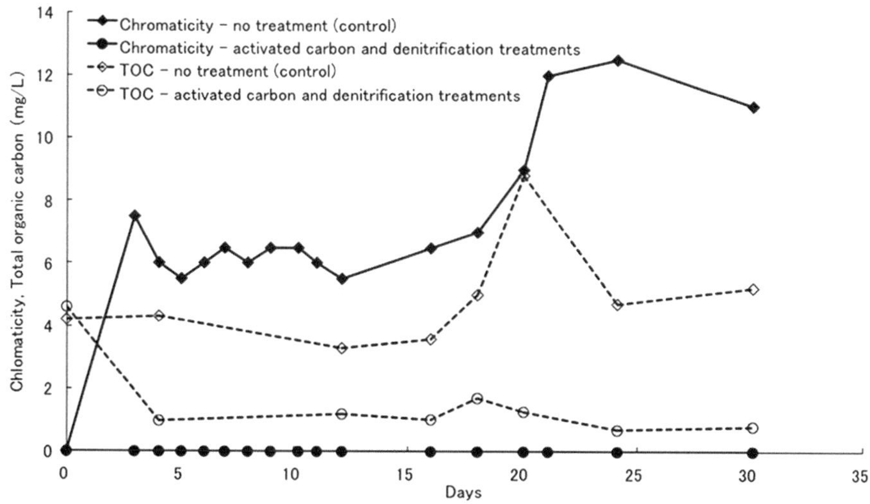

Fig. 8.7 Changes of total organic carbon and chromaticity in the rearing tanks of bloody clam with adding activated carbon and denitrification treatments

the supernatant of the rearing saltwater were measured by total organic carbon analyzer (Shimadzu Co, Kyoto, Japan) and color meter (Abs. 390 nm) (WA-PT-4DG, Kyoritsu Chemical-Check Lab., Corp., Tokyo, Japan), respectively (Fig. 8.7). As a result, survival period in the combination of denitrification and activated charcoal treatment RAS is prominently prolonged compared with the control. Durations for 50% survival in control and both charcoal- and nitrate-added (677 mg-N/L) RAS were 24.4 and 7.6 days, respectively. Survival period in the RAS of added high concentration of nitrate was significantly shortened irrespective of charcoal treatment (Fig. 8.6). TOC and chromaticity in the charcoal-added RAS were maintained lower than those in the control. Chromaticity in charcoal-treated RAS was kept at 0, while that in the control increased up to 12.5, and TOC up to 8.9 mg/L at that time (Fig. 8.7). The change in chromaticity was related to mortality in the control. Bloody clams started to die when chromaticity steeply increased on around 20 days in the control. At this time, nitrite and TAN were detected at 0.2 and 0.6 mg-N/L, respectively. Decay of dead body might trigger mass mortality and decline quality of rearing water. Thus, rapid removal of dead body and water treatment system are important to maintain water quality.

In the case of abalone, similar effects of nitrate concentration and chromaticity on mortality and growth were observed (Fig. 8.5) indicating that this is common among molluscan shellfish. But there was not a significant relationship between chromaticity and TOC (data not shown). Chromaticity might tend to affect directly mortality more than TOC concentration.

Others pH control directly affects shell formation. Natural saltwater has a pH of 8.0–8.5 and is relatively stable by buffer function of bicarbonate and boric acid. pH decreases due to robust nitrification and abalone respiration (Fig. 8.8a). If the pH goes below 7.6, the calcium carbonate of the abalone shell starts to dissolve (Leighton 2008). This must be counteracted by the addition of chemicals containing hydroxide, carbonate, or bicarbonate ions. Sodium carbonate (baking powder) is usually used since it is relatively safe, easy to obtain, and dissolved rapidly and completely in water (Loyless and Malone 1997). To ensure this buffering capacity, the alkalinity should be maintained at 200–300 mg/L as $CaCO_3$ (Table 8.3).

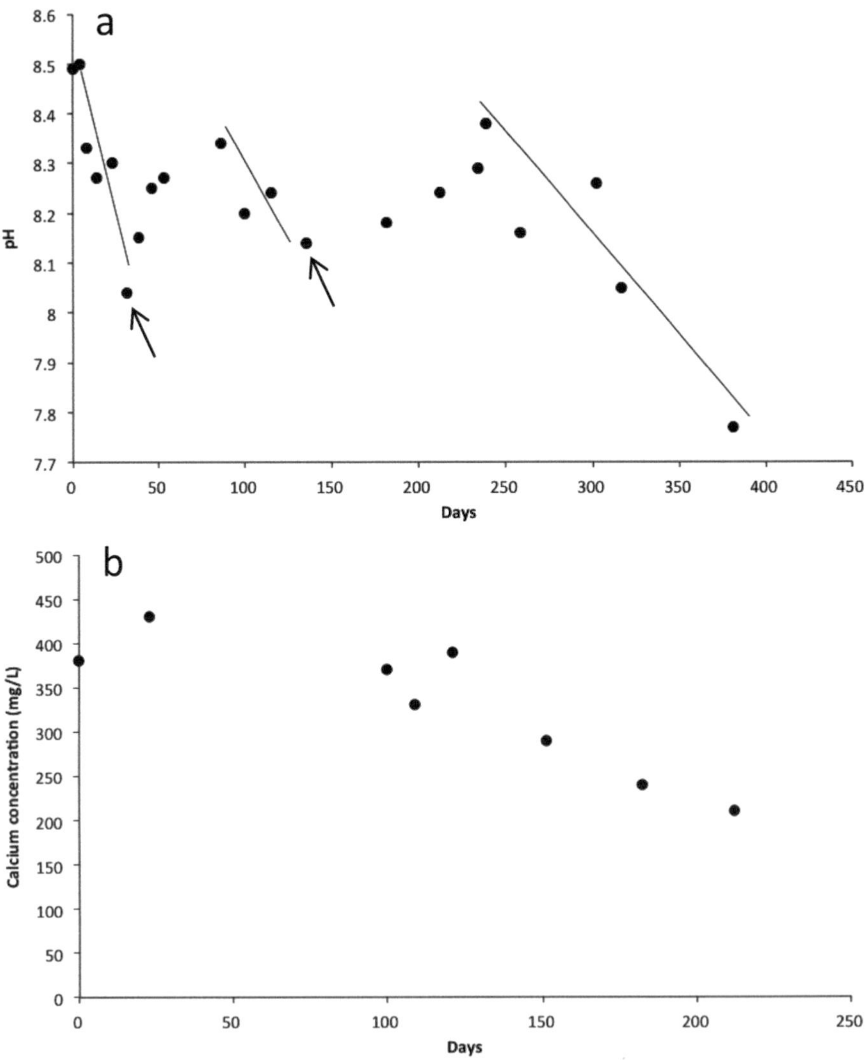

Fig. 8.8 Changes of pH (**a**) and calcium concentration (**b**) in the abalone rearing tank of RAS. Arrows indicate the addition of sodium carbonate

An increase of pH and temperature increases the proportion of the unionized form of ammonia nitrogen. pH and temperature adjustment especially in the case of increase should be gradually carried out with great care.

Calcium concentration in CRAS proportionally decreases without saltwater exchange (Fig. 8.8b). Calcium concentration in natural saltwater is generally of around 410 mg/L and should be maintained at 350–450 mg/L in the RAS (Table 8.3). Shortage of calcium concentration leads shell formation trouble because abalone shell mainly consists of calcium carbonate. This must be counteracted by the addition of chemicals containing calcium such as calcium hydroxide and calcium chloride. The addition of calcium hydroxide could also increase pH. On the other side, the addition of sodium carbonate decreases calcium concentration by precipitation of calcium carbonate which degree of solubility is 0.00015 mol/L at 25 °C. Thus, adjustment of pH, alkalinity, and calcium concentration by adding one or more of a number of chemicals must be conducted as considering chemical interreactions.

Potassium and phosphate are relatively highly contained in diet. For example, 100 g of dried kelp (*Laminariaceae* sp.) contains potassium, calcium, magnesium, phosphate, and nitrogen at 5,300, 760, 540, 240, and 1,280 mg, respectively. Artificial diets contain approximately 30% of protein, which is converted into 4800 mg-N/100 g. Although calcium and magnesium are taken in abalone growing, potassium, phosphate, and nitrogen accumulate in the RAS unless rearing water is periodically exchanged with fresh saltwater. Thus, removal system for those elements would be further needed in order to achieve intensive CRAS. Effects of accumulation of potassium and phosphate on abalone mortality would be also investigated by intergrading concentrations of those elements as shown in Figs. 8.5 and 8.6.

Many analytical instruments are used in various water quality analyses. Periodical calibration and maintenance must be indispensable to obtain accurate data. Another point to keep in mind is that water temperature affects measurements of some parameters such as DO, unionized ammonia, and electric conductivity. For example, an electric conductivity of 37.0 mS/cm (33.0‰) at 10 °C increases to 47.7 mS/cm (35.2‰) and 58.4 mS/cm (36.0‰) at 20 °C and 30 °C, respectively. Thus, attention should be paid to temperature when salinity is measured by electric conductivity.

8.4 Water Treatment Technologies for Maintenance of Water Quality in RAS

Since abalone does not tolerate low salinity of saltwater, they need fresh natural or artificial saltwater. It is very costly and labor intensive to obtain fresh saltwater in the place far from sea. Thus, land-based abalone farming should be established near the sea if natural saltwater, which has an optimum quality for rearing abalone, can be abundantly pumped up. However, appropriate coastal place for abalone farming is restricted in Japan because of geographical condition, water temperature, salinity, red tide occurrence, and so on. Although Wando district in Korea is robustly ongoing abalone farming on the sea, uneaten food, feces, dead body, and removed excrescence accumulate on the bottom of the sea and cause anoxic water leading to mortality by a shortage of oxygen. Semi-closed RAS is abundantly required to prevent from damage caused by natural disaster and water pollution. Moreover, zero-emission land-based aquaculture using CRAS independent of the sea is expected to save marine environment and to recover indigenous abalone stock in the future. Although RAS is advantageous for saving energy via temperature control, feeding control, and rearing management, water treatment technologies for maintenance of water quality are essential. More strict water quality control is needed for abalone than teleost fish. In this chapter, three main technologies for nitrification, denitrification, and dissolved organic matter treatment by electrolysis are introduced.

8.4.1 Nitrification Using Polypropylene Short Tubular Media

There are many types of biofilters that are commonly used in intensive RAS: submerged biofilters, trickling biofilters, rotating biological contractors, floating bead biofilters, and fluidized-bed biofilters (Timmons et al. 2002). Although each biofilter has its own strength and weakness, basic principle of nitrification is common among them.

There are two phylogenetically distinct groups of bacteria (ammonia- and nitrite-oxidizing bacteria) to perform nitrification. They grow slowly because of chemosynthetic autotrophic bacteria. pH of rearing water decreases, and nitrate accumulates as end product in the RAS with biofilters. Coral gravel is normally used for nitrification media because it is cheap and has wide surface area. However, loading a lot of coral gravel into the submerged biofilter tank and washing clogged media are labor intensive because of its heaviness. Thus, light polypropylene (PP) short tubular media (ϕ5 mm $\times$ 5 mm) is preferable and its price is almost same as coral gravel. Weights per unit volume of coral gravel and PP media are 1.07 and 0.28 kg/L, respectively. Ammonia conversion ratio of PP media (21.0 mg/L•day) was less than that of coral gravel (28.5 mg/L•day) calculated under the condition of 10 mg-N/L of ammonia at 20 °C on a basis of per volume of media. Although ammonia conversion ratio of PP media decreased by 26% compared to that of coral, handling of PP media is much easier than that of coral. Nevertheless, coral gravel is used together for weight on PP media and calcium supply.

Rates of nitrification increased by increasing turnover ratio of biological filter but reached a plateau at some point (Fig. 8.3). Independent circulation of nitrification tank contributes to some extent for downsizing. Continuous addition of ammonia chloride accelerated nitrification activities but reached at plateau after 22 times

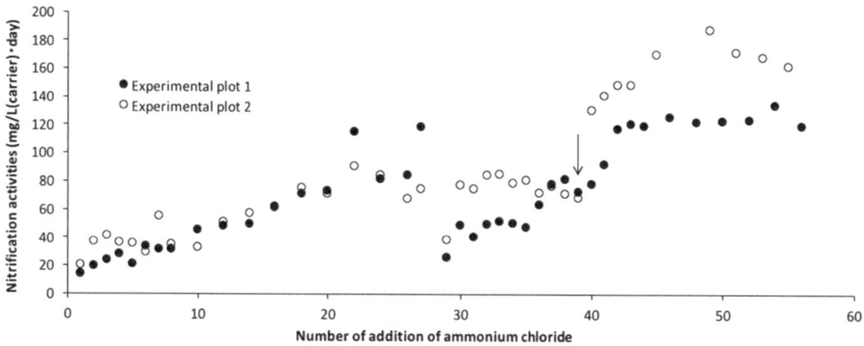

Fig. 8.9 Change in nitrification activities by periodic addition of ammonia chloride and seawater replacement. Arrow indicates saltwater replacement

of addition (Fig. 8.9). But nitrification activities accelerated again after new saltwater was exchanged. This means that accumulation of nitrate inhibit nitrification.

8.4.2 Automated Batch Processing Denitrification System

Nitrate is the end product of nitrification, and its accumulation is inevitable without exchanging saltwater or installing denitrification system. Although nitrate does not show an acute toxicity, it must not be left to accumulate. High concentration of nitrate may cause chronic toxicity (Fig. 8.5) and inhibition of nitrification (Fig. 8.9). Abalone started to die with increase of nitrate concentration and more than 100 mg-N/L of nitrate concentration significantly increased mortality (Fig. 8.5c). Levels of nitrate concentration ideally should not exceed 50 mg-N/L within RAS.

Level of nitrate concentration is determined by "emission speed of nitrogen compounds from abalone and uneaten diet" and "water exchanging ratio" in the case of prevention of nitrate accumulation by water exchange. The more saltwater exchange, the lower nitrate concentration is maintained, but it is more costly in order to gain saltwater and to adjust water temperature, and environmental load increases. There is a trade-off between level of nitrate concentration and cost for fresh saltwater. For example, 630 g-N/day of nitrogen load from abalone, which corresponds to approximately 200,000 abalone (shell length 70 mm) in a RAS of 42 m^3 requires a daily water exchange rate of 25% to maintain the level of nitrate less than 50 mg-N/L (Fig. 8.10a). Under this assumption, saltwater annually will cost 190,000 dollars if 1 m^3 of saltwater can be purchased at 50 dollars. Thus, business of marine-independent and land-based abalone farming is impossible to become profitable unless saltwater can be purchased at an extremely lower price.

Denitrification system has been mainly developed in advanced sewage treatment (Rittmann and McCarty 2001) but could not be applied to aquaculture system without any alteration. Criteria for denitrification control of sewage treatment are

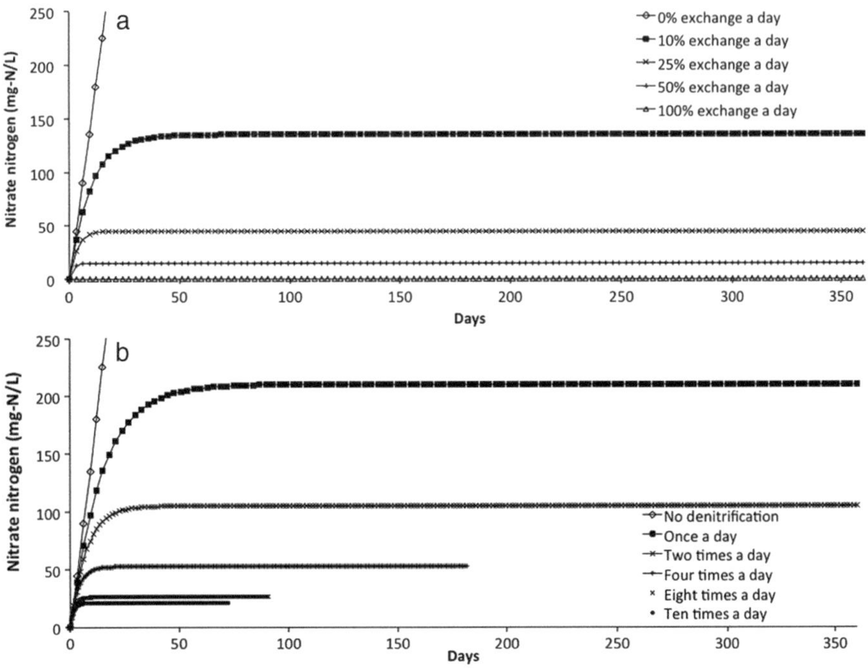

Fig. 8.10 Simulation for balanced nitrate concentration in treatment of exchanging water (**a**) or denitrification (**b**)

not likely to apply to aquaculture because water quality for fish farming is severely required (Colt 2006; Poxton and Allouse 1982). Although there are many studies on application of denitrification in freshwater and saltwater RAS during more than 30 years period, unifying concept for design and operation of these systems has not been developed (van Rijn et al. 2006; Klas et al. 2006). We have newly designed an automated denitrification system specialized for RAS and reared abalone for more than 1 year without water replacement (Fig. 8.11). The biodegradable aliphatic polyester was chosen to be used for a denitrification substrate in our study after classification of turbidity occurrence, hydrogen sulfide production, and denitrification activity among commercially available polymers. According to the computer simulation (Fig. 8.10b), nitrate concentration in the rearing tank reached an equilibrium, effectively balancing the nitrogen emitted by the abalone and batch denitrification treatment. Power approximated curve is fitted to the relationship between the balanced nitrate concentration and the turnover rate of the denitrification apparatus. This simulation indicated that excessive turnover is ineffective in decreasing low nitrate concentration in RAS.

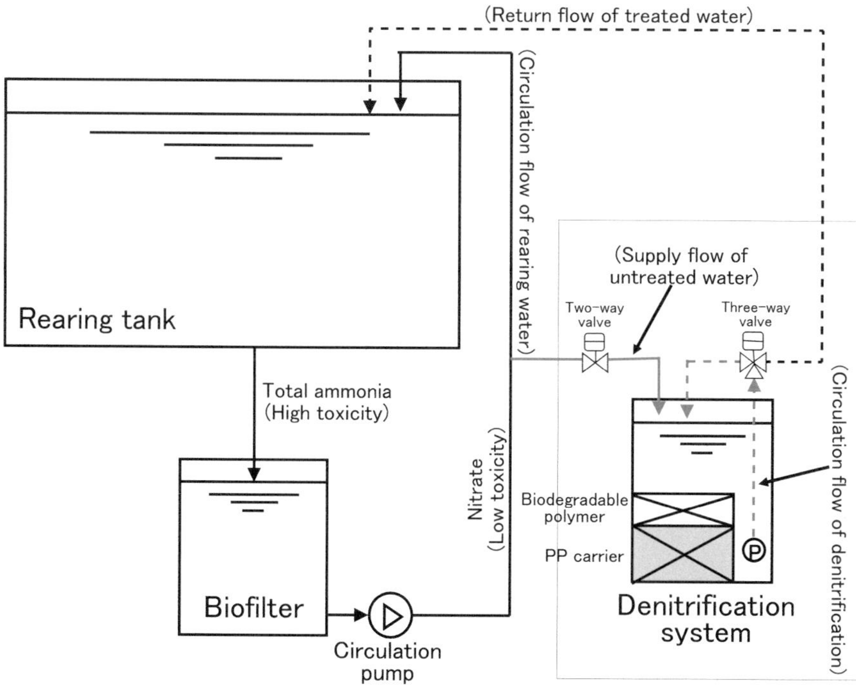

Fig. 8.11 A scheme of automated denitrification system for RAS

8.4.3 Treatment of Recalcitrant Dissolved Organic Matter (DOM)

As described above, rapid removal of settlement, suspended, and dissolved organic matters is very important (§8.3.3 Solids). Treatments of DOM are roughly divided into three groups on a principle basis: physical (form fractionators and activated carbon), photochemical (UV irradiation), and chemical (ozonation, chlorine and hypochlorite). Photochemical and chemical treatments also work as bacterial inactivation. Which treatment systems to be suitable should be decided depending on the situations.

Form fractionator or protein skimmer is commonly used to remove fine solids (<30 μm) and dissolved organic compounds in the aquarium fields. These devices work by generating small bubbles. The air/water interfaces of bubbles and surfactants are important for effective removal of DOM and particles. In the case of abalone culture, foam fractionators are relatively effective because abalone secrete mucous from the foot. Foam fractionation also contributes to increasing DO and degassing of CO_2 in the rearing tanks.

Activated carbon is widely used as a porous material which can physically adsorb fine solids and high molecular organic matters. There is a quick effect for color removal and deodorizing by activated carbon filtration or immersion.

However, adsorption ability is lost after fine polar is saturated with molecules and organic matters (adsorption breakthrough). Since abalone mortality is strongly affected by color of culturing water (Fig. 8.5b, c), activated carbon can be helpful as emergency treatment for decreasing chromaticity.

Flow-through UV sterilizer is commonly used in RAS. The low-pressure mercury-vapor lamp emitting light at a wavelength of 254 nm is used for disinfection. UV irradiation disinfection is very safe because UV light does not leak from the apparatus and produce toxic residuals. UV bulb should be changed at least once a year because output strength gradually decreases at 40% a year (Timmons et al. 2002). Prefilter is needed because organic compounds in water absorb UV irradiation and particles block UV transmission, resulting in reducing the strength of microbial inactivation. When UV irradiation is used for abalone rearing in the RAS, attention of induced spawning should be taken. Spawning in a chain reaction leads to terrible pollution of the rearing water.

Ozonation is utilized in advanced drinking water treatment as well as activated carbon filtration. Disinfections by UV irradiation or ozonation are two methods often applied in aquaculture. Oxygen is a reaction end product of ozonation. It is very helpful for oxygen supply. Non-biodegradable organic molecules that contribute to the yellowing of culture water are rapidly oxidized into smaller and more biodegradable molecules by ozonolysis (Summerfelt and Hochheimer 1997; Christensen et al. 2000). Ozone residuals should be toxic to abalone even at a low concentration although lethal concentration for abalone has not yet been determined. In general, ozone residuals can be lethal to fish at ozone concentration as low as 0.01 mg/L (Summerfelt and Hochheimer 1997).

Hypochlorite has been commonly used in disinfection of drinking water and more than 0.1 mg Cl^-/L of concentration is defined as a standard in Japan. In the case of saltwater RAS, chlorine is economically generated by electrolytic water treatment. Hypochlorite is generated from chlorine by dissolving with water. Chlorine disinfection has not been widely used in RAS because of its toxicity and formation of halogen compounds and chloramines, which are highly reactive and have a long half-life. Nevertheless, utilization and safety of electrolyzed saltwater have been investigated since it provides an effective and economically attractive alternative (Jorquea et al. 2002; Taparhudee et al. 2008 and Katayose et al. 2007). Katayose et al. (2007) revealed that more than 90% of the generated halogen compounds were bromoform but its amount was far less than the reference values for drinking water standards in Japan and the USA. However, application of an electrolytic chlorine water in a Pacific white shrimp (*Litopenaeus vannamei*) closed-hatchery system had a lower survival rate and a greater nitrite-nitrogen level than the control (Taparhudee et al. 2008). In a yellowtail amberjack (*Seriola lalandi*) hatchery, nasal cavity was ruptured by the direct electrolytic treatment of saltwater, though chlorine was removed with activated carbon filter and no detectable free Cl^- ion was confirmed during experiments. Impact upon widening cavity depended on the electrical current for chlorine generation (personal communication). Thus, this technique is not likely to apply abalone hatchery, and further study is needed to determine the effect of an electrolytic water treatment on fish hatchery.

The direct electrolytic water treatment is effective for the yellowing culture water in grow-out culture of abalone with monitoring free and combined Cl^- ion. Extremely small bubbles generated from platinum thin-coated titanium electrodes contain highly concentrated chlorine gas. Thus, recalcitrant DOM might be rapidly oxidized into smaller and more biodegradable molecules in the gas/water interfaces of bubbles with free residual Cl^- ion even at the undetectable level.

8.5 Business Strategy

Abalone culturing business by using RAS is not widespread unless it becomes profitable, though RAS culturing has many advantageous points compared to marine culture. There are three important strategies to make RAS culture profitable: "reduction of capital expenditure and operating expense (engineering)," "increase of abalone productivity (scientific)," and "sales and marketing managements (marketing)." A consistent and cross-sectoral approach is needed for the successful land-based abalone farming by RAS. Management perspective must be indispensable. First of all, a business plan is built as considering income and payout. Once a project starts, business and financial conditions must be perceived from profit-and-loss sheet (P/L), balance sheet (B/S), and cash flow statement (C/F).

8.5.1 Reduction of Capital Expenditure and Operating Expense (Engineering Aspect)

Abalone culturing style has been developed in the order of marine culture, land-based flow-through culture, and CRAS. Marine and flow-through cultures have been already profitable and widespread especially in Korea and China, respectively. Abalone farming remains a matter to grow popular in Japan.

Abalone culture by CRAS is still too costly. Water treatment systems for RAS (§8.4) might contribute to making abalone farming by CRAS profitable because it can largely reduce costs for saltwater replacement, temperature adjustment, and environmental loads. Costs for facility depreciation, seeding, food, labor, and electricity may account for more than 80% of total annual expense. Building is planned to provide optimum environments (temperature and light condition) for abalone culturing. Thermal insulation is helpful for saving energy for temperature control. The cheapest but secure construction is multi-insulating vinyl film greenhouse. Diversion of unused building like closed schools, factories, or storage can contain a construction cost. Arrangement of water treatment systems and rearing tanks in the facilities is important to maximize abalone productivity. Elimination of settleable solids in the rearing tanks by recirculating water flow saves time and labor-intensive work for cleaning, leading to reduction of labor cost. Every effort for growing capability and downsizing of water treatment systems and reduction of production cost should be made to increase abalone productivity.

Utilization of waste heat from industry such as municipal garbage incineration plant and renewable energy such as solar heat, geothermal heat, underground cold water, and so on enormously contribute to saving energy payment for temperature control though the capital expenditure is needed for the installation of heat-utilizing systems. For example, verification study of air conditioning for an Internet data center by a heat exchange system of snow cold energy, so-called White Data Center Project, has started in Bibai city of Hokkaido, a heavy snow area in Japan. Utilization efficiency of mass snow cold energy and waste heat from server devices in abalone culturing by using land-based CRAS will be evaluated. Cross-industrial cooperation might remarkably accelerate commercializing abalone farming by using RAS technologies.

8.5.2 Increase of Abalone Productivity (Scientific Aspect)

Basic data about abalone rearing such as growth curve (shell length, body weight, and body mass index), yield ratio (mortality and growth variability), and rearing density must be obtained in order to make a plan for maximizing abalone production. Body mass index (BMI) largely affected commodity value in abalone markets. BMI is empirically calculated as described below:

$$\mathrm{BMI} = \text{body weight (g)}/(\text{shell length})^3\ \left(\mathrm{cm}^3\right)$$

Mortality directly impacts business profitability. If increase of dead abalone is observed, a specific reason for mortality should be identified as soon as possible (Tables 8.2 and 8.3).

There are two strategies to reduce production cost: “reduction of operating expense” and “increase of productivity.” In the former case, feeding efficiency is important to save production cost. Food for abalone is seaweeds or artificial diet. Wild or cultured seaweed is available but there is picking season. Dried and stocked seaweed can be fed during no picking season. Since dietary value is greatly different depending on species, appropriate species of seaweed should be chosen as diet. Uki et al. (1986) investigated dietary value of seaweeds for growth of the abalone *Haliotis discus hannai* indicating that Order Laminariales, *Desmarestia* spp., *Chondria*, and *Enteromorpha* have superior value. On the other hand, many types of artificial diet are commercially available. Ingredients of artificial diet remain a matter of improvement. Uki et al. (1985) suggested optimum level of sodium alginate, lipid, and the mineral mixture in a casein diet based on the results of growth rate and feed conversion efficiency (FCE). In our study, almost the same growth rates (shell length, body weight, and BMI) of abalone fed with dried *Laminaria angustata* and artificial diet have been achieved in the same rearing density (7.5–13.4%) (Fig. 8.12). Rearing density is defined as occupancy of total

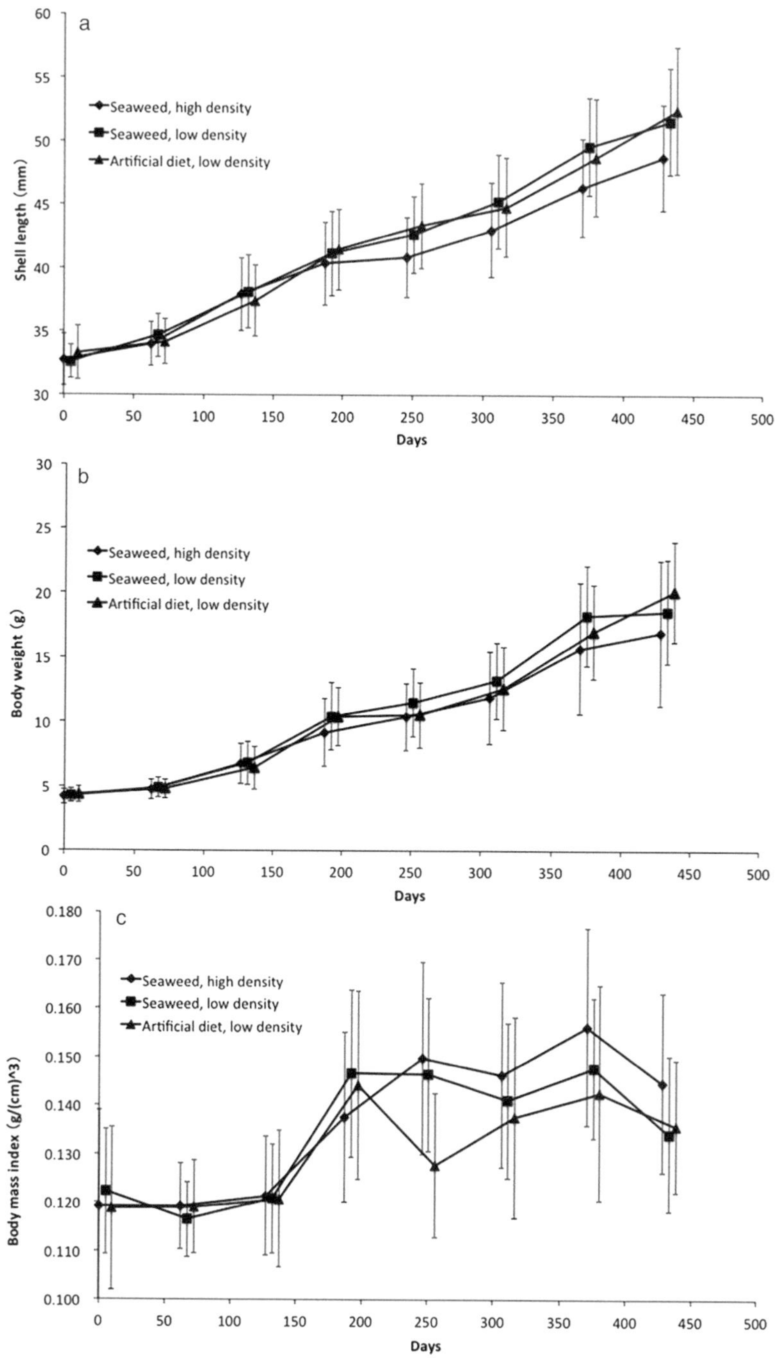

Fig. 8.12 Changes in individual shell length (**a**), body weight (**b**), and body mass index (**c**), comparing among food type and rearing density (Tukey HSD; $P < 0.05$, $n = 40$–110)

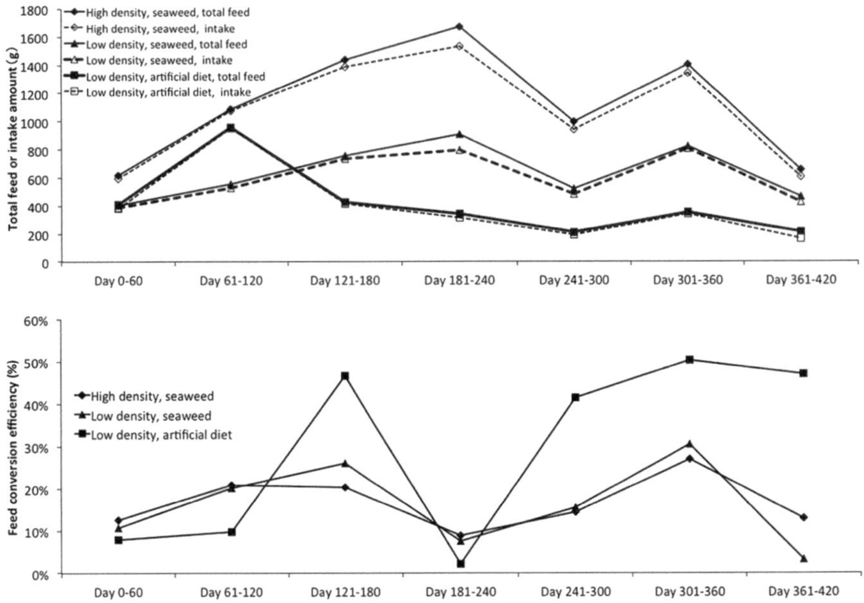

Fig. 8.13 Change in feed amount (**a**) and feed conversion efficiency (**b**), comparing among food type and rearing density

shell area to the area of base. Individual shell area is roughly estimated by 0.7 × (shell length)^2. FCE of artificial diet (approximately 50%) is about two times higher than that of algae during half of the experiment period though FCE irregularly changed (Fig. 8.13). A decision of food type should be made as considering cost-benefit performance. Weight gain of 10 t is achieved by feeding more than 20 t of artificial diet or 40 t of seaweeds (dry weight). Monitoring and management of feed amount and FCE lead to making production more efficient.

In the latter case, maximization of growth rate and rearing density are important to increase productivity. There is a multiple approach to maximize growth ratio: selective breeding, environmental control, diet, and rearing density. Some studies have indicated that selective breeding of abalone could enhance growth rates (Hara and Kikuchi 1992; Kawahara et al. 1997; Kobayashi et al. 2006). A systematic approach to breeding abalone will be required on the basis of breeding and genetics. Optimum environmental conditions such as water temperature, light, and water quality have been well studied and reviewed in many articles (Uki 1989; FAO 1990; Fujinaga et al. 1999; Sasaki 2005; Heasman and Savva 2007; Leighton 2008). Environmental control is much easier by CRAS than that by marine and land-based flow-through cultures. The rearing density largely affects growth properties (Fig. 8.12) (Ishida 1993; Miyauchi et al. 2006; Akimoto et al. 2007). The shell length and body weight in the low rearing density (7.5–13.4%) were significantly higher than those in the high rearing density (15.8–21.9%) (Fig. 8.12a, b). But the BMI in the low rearing density was lower than that in the high rearing density (Fig. 8.12c). Efficient production of well-fed abalone might be achieved by growing

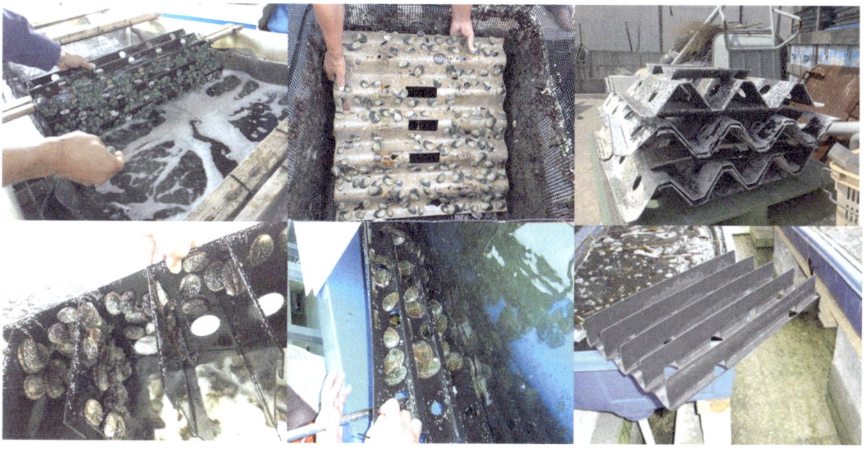

Fig. 8.14 Variety of feeding attachment plate for rearing juvenile and adult abalone

in the low density at the beginning then in the high density before the shipment. Shell length and body weight varied wider as they grew (Fig. 8.12a, b). Thus, optimum density for rearing should be decided in order to maximize absolute population of desired marketable size.

High-density rearing is accomplished by using feeding attachment plate. Feeding plate is normally made with corrugated vinyl chloride material in Japan. Width of peak and valley is decided by size of abalone. There are many kinds of feeding plates which are empirically designed (Fig. 8.14). Structure of feeding plate, water current in the rearing tank and feeding operation greatly influence how abalone distribute in the tank and how dense abalone can be reared.

8.5.3 Sales and Marketing Managements (Marketing Aspect)

Developing sales and sales strategy are important because no income is made unless grown abalones are sold. Furthermore, it is difficult to set a production schedule in accordance with the developments of the future market because it takes at least 3 years for abalones to attain the desired marketable size. However, grown abalone in land-based CRAS has a number of advantages. From an international perspective, environmentally sustainable and socially responsible aquaculture is demanded as addressed by Aquaculture Stewardship Council (ASC). The abalone cultured in RAS, which discharge minimal waste water into the environments, likely meet the demand. Marine-independent culture of abalone by RAS can avoid the risk of natural disaster and disease. Thus, abalone produced in RAS enables more stable supply and market price than wild abalone.

Creating added value in product is also needed for sales promotion. Production and hygiene control and product traceability are responsible. Introduction of third-party certification system such as global GAP (Good Agriculture Practice), ASC certification, and HACCP is advantageous to prove its safety and add value. Setting a tough standard in size and body weight of marketable abalone would be valuable since uniform quality of ingredients is desired in food-processing and restaurant industry.

Analytical evaluation of marketable abalone might support sales promotion. For example, seafood authentication is recently a growing issue in global marketplace. Wong and Hanner (2008) investigated market substitution in North American seafood by DNA barcoding and indicated that 25% of the samples were potentially mislabeled. Japanese abalone is difficult to discriminate species because morphological features are very similar. Hamaguchi et al. (2006) developed discrimination methods by using molecular techniques. Species identification and certification of origin in an objective manner such as molecular techniques would be an important judgmental standard for consumers.

Free amino acid (FAA) analysis revealed that abalone fed with artificial diet was higher in some FAA constituents than that fed with seaweed (Fig. 8.15). Nutritional and physiological research on FAA has been intensively carried out, and some FAA constituents affect taste. Thus, FAA rich abalone fed with artificial diet might have an advantage in human health and taste. Rupture strength was significantly different between cultured (11.2 ± 1.8, $n = 7$) and wild (15.6 ± 2.0, $n = 8$) abalone (mean ± SD, Tukey HSD; $p < 0.05$). Our hearing survey revealed that cultured abalone is preferred rather than wild one due to its softness when raw abalone is eaten as Sashimi. These analytical evaluations will corroborate the feature of the cultured abalone.

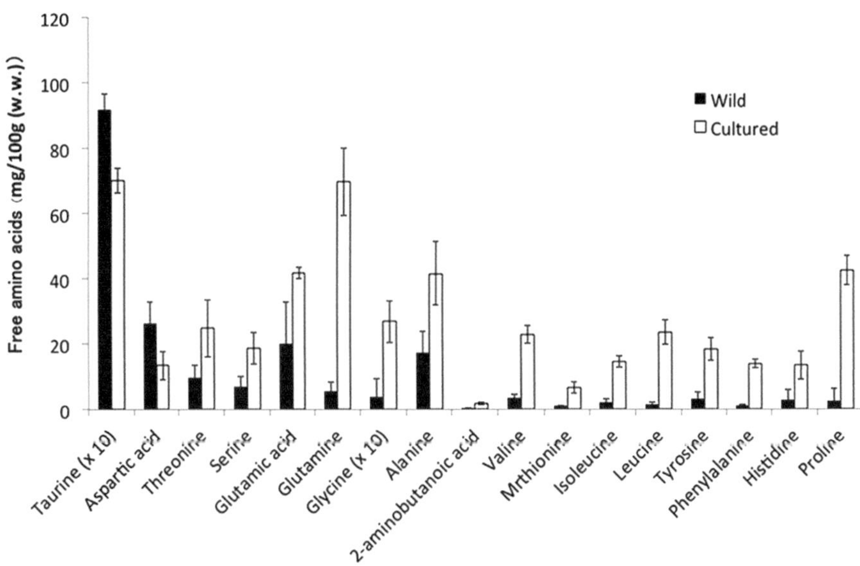

Fig. 8.15 Comparison of free amino acid concentration in body of abalone fed with artificial diet and natural seaweed. All contents in the bar chart showed significant difference between wild and cultured abalone (Tukey HSD; $P < 0.05$, $n = 3$)

8.6 Conclusions

Abalone is considered as one of the most valuable fishery products elsewhere in the world. However, landings from legal fisheries are declining year by year because of overexploitation, illegal harvesting, and habitat degradation. Thus, abalone farming increasingly plays an important role in maintaining abalone supply.

Methods of abalone production vary country by country are still being developed. Sea-independent land-based abalone aquaculture by using closed recirculating systems may be most developed and more advantageous than conventional abalone production: avoidance of risk for natural disaster and poaching, easiness of harvesting, high productivity, scheduled production, intensive traceability, much less environmental load, and so on. The breeding and culture technologies for abalone have already well developed. Thus, it is possible to achieve a land-based recirculating culture of abalone as long as water quality is maintained.

Abalone requires more strict water quality than teleostean fish. In the case of the flow-through aquaculture system, water in the rearing tanks is exchanged with pumped saltwater to maintain water quality. A larger amount of fresh saltwater is necessary when farming facilities become huge. Profitable abalone culturing is impossible in locations where ambient saltwater is not available.

Thus, water treatment systems are essentially required to maintain water quality in a land-based CRAS. Water quality standards and water treatment technologies for abalone rearing were have been acquired through various experiments. The key to success of abalone culture by CRAS is to avoid accumulation of solids, nitrogen compounds, and recalcitrant dissolved organic matters. Moreover, calcium supplement and adjustment of pH, alkalinity, and salinity by the addition of some chemicals are required as considering chemical interreaction.

Reduction of capital expenditure and operating expense, increase of abalone productivity, and sales and marketing management must be intensively promoted to disseminate the technology of CRAS. For the successful land-based abalone farming by CRAS, a consistent and cross-sectional approach is required from the aspects of engineering, scientific, and marketing. Research and development for successful abalone farming by RAS would be pursued to increase diversity of fishery production methods and contribute to recovering wild fishery resources and to strengthening food security.

References

Akimoto T, Sato H, Ushirokawa T (2007) Reasonable accommodation density by young abalone, *Haliotis discus*, culture technique. Bull Fukuoka Fisheries Mar Technol Res Cent 17:15–20. (In Japanese with English abstract)

Anonymous (2011) The total amount of damage on fish farming caused by the Tohoku tsunami reached 10 billion yen, Nikkei. Evening paper 18 May, p 14. (In Japanese)

Anonymous (2012) The total losses in abalone farming caused by red tide were 2.7 billion yen in Fujian province, China, Xinhua Japan, 30 May, http://www.asahi.com/business/news/xinhuajapan/AUT201207060135.html. Accessed 21 Nov 2014. (In Japanese)

Anonymous (2014) The mass mortality in abalone farming caused by continued low water temperature in Hokkaido, Japan, Hokkaido Shimbun. Morning paper 29 May, p 6. (In Japanese)

Christensen JM, Rusch KA, Malone RF (2000) Development of a model for describing accumulation of color and subsequent destruction by ozone in a freshwater recirculating aquaculture system. J World Aquacult Soc 31:167–174

Colt J (2006) Water quality requirements for reuse systems. Aquacult Eng 34:143–156

Cook PA, Gordon HR (2010) World abalone supply, markets, and pricing. J Shellfish Res 29:569–571

FAO (1990) Training manual on artificial breeding of abalone (*Haliotis discus hannai*) in Korea DPR. FAO Corporate Document Repository

FAO Fishery and Aquaculture (2017) http://www.fao.org/fishery/statistics/en. Accessed 7 Oct 2017

Fishtech Inc. (2012) http://www.fishtech.com/2012abalonemarket.html. Accessed 21 Nov 2014

Fujinaga A, Iwata N, Sakaguchi I (1999) Effects of water temperature on feeding and growth of Japanese abalones, *Haliotis gigantea* and *Haliotis discus discus*. CRIEPI report U98030. (In Japanese with English abstract)

Hamaguchi M, Sasaki M, Horii T, Kiyomoto S, Ohashi S, Fujii A, Takiguchi N, Hashimoto K, Takenouchi T (2006) Development of a discrimination methods by using molecular and monoclonal antibodies techniques for *Haliotis discus discus, H. madaka* and *H. gigantia* during their early life stages. Bull Fish Res Agen suppl 5:75–83. (In Japanese with English abstract)

Hara M, Fujio Y (1992) Genetic relationship among abalone species. Suisan Ikusyu 17:55–61. (In Japanese with English abstract)

Hara M, Kikuchi S (1992) Increasing the growth rate of abalone, *Haliotis discus hannai*, using selection techniques. NOAA Technical Report NMFS 106:21–26

Hara M, Sekino M (2005) Genetic difference between Ezo-awabi *Haliotis discus hannai* and Kuro-awabi *H. discus discus* populations: microsatellite-based population analysis in Japanese abalone. Fish Sci 71:754–766

Heasman M, Savva N (2007) Manual for intensive hatchery production of abalone -Theory and practice for year-round, high density seed production of blacklip abalone (*Haliotis rubra*)- NSW Department of Primary Industries. http://www.dpi.nsw.gov.au/__data/assets/pdf_file/0013/222331/Manual-for-intensive-hatchery-production-of-abalone-section1.pdf. Accessed 21 Nov 2014

Ino T (1952) Biological studies on the propagation of Japanese abalone (genus *Haliotis*). Bull Tokai Reg Fish Res Lab 5:1–102. (In Japanese with English abstract)

Ishida O (1993) Effect of population density on the growth of juveniles of the abalone, *Haliotis discus*. Suisan Zousyoku 41:431–433. (In Japanese with English abstract)

Jorquea MA, Valencia G, Eguchi M, Katayose M, Riquelme C (2002) Disinfection of seawater for hatchery aquaculture systems using electrolytic water treatment. Aquaculture 207:213–224

Katayose M, Yoshida K, Achiwa N, Eguchi M (2007) Safety of electrolyzed seawater for use in aquaculture. Aquaculture 264:119–129

Kawahara I, Noro T, Omori M, Hasekura O, Kijima A (1997) Genetic progress for growth in different selected populations of abalone, *Haliotis discus hannai*, at different hatcheries. Suisan Zosyoku 25:81–90. (In Japanese with English abstract)

Kikuchi S, Uki N (1974) Technical study on the artificial spawning of abalone, genus *Haliotis*. Part II: effect of irradiated sea water with ultraviolet rays on inducing to spawn. Bull Tohoku Reg Fish Res Lab 33:79–86. (In Japanese with English abstract)

Klas S, Mozes N, Lahav O (2006) A conceptual, stoichiometry-based model for single-sludge denitrification in recirculating aquaculture systems. Aquaculture 259:328–341

Kobayashi M (2002) Manual for marine cage culture of abalone (*Haliotis discus hannai*). Iwate Fish Tech Cent, Kamaishi p 1. (In Japanese)

Kobayashi T, Hara M, Kobayashi M, Sekino M (2006) Evaluation of growth performance of pacific abalone *Haliotis discus hannai* selected for juvenile size for 4 generations. Suisan Zosyoku 54:209–215. (In Japanese with English abstract)

Leighton P (2008) Abalone hatchery manual. Series: BMI Aquaculture Explained Issue 25. http://www.bim.ie/media/bim/content/publications/bimno_25_Abalone_Hatchery_Manual_2008.pdf. Accessed 21 Nov 2014

Liao PB, Mayo RD (1974) Intensified fish culture combining water reconditioning with pollution abatement. Aquaculture 3:61–85

Loyless CL, Malone RF (1997) A sodium bicarbonate dosing methodology for pH management in freshwater-recirculating aquaculture systems. Prog Fish Cult 59:198–205

Mie Prefecture Research Institute HP (2013) Manual for releasing abalone juvenile. http://www.mpstpc.pref.mie.lg.jp/SUI/inform/abalonmannalH24.pdf. Accessed 21 Nov 2014. (In Japanese)

Miyauchi M, Shibata T, Nagashima T (2006) Production of 60 mm or more large-scale abalone, *Haliotis discus discus*. Bull Fukuoka Fish Mar Technol Res Cent 16:41–49. (In Japanese with English abstract)

Nonaka T (2011) History of abalone propagation. Seibutsu Kenkyusya, Tokyo. pp 7–9, pp 11–12, pp 67–68, p 107. (In Japanese)

Poxton M (2003) Water quality. In: Aquaculture -farming aquatic animals and plants. Blacwell publishing, Oxford, pp 47–73

Poxton MG, Allouse SB (1982) Water quality criteria for marine fisheries. Aquacult Eng 39:153–191

Rittmann BE, McCarty PL (2001) Environmental biotechnology: principles and applications. McGraw-hill International Editions, Boston, pp 470–474. pp 497–500

Sasaki R (2005) Abalone *Haliotis* spp. In: Aquaculture system 3. Kouseisya-kouseikaku, Tokyo, pp 89–90. (In Japanese)

Summerfelt ST, Hochheimer JN (1997) Review of ozone processes and application as an oxidizing agent in aquaculture. Prog Fish-Cult 59:94–105

Taparhudee W, Suksamran S, Srisapoome P, Chuchird N (2008) Application of an electrolytic water treatment technique in a *Litopenaeus vannamei* (Boone, 1931) closed-hatchery system. Kasetsart J (Nat Sci) 42:503–512

Taya K (2007) Current situation and problem of abalone poaching in Japan. Yousyoku 7:80–83. (In Japanese)

Timmons MB, Ebeling JM, Wheaton FW, Summerfelt ST, Vinci BJ (2002) Recirculating aquaculture systems, 2nd edn. Cayuga aqua ventures, New York. pp 19–50, pp 157–203, pp 212–223, p. 395

Trade Statistics of Japan, Ministry of Finance HP (2013) http://www.customs.go.jp/toukei/info/index_e.htm. Accessed 21 Nov 2014

Uki N (1989) Abalone seedling production and its theory (1). Int J Aqua Fish Technol 1:3–15

Uki N, Kikuchi S (1975) Oxygen consumption of the abalone, *Haliotis discus hannai* in relation to body size and temperature. Bull Tohoku Reg Fish Res Lab 35:73–84. (In Japanese with English abstract)

Uki N, Kikuchi S (1984) Regulation of maturation and spawning of an abalone, *Haliotis* (Gastropoda) by external environmental factors. Aquaculture 39:247–261

Uki N, Kemuyama A, Watanabe T (1985) Development of semipurified test diets for abalone. Bull Jpn Sci Fish 51:1825–1833. (In Japanese with English abstract)

Uki N, Sugiura M, Watanabe T (1986) Dietary value of seaweeds occurring on the pacific coast of Tohoku for growth of the abalone *Haliotis discus hannai*. Bull Jpn Soc Sci Fish 52:257–266. (In Japanese with English abstract)

van Rijn J, Tal Y, Schreier HJ (2006) Denitrification in recirculating systems: theory and applications. Aquacult Eng 34:364–376

Wong EH-K, Hanner RH (2008) DNA barcoding detects market substitution in North American seafood. Food Research International 41:828–837

Yamakawa H (2006) Estimation of abalone illegal exploitation quantity in Miyagi prefecture, Japan. Suisan Zousyoku Kenkyu Kaihou 36:23–41. (In Japanese)

Part IV
Applications and Other Consideration for Recirculating Aquaculture Systems

Chapter 9
Seed Production Systems

Yoshihisa Yamamoto

Abstract A practical closed recirculation aquaculture system (CRAS) for seed production has been developed by Fisheries Research Agency (FRA) in Yashima Laboratory. This system consists of a culturing tank, reservoir tank (with filter treatment by collection nets), foam separation unit, biofiltration unit, UV disinfection unit, water level control tank, and recirculation pump. The foam separation unit is the most important equipment in CRAS for seed production. This is because CRAS has a high ability to remove suspended matter in the water by using microbubbles. Optimum control conditions of CRAS for seed production have been shown efficient biofilter materials and recirculation rate in biofilter and total system. Seed production using CRAS is more advanced than the flow-through system. Using CRAS, the advantages of low salinity seed production of tiger puffer, *Takifugu rubripes*, and high density culturing (20,000 fishes/kL of 25 mm in TL) for red sea bream, *Pagrus major*, seed production were demonstrated. Preventing diseases such as VNN was proved in seed production for red-spotted grouper, *Epinephelus akaara*, using CRAS in Kagawa Prefectural Fisheries Experimental Station.

Keywords Closed recirculation system • Seed production • Form separation unit • Recirculation rate • High rearing density • Disease preventing • Low salinity

9.1 Introduction

Given the characteristics of environment preservation, energy saving, disease preventing, high productivity, and availability to provide traceability, CRAS aquaculture has been in the spotlight for providing security and safety for seafood supply. Therefore industrial promotion of CRAS can be expected in Japan. Various accomplishments have been made in this field; however, it still has not been promoted much in Japan. The problem with seeds and seed production is thought

Y. Yamamoto (✉)
Japan Fisheries Research and Education Agency, National Fisheries University,
2-7-1 Nagata-Honmachi, Shimonoseki, yamaguchi 759-6595, Japan
e-mail: yama1215@fish-u.ac.jp

T. Takeuchi (ed.), *Application of Recirculating Aquaculture Systems in Japan*,
Fisheries Science Series, https://doi.org/10.1007/978-4-431-56585-7_9

to play a large part. Prevention of diseases that originate from seedlings is a required condition for establishing a CRAS (Maruyama and Suzuki 1998).

However, under the current conditions of seed production using a flow-through system, the disease problem clearly exists. Diseases which originate from seedlings such as VNN cause mass mortality and cause damage in aquaculture (Yamamoto and Hayase 2008). Therefore, technical advancements in CRAS to prevent diseases are extremely important (Yamamoto and Hayase 2008).

On the other hand, many CRAS has been designed and experiments have been conducted. However, the number of experiments is still insufficient for constructing a full CRAS for marine fish seed production with high performance and no maintenance at a low cost (Yamamoto 2013).

Since the year 2000, FRA has started a systematic research on CRAS for seed production at National Research Institute of Fisheries and Environmental Inland Sea (NRIFEIS) (Yamamoto 2013). This section will introduce the current status of developing CRAS for seed production, as well as case studies along with experiments.

9.2 The Characteristic of Seed Production in CRAS

Fish culturing can be divided into three major stages: broodstock management (the process of bringing adults into spawning condition and collecting the fertilized egg), seed production (the process of hatching eggs and nursing hatchlings into juveniles), and grow-out (the process of culturing juveniles to commercial size). The culturing method for each stage is different. In the case study explained below on marine fish seed production, the water management is conducted in a unique way. First, the culturing tank is filled up with filtrated and disinfected water, and then the fertilized eggs or hatched larvae are set in the culturing tanks. Second, the water is not exchanged for several days and after that period; the water exchange rate is gradually increased to maintain water quality. Another characteristic of this case study is that the culturing period for seed production is 1–3 months, which is significantly shorter than the conventional aquaculture method which takes about 1–3 years.

Feed for larvae in this seed production is also supplied in a unique way. In marine fish seed production, larvae are usually fed rotifer in the first stage, then nauplius of *Artemia salina* and finally artificial feed. In the early stages of larvae, the zooplankton (rotifer) is fed every day, and the density of rotifer in seed production tank is kept at 5–10 rotifer/mL. This is to increase the survival rate of larvae because its feeding ability during this stage is low. Meanwhile, management is necessary to maintain an environment where rotifer can survive in and increase the density in the seed production tank. Therefore the addition of phytoplankton (*Chlorella* sp. or *Nannochloropsis* sp.), which is food for rotifer is important for growing rotifer with high nutritional value. Thus the water in the seed production tank for larvae at the early stage contains phytoplankton, zooplankton, larvae, and bacteria. In other words, there are a lot of minute organic suspended matters in the water (Fig. 9.1).

Since the larvae do not have a completely developed bone formation and swimming ability, the water current and water recirculation need to be lowered during seed production. There is a high risk of mortality for larvae in the early stage when it is cultured in a flow-through system due to the rapid current that occurs from the high rate of water exchange. Therefore, water exchange (or recirculating) rate needs to be low during seed production (Yamamoto 2013).

Usually in grow-out, water exchange (or recirculating) rate is set to be high in both the flow-through system and CRAS. Since the high water exchange (or recirculating) rate leads to high water quality for fish culturing in grow-out, it is necessary to use a high-powered pump to achieve a high water exchange rate. In this aspect, grow-out is different from seed production (Yamamoto 2013). Also, compared with seed production, the quantity of organic discharge including urine and feces from fish and the leftover feed are more in the grow-out.

There are obvious differences between seed production and the grow-out such as culturing period, condition of suspended matters, water exchange (or recirculating) rate, and quantity of organic discharge matters. In seed production, the key points of the culturing conditions are as follows: (1) short culturing period, (2) large amounts

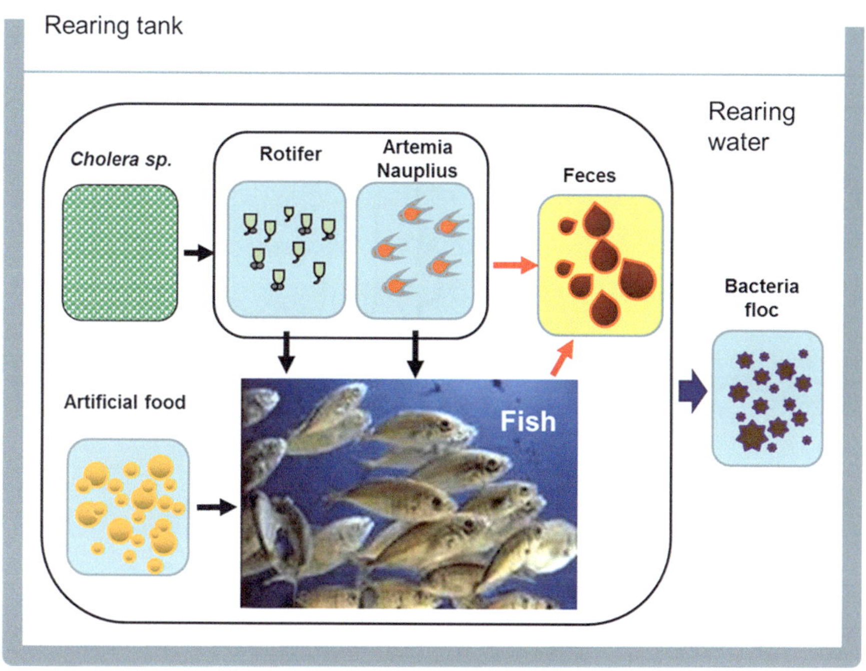

Fig. 9.1 Organic matters and organism in rearing water of seed production tank

of minute organic suspended matters in the water, (3) low water exchange (or recirculating) rate, and (4) low biomass in culturing space (Yamamoto 2013).

9.3 The Characteristic of CRAS for Seed Production

As mentioned above, there are different conditions for seed production and grow-out. CRAS for grow-out is not completely applicable to CRAS for seed production. CRAS for grow-out is constructed with the following units: culturing tank, settling tank, physical treatment units (for example, drum filter unit and foam separating units), biofiltration units, disinfection units, denitrification unit, oxygen supply unit, wastewater treatment units, recirculation pump, CO_2 removal units, temperature control units, etc. (Fig. 9.2).

However, CRAS for seed production may not include the following units: deposit tank and large filtering unit for removing feces, oxygen supply units which is necessary in high density culturing, CO_2 removal units which is necessary for removing dissolved CO_2 in the water from fish, and waste treatment units for removing waste from the system. These units are not necessary in CRAS for seed production because oxygen is supplied by aeration units and foam separation unit, while CO_2 is removed by aeration and foam separation unit. The other units can also be substituted with simpler methods since the scale of the system is smaller for seed production (Yamamoto 2013).

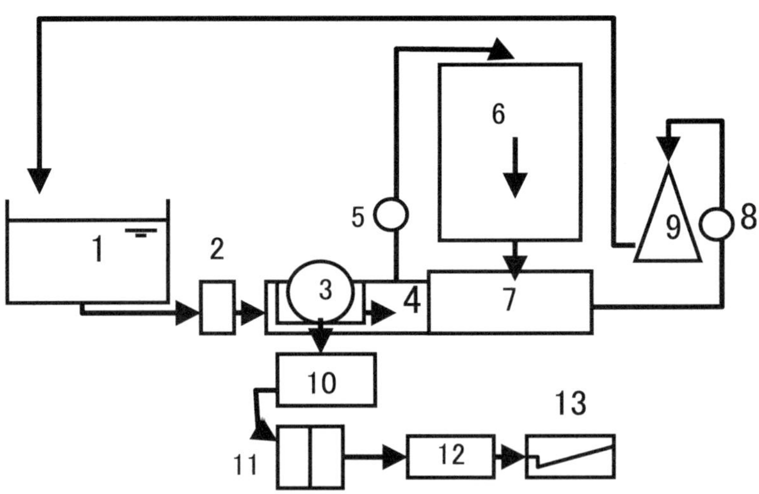

Fig. 9.2 Formation of CRAS for aquaculture in Wageningen University in Netherlands: *1*, rearing tank; *2*, settling tank; *3*, drum filter unit; *4*, UV sterilization unit; *5*, recirculation pump; *6*, biofilter unit; *7*, reservoir tank; *8*, recirculation pump; *9*, oxygen supply unit; *10*, wastewater storage tank; *11*, denitrification unit; *12*, dephosphorylation unit; *13*, solid-liquid separation unit after flocculants treatment

9.4 Construction of CRAS for Seed Production

At FRA, a simple CRAS for seed production under the basic concept of fulfilling all the conditions for seed production with the bare necessities and minimal initial cost was constructed. This fundamental system consisted of the culturing tank, reservoir tank (with filter treatment by collection nets), foam separation unit, biofiltration unit, UV disinfection unit, water level control tank, and recirculation pump (Figs. 9.3 and 9.4) (Yamamoto 2013).

The system is easily adaptable to the varying conditions of the culturing methods in seed production. Given the differences from the grow-out system, the total recirculation rate in the system is set at low level. The distinctive point is that the recirculation rate in the biofilter unit is kept high. Generally, the low total recirculation rate will lead to the accumulation of ammonia in the system. In order to divert this risk, the water exchange rate of the biofilter is kept high. In addition, the newly developed foam separation unit works efficiently and assists the biofilter unit in keeping the nitrification performance at a high level. Therefore this

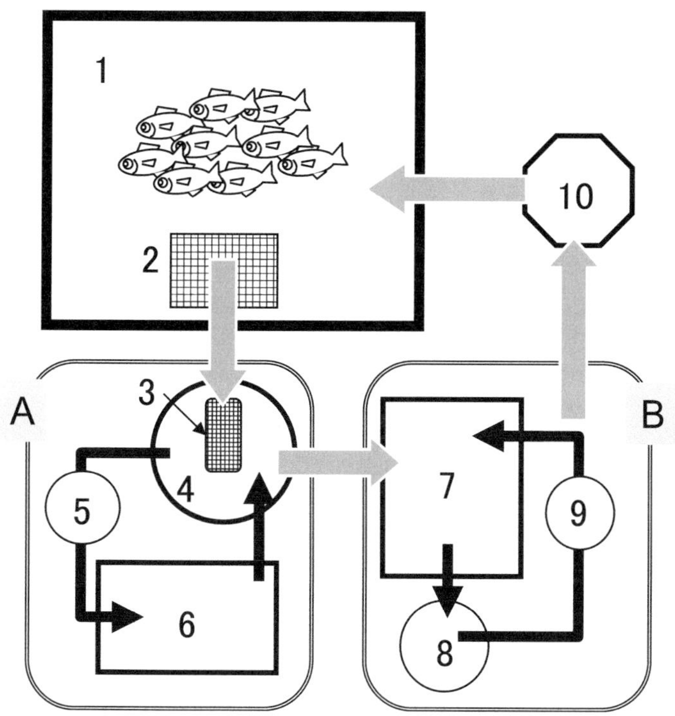

Fig. 9.3 Schema of formation of CRAS for seed production of Yashima station in FRA: *1*, rearing tank; *2*, strainer; *3*, collection nets; *4*, reservoir tank; *5*, recirculation pump; *6*, foam separation unit; *7*, biofilter unit; *8*, water level control tank; *9*, recirculation pump; *10*, UV sterilization unit; *A*, part of physical filter; *B*, part of nitrification

Fig. 9.4 Setting of prototype CRAS for seed production of Yashima station in FRA: *1*, reservoir tank; *2*, collection nets; *3*, foam separation unit; *4*, biofilter unit; *5*, water level control tank; *6*, UV sterilization unit

system will provide a stable operation with high and efficient performance for seed production.

The purification unit in this system is the foam separation unit, which is commonly used for sewage disposal. It is especially effective at removing of minute suspended matters. This function is suitable for removing and treating mass organic suspended matters in the water for seed production. In the case of inland aquaculture in Japan, Maruyama et al. demonstrated the effectiveness of foam separation unit in CRAS for food fish culturing (Maruyama et al. 1999; Suzuki et al. 1999).

Also, in seed production in Japan, Dr. Yoshihisa Yamamoto of Yashima Laboratory developed and designed a practical type of CRAS for seed production (Yamamoto 2013). It was constructed at Yashima station of NRIFEIS in FRA and Kagawa Prefectural Sea-Farming Center. It has been very successful and has gotten actual results in producing high-quality seed (Fig. 9.5). And another practical type of CRAS for mass seed production (rearing tank volume is 40 kL) was designated by Dr. Yamamoto of Yashima Laboratory (Fig. 9.6) and constructed at Kagawa Prefectural Fisheries Experimental Station which the fundamental detail of characteristic and function in several unit and system has been described in Chap. 2.

Fig. 9.5 Facility of CRAS for seed production of Yashima station in FRA

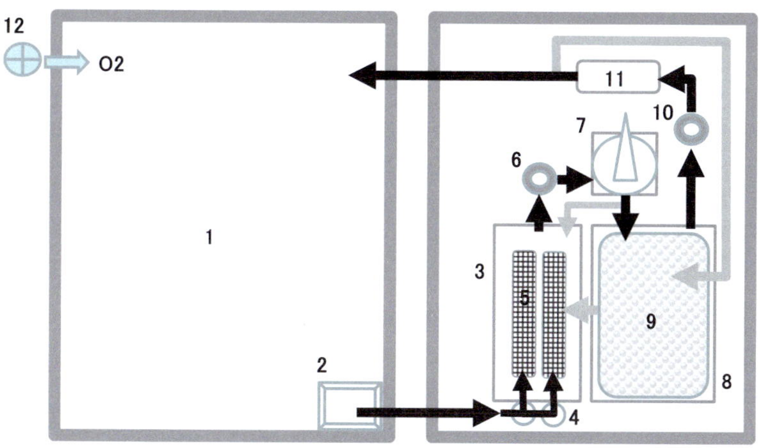

Fig. 9.6 Diagram of practical CRAS for seed production at Kagawa Prefectural Fisheries Experimental Station (practical type system was developed by Yashima Laboratory in FRA): *1*, seed production tank; *2*, screen net for recirculation; *3*, reservoir tank; *4*, water level control pipe; *5*, filtration net; *6*, foam separation pump; *7*, foam separation unit; *8*, biofilter unit; *9*, biofilter materials; *10*, recirculation pump; *11*, UV disinfection unit; *12*, oxygen generator

9.5 The Fundamental Units, Their Function, and Suitable Conditions in Operation

9.5.1 *Reservoir Tank and Net Filtration Unit*

The function of reservoir tank in this system is stocking the water which is drained out from the culturing tank. Another function is setting the net filtration unit for collecting the suspended matters in the water. The suspended matters are rotifer, *Artemia*, bacteria flock, feces, etc. One or two reservoir tanks are needed in one system. In the case of a system with one tank, a bypass needs to be set between the biofilter and this reservoir tank to prevent overflow and control the water level.

The volume of reservoir tank is about 5% for total water volume in CRAS, and it is necessary for treatment space for changing the filtration nets by maintenance.

9.5.2 *Foam Separation Unit*

The first function of foam separation unit is the removal of minute organic suspended matters from the water. The foam separation unit is set to remove the minute suspended solids which the net filtering could not remove by sticking the suspended matter in the water to bubbles and collect it by foaming. It also has a function to degas CO_2 and dissolve O_2 using mass bubbles generated in the unit.

The CRAS for grow-out is a one-pass treatment. However for seed production, the water needs to be pumped back to the reservoir tank for foam separation treatment at a low recirculation rate. This method is recognized as the best way to purify the water.

The important function of foam separation units for seed production is separation of minute organic matters which are bacteria flock, *Chlorella* sp. for additional rotifer food in rearing water and face of organism, from rearing water. The ability of this unit has been described in Figs. 2.6 and 2.9 in Chap. 2. Foam separation unit is most important in CRAS for seed production.

9.5.3 *Biofilter Unit*

The biofilter unit is for the nitrification process to convert ammonium nitrogen (NH_3-N) into nitrite nitrogen (NO_2^--N) and nitrate nitrogen (NO_3^--N) by nitrification bacteria. In seed production, the amount of organic matter is relatively less than that of in grow-out. Therefore the volume of biofilter media to the total water volume in the culturing tank for seed production is about 5%. For grow-out, this percentage is more than 20%. A downflow-type biofilter is used in this system since it is easy to set up (Yamamoto 2013).

Using this system, it was confirmed that the higher the water exchange rate in the biofilter, the higher the nitrification performance (Yamamoto 2013). In this system, some water goes through the biofilter and UV disinfection treatment and then back to the culturing tank, while most of the water goes back into the biofilter unit. In general, nitrification performance of the biofilter will increase when it does 100 cycles per day (Fig. 9.7) (Yamamoto 2013). In addition to the pump, adding a simple water cycling unit with an air-lift system may lead to saving cost and energy. Furthermore, introducing an intermittent type to the biofilter can improve the nitrification performance and make the unit more compact.

9.5.4 *UV Disinfection Unit*

The function of UV disinfection unit is to decrease bacteria in the water by strong UV rays. There is a fixed value for operating the UV unit. When the speed of the water is slow, the time for UV disinfection becomes longer, whereby oxidants with high toxicity is generated. In the process of seed production, the recirculation rates

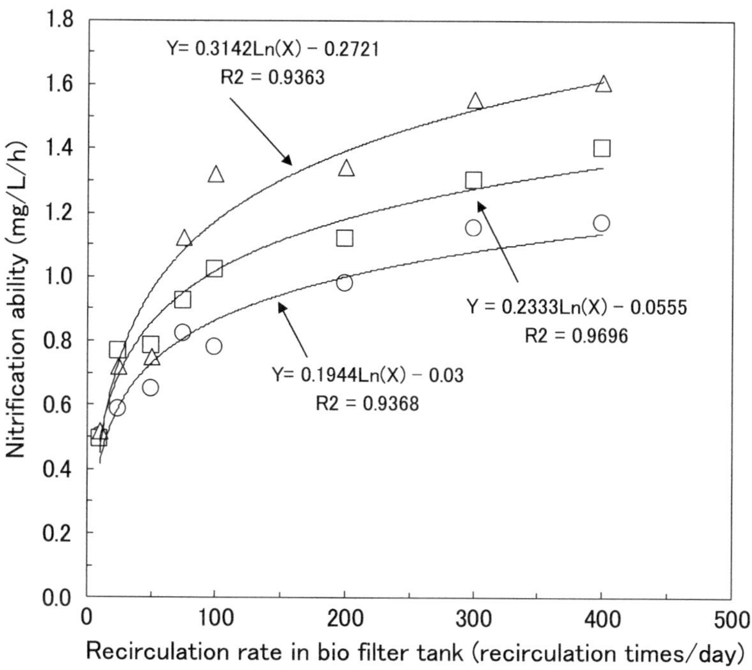

Fig. 9.7 Relationship between nitrification ability and recirculation rate in biofilter unit using pump: □ (open triangle), tank 1; □ (open square), tank 2; ○ (open circle), tank 3

gradually increase from 0%/day to 600%/day. Therefore, during the early stages of seed production, it is important to be aware of the oxidants which are produced. The solution to this problem is to distribute the UV unit and pipeline in CRAS like Fig. 9.3.

9.5.5 Recirculation Pump

The function of the recirculation pump is transferring the water among several units and culturing tank. Once the water is pumped up to the highest position, the water can flow downward using gravity. The standard output for recirculation pump is 0.25–0.4 kw (for 5–10 kL culturing tank) or 0.75–1.5 kw (for 20–50 kL culturing tank).

9.6 Cases of Research on CRAS for Seed Production in Japan

There are few research reports on CRAS for seed production in Japan. Mutsutani et al. (2001) reported a trial of CRAS in seed production, using the simple system which consisted of a net filtering unit, biofilter unit, and recirculating pump. This study showed the possibility of culturing of devil stinger, *Inimicus japonicus*, from hatched larvae to the juveniles of 15 mm in total length (TL).

Since 2000, Yashima station, NRIFEIS in FRA, started a research specifically on CRAS for seed production of marine fish and shellfish. Before this, such an extensive research had not been done in Japan. The research team constructed a practical system for seed production. This system consisted of the main unit and the foam separation unit (Yamamoto 2013).

The research was conducted with a prototype of CRAS in Yashima station, FRA for red sea bream seed production. The results were reported by Tomoda et al. (2005) and Kamoshida et al. (2006). Both results revealed the effectiveness of the CRAS for seed production. The first study was done with 15%/day water exchange rate, and the water exchange rate for the second study was 1.1%/day (Kamoshida et al. 2006). The research has so far been successful in developing CRAS for seed production of juveniles which are 30 mm in TL when the water exchange rate is under 0.5%/day. In these cases, stable results with high survival rate (45–70%) and high density culturing (5000–10,000 juveniles/kL) were achieved. So, the focus was shifted to the technical development of CRAS for seed production (Yamamoto 2013). Along with increasing the productivity, the technical development allow for a higher culturing density (20,000 juveniles/kL) in red sea bream seed production that are 25 mm in TL (Table 9.1, Fig. 9.8).

Table 9.1 Results of the experiment on high density seed production for red sea bream (*Pagrus major*)

Experiment no.	Lots	Initial		Final					
		Number of hatched larva (ind.)	Rearing density (ind./kL)	Days after hatching (days)	Total length (mm)	±SD	Number of juveniles (ind.)	Survival rate (%)	Rearing density (ind./kL)
1	Control	64,400	16,100	45	25.4	3.7	28,640	44.5	7160
	Experiment	214,500	53,625	45	24.1	4.0	80,830	37.7	20,208
2	Control	65,000	16,250	42	24.8	3.2	31,250	48.1	7813
	Experiment	205,500	51,375	42	23.5	4.6	79,210	38.5	19,803

*Control: Usual density of egg stocking in seed production tank
*Experiment: High density of egg stocking in seed production tank

Fig. 9.8 Photo in the case study of seed production of higher rearing density over 20,000 juveniles/kL (rearing water) in red sea bream *Pagrus major*

For system development, an original foam separation unit (patent number; 5130428 in Japan) and a new intermittent biofilter unit (patent number; 4670087 in Japan) which require less maintenance and is high in performance was developed. To improve the performance of the biofilter unit, nitrification ability according to the recirculation rate of the biofilter unit, water temperature, dissolved oxygen, and biofilter media were examined (Yamamoto 2013). Also, the effectiveness of foam separation unit in CRAS for seed production was examined. In this section, detailed subjects about biofilter media and foam separation unit are explained (Fig. 9.9).

9.6.1 Optimum Biofilter Media and Its Relationship with Foam Separation Unit

The concept of the selection for the optimum biofilter media for seed production is different from that of grow-out. The major differences between grow-out and seed production are the biomass and the amount of organic matters. In CRAS, the biofilter media get covered by biofilm and organic matter. When it is too covered, the nitrification ability decreases. However, if the foam separation unit can remove organic matter in the water, it will prevent the biofilter media from losing its nitrification ability. Therefore combining a foam separation unit in the CRAS and the selection of biofilter media is most important since they remove ammonia.

The effectiveness of the foam separation unit was examined via comparative experiment. As a result, in the system without a foam separation unit, organic matter accumulated in the biofilter and blockage occurred in the biofilter unit. Consequently the biofilter overflowed 30 days after hatching. In the biofilter unit,

Fig. 9.9 Newly developed foam separation unit of Yashima station in FRA (Yamamoto and Doi 2012, patent number; 5130428)

organic matters accumulated 2–3 times more when the system did not have a foam separation unit. In addition, the survival rate of the system without foam separation unit was lower than the system with foam separation unit. These results showed that the foam separation unit is important in CRAS, especially for seed production (Table 9.2, Fig. 9.10).

The search for the optimum biofilter media for seed production was conducted. After being matured for 3 months, the biofilter media had significantly higher nitrification ability, and coral showed the best results. The nitrification ability of the different types of media exhibited in the following order: coral > ceramic = carbon > fiber (Fig. 9.11) (Yamamoto 2013).

Considering all of the results, the most important unit in CRAS for seed production is the foam separation unit. It maintains a clean condition in the biofilter by removing the organic matters. In addition, the most effective biofilter media was the porous media such as coral.

Table 9.2 Result of comparative experiment for red sea bream seed production with and without foam separation unit

		Estimation of survival												
		Initial of experiment		6 DAH[a]		15 DAH[a]		Final						
Experiment no.	Lots	Number of hatched larva (ind.)	Rearing density (ind./kL)	Number of survival fish (ind.)	Survival rate (%)	Number of survival fish (ind.)	Survival rate (%)	Days after hatching (days)	Total length (mm)	±SD	Number of juveniles (ind.)	Rearing density (ind./kL)	Survival rate (%)	Occurrence of overflowing trouble in biofilter unit
1	Control (without use)	80,000	20,000	67,700	84.6	38,550	48.2	43	23.4	2.8	23,946	5987	29.9	○
	Experiment (use)	80,000	20,000	64,700	80.9	40,730	50.9	43	24.5	3.0	33,411	8352	41.8	–
2	Control (without use)	75,200	18,800	65,170	86.7	48,970	65.1	46	22.0	4.1	28,800	7200	38.3	○
	Experiment (use)	75,200	18,800	70,850	94.2	54,260	72.2	46	22.6	3.1	39,030	9758	51.9	–

[a]DAH: days after hatching

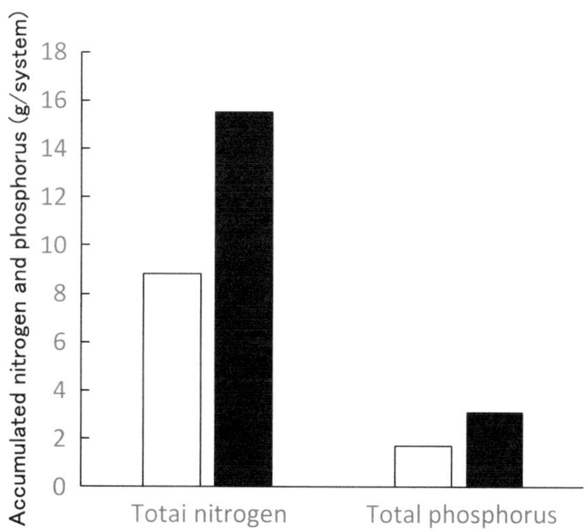

Fig. 9.10 Comparison of accumulated TN and TP in biofilter after seed production of red sea bream by systems with a foam separate unit or without it. □, system with foam separation unit; ■, system without foam separation unit

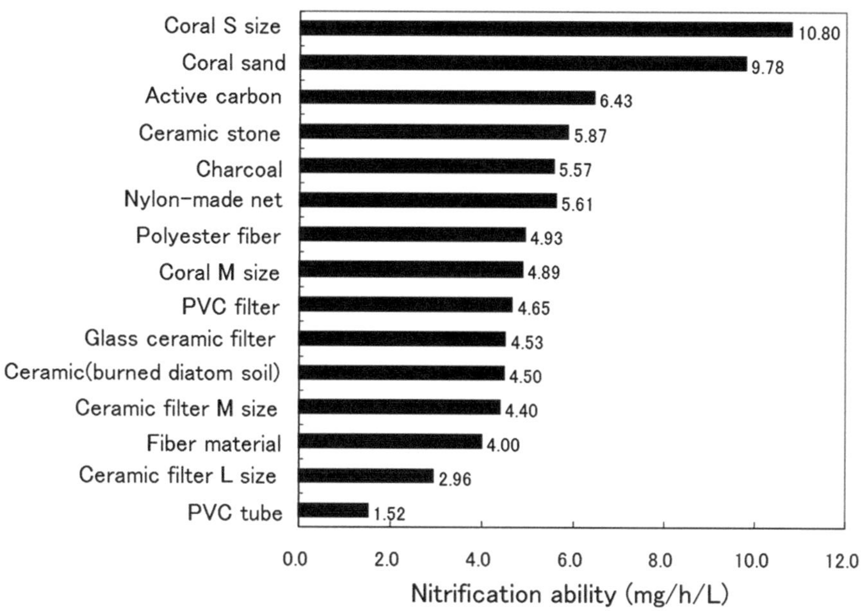

Fig. 9.11 Comparison of nitrification ability in several biofilter materials by nitrification test

9.6.2 Culturing Water; Artificial Seawater, Low Salinity

For the water, the possibility of using artificial seawater for red sea bream (Yamamoto 2013) and the effect of using low salinity for culturing tiger puffer (Imai et al. 2010) and red-spotted grouper were examined. Artificial seawater for CRAS is effective at maintaining stable salinity and preventing disease. Artificial seawater is expected to lead to stable production, especially for seed production. The culturing results between filtrated seawater and commercially sold artificial seawater were compared, and the comparison showed that there were no differences in growth or survival in seed production for juveniles of 25 mm in TL (Table 9.3).

In addition, artificial seawater is highly effective for preventing disease, so it can be used in seed production. When the effect of salinity on tiger puffer, *Takifugu rubripes,* was studied with 50% diluted seawater, it improved the growth and survival rate significantly (Table 9.4) (Imai et al. 2010). The effect of low salinity in CRAS for large-scale seed production has been conducted and has been proven to be effective (Katayama et al. 2013).

9.6.3 Suitable Recirculation Rate in CRAS for Seed Production

The optimum recirculation rate for red sea bream seed production in CRAS was examined (Yamamoto 2013; Yamamoto et al. 2013). The larvae are sensitive to water current and water quality. A strong current during the early stages will have negative effects on the larvae, especially those with low swimming ability. Therefore in the flow-through culturing system for seed production, the water exchange method is important. During the first 5–10 days after hatching, the water is not exchanged, but after that period the water exchange rate is gradually increased. On the other hand, in CRAS for seed production, low recirculation leads to low quality water due to insufficient water purification. Therefore the optimum recirculation rate in CRAS for red sea bream seed production was examined. The results of the experiments showed that the optimum condition of recirculation rate was between 3 and 6 cycles/day. This recirculation rate allows for a stable clean water condition and high survival rate (Fig. 9.12) (Yamamoto et al. 2013).

Thus, Yashima station in FRA will promote the systematic research on CRAS specializing in seed production and work to spread the technology. However, it is necessary to unify separate units into one compact system, as suggested by Yoshino (2003). Making it into one compact system will save cost and lead to a more efficient system. It is also important to arrange the system according to each site. In the near future, consulting business will be necessary for promoting CRAS in Japan.

Table 9.3 Comparison between natural seawater treated by sand filter and artificial seawater in seed production for red sea bream (*Pagrus major*)

Experiment no.	Lots	Initial		Final					
		Number of hatched larva (ind.)	Rearing density (ind./kL)	Days after hatching (days)	Total length (mm)	±SD	Number of juveniles (ind.)	Rearing density (ind./kL)	Survival rate (%)
1	Control	58,600	14,700	40	29.3	3.5	29,300	7325	50.0
	Experiment	60,900	15,200	40	29.7	3.6	30,100	7525	49.4
2	Control	61,000	15,300	40	29.9	3.4	28,100	7025	46.1
	Experiment	56,200	14,100	40	29.1	3.9	29,900	7475	53.2

*Control: Rearing water is used natural seawater treated by sand filter
*Experiment: Rearing water is used artificial seawater

Table 9.4 Body weight and survival rate of tiger puffer (*Takifugu rubripes*) cultured in three different salinity for 40 days in CRAS for seed production at Yashima station in FRA

Designed salinity (psu)	Total body weight (g)	Mean body weight* (g)	Survival rate (%)
16	7040	0.471 ± 0.016^{ab}	24.9
24	6370	0.524 ± 0.034^{a}	20.3
32	3930	0.444 ± 0.057^{b}	14.8

*Values are mean ± standard deviation, and those having different superscript letters are significant difference at $P < 0.05$ (Tukey-Kramer multiple comparison test)

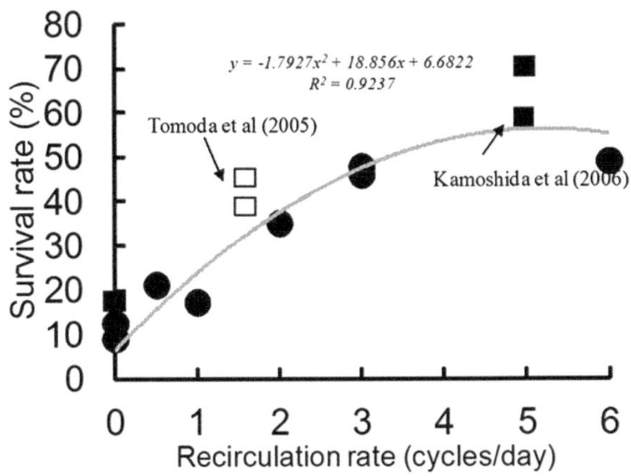

Fig. 9.12 Relationship between survival rate of seed production in red sea bream and recirculation rate in CRAS for seed production in Yashima station, in this study (●), in data of Tomoda et al. 2005 (□), in data of Kamoshida et al. 2006 (■)

9.7 Cases on Demonstration of CRAS for Seed Production

This section will introduce some cases of demonstration tests using the practical type of CRAS at Kagawa Prefectural Fisheries Experiment Station and Hiroshima Prefectural Technology Research Institute of Fisheries Marine Technology Center. Yashima station in FRA has been collaborating with both institutions since 2009.

The purpose of the collaboration at Kagawa Prefectural Fisheries Experiment Station was to prevent viral nervous necrosis (VNN) disease during seed production for red-spotted grouper, *Epinephelus akaara*. The purpose of the test at Hiroshima Prefectural Technology Research Institute of Fisheries Marine Technology Center was to demonstrate high productivity in low salinity.

The concept was to make the system practical, effective, and compact. The principle was to maximally utilize existing materials at each work site. The system also needed to be flexible to change so improvements could be made accordingly.

9.7.1 The Case of Disease Prevention; Kagawa Prefectural Fisheries Experiment Station

The VNN virus in marine fish is a threat in grow-out and seed production. In particular, VNN is the most deadly virus for grouper at the larval stage. Before 2009, there were several cases of the mass mortality caused by VNN at Kagawa Prefectural Fisheries Experiment Station during seed production of red-spotted grouper. Kagawa Prefectural Fisheries Experiment Station took various preventive measures against VNN, and the application of CRAS was their final measure.

The most possible source of infection was considered to be the seawater intake by water pump. Therefore incorporating a dissociation unit in the CRAS was the perfect disinfection treatment. For a 40 kL culturing tank, the CRAS for seed production consisted of a 5 kL reservoir tank, a II type foam separation unit (patent number; 5130428 in Japan), a 1.5 kw recirculation pump, a 5 kL biofilter tank (the downflow submerged type), a 0.75 kw recirculation pump, and a UV disinfection unit. The units and tanks were joined with pipes or hoses (Fig. 9.13).

The effect of preventing VNN using this system in Kagawa Prefectural Fisheries Experiment Station has been conducted every year since 2009. In the 18 seed production trials in CRAS, diseases caused by VNN were not observed until the juveniles of 50–60 mm in TL after hatching. The survival rate in CRAS seed production increased four to five times more than that of the flow-through system. Therefore these results demonstrated the effect of CRAS in preventing diseases. Specifically, CRAS for seed production is effective at preventing VNN which lead to stable of seed production (Fig. 9.14).

9.7.2 The Case of High Productivity in Low Salinity; Hiroshima Prefectural Technology Research Institute of Fisheries Marine Technology Center

Hiroshima prefecture is located in Seto Inland Sea, where inshore fisheries are prosperous. The project to increase stock of coastal fish, e.g., marbled rock fish *Sebastiscus marmoratus*, red-spotted grouper *Epinephelus akaara*, devil stinger *Inimicus japonicus*, etc., is conducted on this site. It is necessary to accomplish the project to produce the mass seeds for releasing. Several coastal fish are adapted to the low salinity condition at specific periods in their life cycle. The growth and survival rate in 25–75% diluted seawater are better than those in 100% seawater. There are some studies which report this kind of phenomenon for rock fish, tiger puffer, etc. Therefore Hiroshima Prefectural Technology Research Institute of Fisheries Marine Technology Center examined the effect of low salinity during culturing. In CRAS it is necessary to keep a certain salinity level. In Hiroshima Prefecture, it was demonstrated that in CRAS it is easy to keep the optimum condition that leads to high productivity with higher growth and survival rate.

Fig. 9.13 Setting of practical type CRAS for seed production at Kagawa Prefectural Fisheries Experimental Station (practical type system was developed by Yashima station in FRA): *1*, rearing tank; *2*, reservoir tank; *3*, foam separation unit; *4*, biofilter unit; *5*, UV sterilization unit

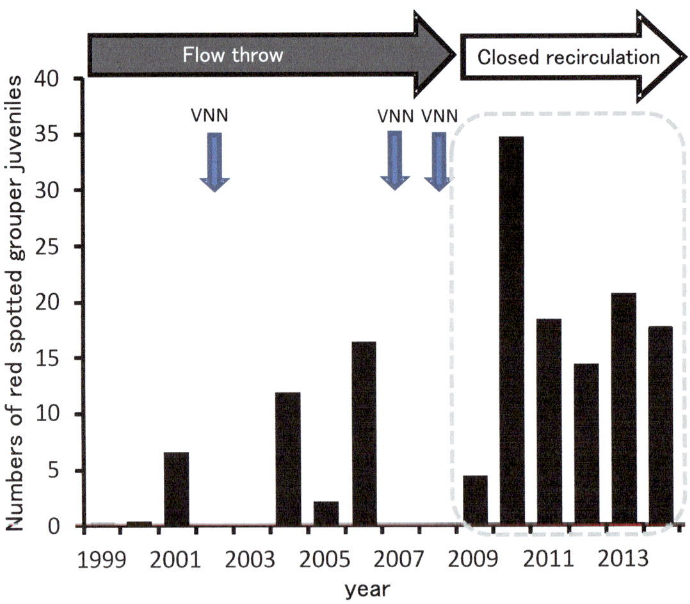

Fig. 9.14 Case study of avoidance against external disease risk of VNN in seed production of red-spotted grouper *Epinephelus akaara* using CRAS for seed production at Kagawa Prefectural Fisheries Experimental Station, its system design by Yashima station in FRA

Fig. 9.15 Setting of practical type CRAS for seed production at Hiroshima Prefecture, Fisheries and Marine Technology Center (practical type system was developed by Yashima station in FRA): *1*, reservoir tank; *2*, recirculation pump; *3*, foam separation unit; *4*, biofilter unit; *5*, UV sterilization unit; *6*, rearing tank

For a 5 kL culturing tank, CRAS for seed production consisted of a 0.5 kL reservoir tank, a I type foam separation unit (patent number; 5130428 in Japan), a 0.4 kw recirculation pump, a 1 kL biofilter tank (the downflow submerged type), a 0.4 kw recirculation pump, and a UV disinfection unit (Fig. 9.15). The units and tanks were connected with pipes and hoses. Using this system, the effect of 50% diluted seawater was compared with the flow-through system. The survival rate of larvae in seed production using CRAS was 1.6 times as higher than that of the flow-through system. This result showed the effectiveness of CRAS for seed production. In seed production, it is possible to achieve high productivity with CRAS.

9.8 Conclusion

As mentioned in the beginning, it is essential to produce healthy seedlings in order to properly conduct land-based aquaculture and spread the technology. It is also important to produce seedling without disease and to create the supply system in Japan because many inland aquaculture businesses closed down due to mass mortality caused by disease-infected seedling. In CRAS, there is a higher risk of disease spread once a pathogen enters the system since it is a closed system. Therefore it is necessary to prevent the disease with an ironclad treatment. Except for a few companies conducting seed production in semi-closed recirculation systems in Japan, there are not many private companies that use CRAS for seed production. Therefore, it is necessary to encourage private seed production companies to use CRAS in order to promote inland aquaculture. In Japan there are over 100 fish farming centers so it is necessary to construct a cooperation system between the public institutions and the private sector for technological support.

References

Imai T, Arai D, Morita T, Kogane T, Yamamoto Y, chida N, Endo M, Takeuchi T (2010) Effect of low salinity on growth and survival of tiger puffer *Takifugu rubripes*, and the purification of rearing water in closed recirculating seed production. Aquac Sci 58(3):373–380

Kamoshida M, Yamazaki H, Yamamoto Y (2006) Seed production of red seabream, *Pagrus major*, using a closed recirculating system. Saibai Giken 33(2):67–76

Katayama T, Morita T, Imai T, Yamamoto Y (2013) Mass seed production of ocellate puffer *Takifugu rubripes* in closed Recirculating system. J Fish Tech 5(2):165–169

Maruyama T, Suzuki Y (1998) The present state of effluent control in Japan and pollutant load from fish culture to environment - possibility of intensive recirculating fish culture systems. Nippon Suisan Gakkaishi 64:216–226

Maruyama T, Suzuki Y, Sato D, Kanda T, Michishita T (1999) Performance of a closed recirculating system with foam-separation and nitrification units for intensive culture of Japanese flounder. Nippon Suisan Gakkaishi 65(5):818–825

Mutsutani K, Nakase G, Ryu S-L (2001) Seed production of Scorpaenid fish Inimicus japonicas-II running water rearing and filter circulation rearin. Saibai Giken 29(1):1–7

Suzuki Y, Maruyama T, Takemoto S, Oda R (1999) Performance of a closed recirculating system with foam separation, nitrification and denitrification units for the intensive culture of Eel. J Jpn Soc Water Environ 22(11):896–903

Tomoda T, Fushimi H, Kurokura H (2005) Performance of a closed recirculation system for larviculture of red sea bream, *Pagrus major*. Fish Sci 71:1179–1181

Yamamoto Y (2013) Research of development for closed recirculation system for seed production in marine fishes. Doctoral dissertation of graduated school of marine science and technology, Tokyo University of Marine Science and Technology, p 206

Yamamoto Y, Hayase S (2008) Japan, in thematic regional reviews, the future of mariculture; a regional approach for responsible development in the Asia-Pacific region. FAO Fish Proc 11:189–198

Yamamoto Y, Doi K (2012) Water treatment method and foam separation unit using this method, patent number 5130428

Yamamoto Y, Kamoshida M, Takeuchi T (2013) Examination of suitable condition of recirculation rate in a closed recirculating system for larviculture of red sea bream, *Pagrus major*. Eco-Engineering 25(2):49–58

Yoshino H (2003) Research of closed recirculating aquaculture system. Doctoral dissertation of graduate school of fisheries Science, Hokkaido University, p 216

Chapter 10
Aquarium Recirculation System

Naoyuki Kato and Mutsumi Kawamata

Abstract This chapter introduces a closed recirculating system that includes the latest water treatment system recently introduced to aquaria in Japan. Conventional aquaria have been constructed in coastal areas where large quantities of natural saltwater can be supplied as needed. Due to the need for increasing profitability, there has been growing demand for inland aquaria that can attract large numbers of visitors, but this will require the resolution of operational issues not faced by aquaria in coastal areas. A closed-circulation system has been developed to resolve these issues. This chapter describes the three essential components constituting a closed recirculating system: an artificial saltwater manufacturing system, a high-performance filtration system, and a saltwater reuse system. The performance of the denitrification system included in the high-performance filtration system will also be covered.

Keywords Aquarium • Artificial saltwater • Denitrification (system) • Filtration • Nitrification • Recirculation • Water treatment

10.1 Closed-Circulation Systems in Aquaria

10.1.1 Background

Aquaria require large quantities of saltwater, supplied continuously, for maintaining and breeding the exhibited creatures and for cleaning the filtration equipment. For this reason, until now aquaria have typically been built in locations where natural saltwater could be obtained. Because these locations are often far from cities with large populations, the inability of such facilities to attract large numbers of visitors has been a problem, and aquarium operators have desired to locate aquaria in inland regions. But unlike coastal aquaria, such inland aquaria would not have access to saltwater from a nearby ocean, so saltwater would have to be transported to them by truck. The need to transport saltwater dramatically

N. Kato (✉) • M. Kawamata
Taisei Corporation, 1-25-1 Nishishinjuku, Shinjuku, Tokyo 163-0606, Japan
e-mail: kato@eng.taisei.co.jp

T. Takeuchi (ed.), *Application of Recirculating Aquaculture Systems in Japan*, Fisheries Science Series, https://doi.org/10.1007/978-4-431-56585-7_10

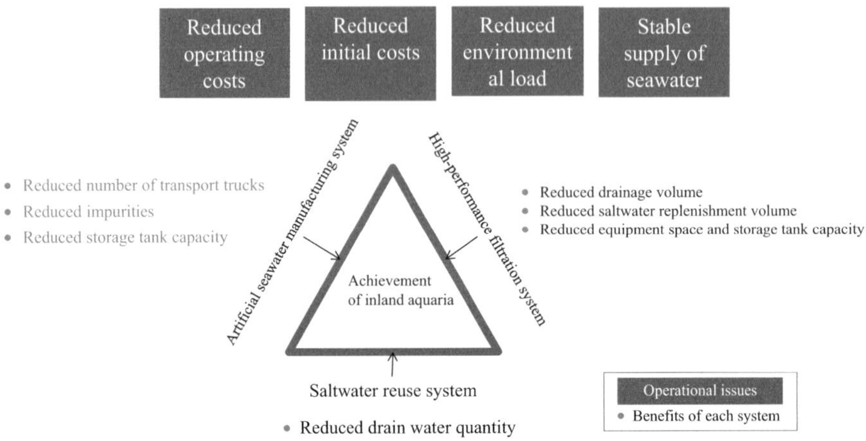

Fig. 10.1 Features of inland aquarium water treatment system

increases the operating costs of inland aquaria and also creates other operational problems such as the instability of saltwater supply and the need to reduce the CO_2 emissions generated by truck transport. For these reasons, a closed-circulation system that could resolve these problems was thought necessary to achieve the vision of inland aquaria.

The closed recirculating type water treatment system described in this chapter is composed of three components: an artificial saltwater manufacturing system, a high-performance filtration system, and a saltwater reuse system. Figure 10.1 shows the issues that could be resolved with successful use of the water treatment system and the objectives of the introduction of each of the component systems.

10.1.2 Overall Configuration of the Latest Water Treatment System

The water treatment system for an inland aquarium must provide the healthful saltwater environment needed for the maintenance and breeding of marine animals, in addition to the lack of turbidity needed for the exhibition tanks. Figure 10.2 shows a simplified view of the water treatment process in the three-component systems that have been designed to meet these requirements.

Instead of using natural saltwater transported by truck, artificial saltwater manufactured by the artificial saltwater manufacturing system is sent to each exhibition tank. The water in the exhibition tanks is constantly purified by the high-performance filtration system. The backwash wastewater from the filters and the wastewater from each of the breeding or exhibition tanks is purified and recycled by the saltwater reuse system and then used for backwashing the filters. A portion of the saltwater that has been reused repeatedly is mixed with and diluted

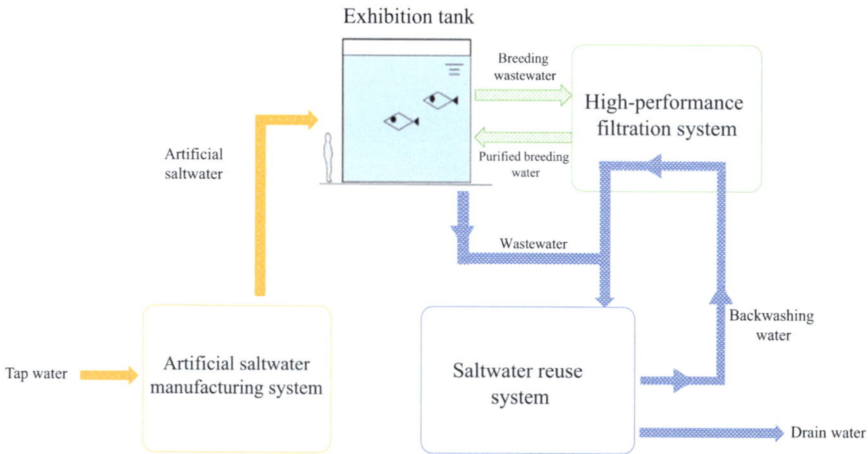

Fig. 10.2 Overall configuration of water treatment system (simplified)

by ordinary drain water and finally discharged into the sewer system. See 4.2.2 "Three components of the latest water treatment system" for a detailed description of each component of the system.

10.2 Three Components of the Latest Water Treatment System

10.2.1 Development of an Artificial Saltwater Manufacturing System

Background Inland aquaria in many countries generally use artificial saltwater. However, because there are no artificial saltwater concentrate products for aquarium use that are manufactured by domestic manufacturers in Japan, it has been difficult to operate Japanese aquaria using artificial saltwater. Nevertheless, given the unstable supply of materials from overseas manufacturers, a unique artificial saltwater concentrate that can be used to manufacture artificial saltwater has been developed for aquarium use. The Kyoto Aquarium, which opened in March 2012, became the first aquarium in Japan to use artificial saltwater made using the locally produced artificial saltwater concentrate and is therefore the first aquarium in Japan that does not require replenishment of its water supplies with natural saltwater.

Objectives The objectives of this system are not only to reduce the cost of transporting natural saltwater but also to reduce the CO_2 emissions that result from transporting the saltwater by truck. In addition, dissolving a concentrated solution to manufacture the artificial saltwater makes it possible to reduce the

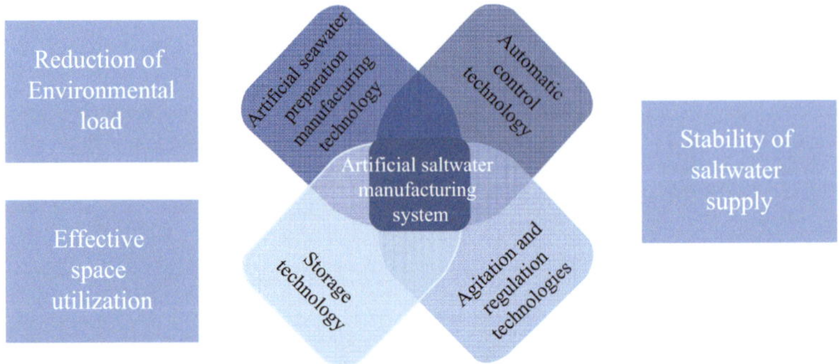

Fig. 10.3 Overview of artificial saltwater manufacturing system

capacity of the saltwater storage tanks and thereby reduces the space needed for installation of the dissolution tanks.

System Functions and Technologies As shown in Fig. 10.3, the artificial saltwater system combines the following four fundamental technologies:

I. Preparation manufacturing technology: a technology for manufacturing artificial saltwater concentrate
II. Agitation and regulation technologies: technologies for dissolving the preparation and regulating it to a suitable concentration
III. Storage technology: a technology for regulating and storing saltwater
IV. Automatic control technology: a technology for automatic control of the system

Production of Artificial Saltwater Figure 10.4 shows the process used to manufacture artificial saltwater.

① Insertion port
② Agitator
③ Water level gauge
④ Freshwater supply motor valve
⑤ Salinometer
⑥ Aeration pipe
⑦ Blower unit
⑧ Flow meter
⑨ Motor valve
⑩ Control panel

Step I Addition of the artificial saltwater concentrate

First, the saltwater concentrate preparation is added through the insertion port ① in (Fig. 10.4) to create highly concentrated artificial saltwater. Different preparations are provided for various species of fish and marine animals. The preparations

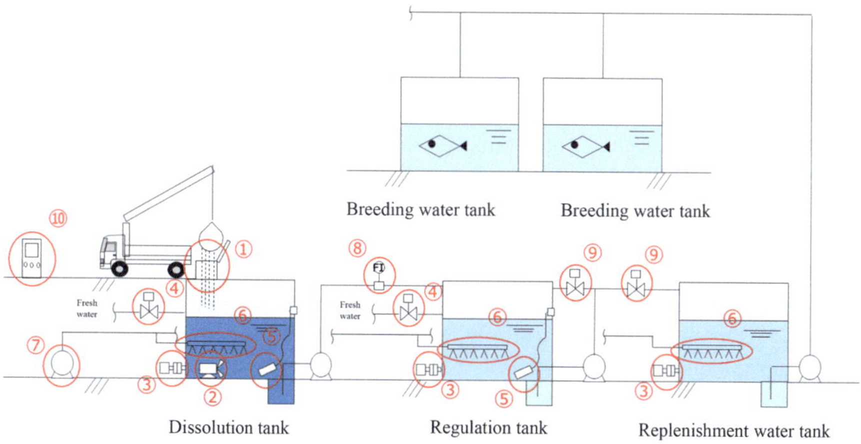

Fig. 10.4 Overview of artificial saltwater manufacturing system

are stored in flexible container bags which are transported to the site by truck and are added to the dissolution tanks using vehicle-mounted cargo-loading cranes. During this process, the vehicle-mounted cargo-loading crane moves the flexible container bag directly above the insertion port ①. The bottom of the bag is then opened, and the preparation is emptied from the bag directly into the tank. This process eliminates the need to keep the saltwater preparation in stock.

Step II Dissolution of the saltwater preparation

Next, the operator uses the control panel ⑩ to check the remaining capacity in the dissolution tank as detected by the water level gauge ③. If there are no problems with the remaining capacity, the operator manipulates the operation of the agitator ②, freshwater supply motor valve ④, and blower unit ⑦. The interior of each tank has an aeration pipe (equipped with multiple blow ports) that delivers compressed air from the blower unit. The agitator in the tank and the compressed air blown from the aeration pipe agitate the water and the saltwater preparation in the tank to accelerate the dissolution of the saltwater preparation. In addition, the agitator can agitate the contents of the tank more vigorously than the blower unit and thereby effectively prevent the heavier saltwater preparation from settling to the bottom of the tank. Furthermore, the salinity of the dissolved saltwater is checked automatically during agitation by the salinometer ⑤. Subsequently, when the salinity concentration reaches the set level, the freshwater supply motor valve ④ is closed by the control panel ⑩ to stop the supply of freshwater and the operation of the agitator.

Step III Regulation of the saltwater to standard concentration and replenishment

When the remaining capacity in the regulation tank is sufficient, the highly concentrated saltwater that has been generated is automatically transferred to the regulation tank. After the saltwater salinity in the regulation tank has been checked by the salinometer ⑤, the freshwater supply motor valve ④ is activated to supply freshwater for further diluting the saltwater to the concentration for actual use. Subsequently, by the controlled action of the motor valve ⑨, the artificial saltwater (which has been regulated to the standard concentration in the regulation tank) is sent to the replenishment water tank in accordance with its remaining capacity, where the saltwater remains on standby to be sent to the exhibition tanks.

However, if there is sufficient saltwater in the replenishment water tank, the artificial saltwater in the regulation tank (which has been regulated to the standard concentration) is sent back to the regulation tank. But, if there is space in the replenishment water tank, the artificial water in the regulation tank can be sent to the replenishment water tank. Similar to the dissolution tank and the regulation tank, the replenishment water tank is equipped with an aeration pipe for agitating the saltwater to maintain water quality while the water is being stored.

10.2.2 Development of a High-Performance Water-Conserving Filtration System

Background In aquaria, approximately 5–10% of the tank volume of saltwater is generally replenished each day to maintain water quality. In the filtration systems that have been used to date, biofiltration was performed to nitrify the ammonia and other organic matter into low-toxicity nitrate nitrogen (NO_3^--N) and prevent the water quality from becoming degraded by the excess feed, excreta, and other organic matter generated by the housed organisms. However, when this nitrate nitrogen accumulates in the water, it can have an adverse effect on the aquarium organisms (Nishi and Sawatari 2007). For this reason, it is necessary to replenish the aquarium water with large quantities of fresh saltwater to dilute the organic matter that would accumulate in the tank water. From the standpoint of operating costs, it is critical to reduce the amount of replenishment water needed. Accordingly, three technologies have been employed to construct a high-performance filtration system capable of reducing the quantity of both replenishment water and drain water.

Objectives One of the objectives was to reduce the daily quantity of replenishment saltwater required from 5–10% to approximately 1% of the tank volume and also to reduce the quantity of drain water. These water volume reductions were also expected to reduce the burden on the aquarium infrastructure and also reduce the amount of energy needed for water temperature regulation. Another objective was to reduce the amount of space occupied by the equipment used in the system and thereby make the facility more compact.

10.2.2.1 System Functions

1. Foam separation technology for removing protein and other organic matter

Tiny air bubbles to which the pollutants adhere are released into the water, which allows them to be removed from the water. The type of foam separator used in this system has been used at aquaria in Europe and the USA since the latter half of the 1960s and is considered effective for efficiently discharging feces, excess feed fragments, exfoliated mucous membranes, and other suspended solids from the tank water. This type of foam separator has been used in Japan since the 1990s and has helped to reduce turbidity and improve transparency of water in aquarium tanks (Suzuki and Nishi 2005).

2. Biofiltration technology including nitrification treatment technology to convert high-toxicity ammonia nitrogen (NH_3-N) and non-toxicity ammonia nitrogen (NH_4^+-N) into low-toxicity nitrate nitrogen (NO_3^--N)

The hazardous soluble substances that arise from suspended solids, excreta, and other residues in the tank water are removed and the treated water is recirculated. The sand filtration equipment used in this process removes the suspended solids and the hazardous nitrogen components from the tank water (Honma 1990). Besides its role in physical filtration, the sand bed plays a crucial role in nitrification treatment via the action of nitrifying bacteria. Highly toxic NH_3-N is converted at the first step into nitrite nitrogen (NO_2^--N) and then converted at the second step into NO_3^--N with lower toxicity. Percentage of NH_3-N into the ammonia nitrogen (NH_3-N, NH_4^+-N) is dependent on pH and temperature in the water.

3. Denitrification technology to remove nitrogen gas from nitrate nitrogen (NO_3^--N)

Although the toxicity of the excretory NH_3-N has been decreased through physical filtration and nitrification treatment, the high concentrations of NO_3^--N that accumulate in the tank water may become toxic if it undergoes reduction. To prevent the NO_3^--N from becoming hazardous again, nitrogen gas is removed by reduction of the NO_3^--N, rather than being diluted with fresh saltwater.

4. Water treatment by means of high-performance filtration

In the high-performance filtration system, the water in the exhibition tanks is pretreated to continuously remove organic matter by using a foam separator (protein skimmer). Then biofiltration is conducted using the same biofiltration technology as in the conventional system, but denitrification is performed simultaneously (Fig. 10.5).

5. Denitrification system performance

Denitrification treatment can be generally divided into two types: physicochemical treatment (e.g., ion exchange, chlorine treatment, electrodialysis, reverse osmosis, or coagulating sedimentation) and biological treatment (e.g., the activated

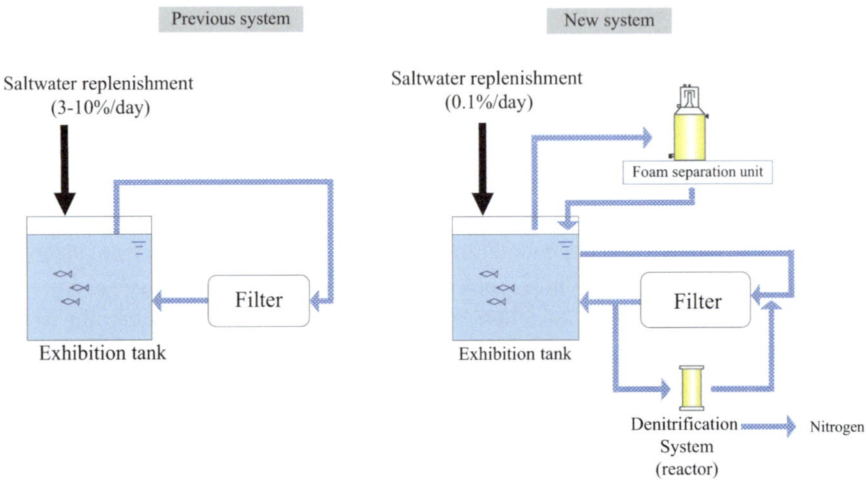

Fig. 10.5 Process of high-performance filtration water treatment

sludge method, the biofilm method, or the anammox method). However, there are few examples of using the former methods to treat large quantities of water due to their high cost. In contrast, biological treatment has been successfully used at many sewage-treatment plants, and its cost is generally low. The operation of denitrification systems based on biological denitrification will be discussed in detail below.

First, various types of tests were conducted to confirm the basic performance of biological denitrification using marine microorganisms (denitrifying bacteria) (Kurabe et al. 2010; Hamaguchi et al. 2010b, c; Ono et al. 2010a, b). These microorganisms were then introduced into an actual aquarium tank, in which saltwater fish had been raised, to revalidate their performance.

10.2.2.2 Optimization of Denitrification Reaction Conditions

1. Selection of electron donor

To develop a denitrification system using marine microorganisms to remove the accumulated NO_3^--N from tank water, a study was conducted to select the electron donor needed for the denitrification reaction that would be carried out by denitrifying bacteria in an anaerobic environment (Hamaguchi et al. 2010a).

The denitrification reactor was a fixed upflow denitrification tank with a capacity of 2 L, connected in series to three cylindrical columns each measuring 1 m in height. The columns were filled with sponge carriers as fixed-bed carriers (downflow hanging sponge; DHS G3: ϕ 33 mm × 33 mm). Two types of electron donors, either sodium acetate (Osaka et al. 2008) (organic additive) or elementary sulfur (sulfur additive), were used. For the initial test conditions, artificial saltwater

Table 10.1 Test conditions

	Organic reactor			Sulfur reactor	
	Elapsed time (days)	HRT (h)	C/N ratio	HRT (h)	Elementary sulfur (g-S/g)
RUN 1	0–44	10	2	10	120
RUN 2	45–62	6			
RUN 3	63–92		3		
RUN 4	93–115	10			

with a salinity of 3.0% and a NO_3^--N concentration of 40 mg-N/L from sodium nitrate was used as simulated tank water. Dissolved oxygen content of greater than 6 mg/L was generated in the simulated tank water through aeration, and the water temperature was controlled at 25 °C. The organic additive denitrification tank was operated with a hydraulic retention time (HRT) of 6–10 h and a carbon/nitrogen (C/N) ratio of 2–3 (RUN 1–RUN 4). For the sulfur additive denitrification tank, each column was filled with granulated elementary sulfur (S), equivalent to 120 g-S, with an HRT of 10 h. Its denitrification performance was then compared with that of the organic additive denitrification tank. Table 10.1 summarizes the test conditions. When continuous water flow tests were conducted using the organic additive and sulfur additive denitrification tanks, the NO_3^--N removal efficiency was 96% and 35%, respectively. Although in some cases the NO_3^--N removal efficiency dropped due to insufficient supply of electron donors, in general, through appropriate regulation of the HRT and C/N ratio, a nitrate nitrogen removal efficiency of 96% could be achieved with this system, resulting in a NO_3^--N concentration of 1.6 mg-N/L in the treated water. This test confirmed that the organic additive system performed better than the sulfur additive system.

2. Selection of fill carrier

To select the ideal fill carrier, the columns were filled with one of three types of fill carrier to determine the differences in treatment performance for different carriers (Hamaguchi et al. 2010a). The HRT was shortened in stages: (1) to 10 h and 6 h for the fixed-bed denitrification tank filled with sponge carriers; (2) to 12 h, 8 h, 6 h, 4 h, 2 h, and 1.5 h for the upflow sludge blanket (USB) denitrification tank filled with granules; and (3) to 8 h and 4 h for the coral sand denitrification tank filled with coral sand sampled from the sand filtration tank (nitrification tank) in an aquarium.

The oxidation-reduction potential (ORP) for each denitrification tank was in the range of −200 to −400 mV. In each case, the NO_3^--N removal efficiencies were high and stabilized at 95–97% during the test period. Figure 10.6 shows an example of the denitrification performance of the USB denitrification tank filled with granules under the test conditions shown in Table 10.2.

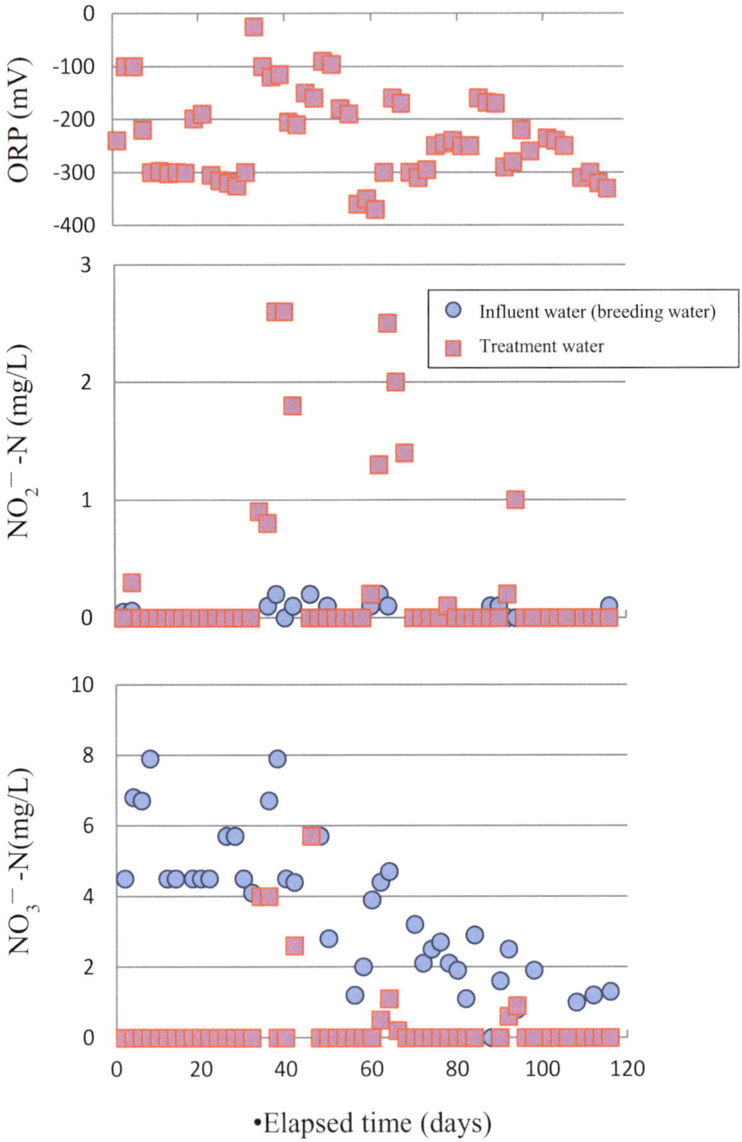

Fig. 10.6 Changes in ORP, NO_2^--N, and NO_3^--N over time

10.2.2.3 Verification

1. Verification of effect on saltwater fish

The denitrification reactor that had been developed above was introduced to a tank in which saltwater fish had actually been raised to check whether it could attain the

Table 10.2 Test conditions

RUN		1	2	3	4	5	6
Period (days)		1–33	34–69	70–80	81–84	85–94	95–116
HRT (h)	DHS nitrification tank	0.09					
	USB denitrification tank	5.0	2.5		1.2		
	(treatment water rate %)	[24]	[48]		[100]		
Aeration(simulated tank without fish)		Yes		No		Yes	
C/N ratio		3.0					5.0

same denitrification treatment performance as in the artificial stimulating water and whether it would affect the growth of saltwater fish. The test system was composed of a breeding tank (120 × 60 × 60 cm) containing 300 L of artificial saltwater, the USB denitrification tank (10 L) used in the previous section, a water temperature regulating cooler, disinfection device, and a filtration tank (nitrification tank). A total of approximately 900 g of marine specimens including saltwater fish—blue damselfish *Chrysiptera cyanea*, striped beakfish *Oplegnathus fasciatus*, and Hong Kong grouper *Epinephelus akaara*—and invertebrates (sea anemone) were held in the tank to achieve an accommodation density of 3 kg/m^3. The feed quantity was approximately 1% of total fish body weight.

To maintain a preferred water environment for marine organisms in this test, a high target was set to attain a low NO_3^--N concentration of less than 10 mg-N/L. The HRT was 6.7 h and the C/N ratio was set at 3. The other breeding conditions in a tank were set at a water temperature of 25 °C and with dissolved oxygen of more than 6 mg/L using aeration.

The ORP for the denitrification-treated water was approximately −350 mV, and denitrification progressed smoothly under these anaerobic environmental conditions. Table 10.3 shows the mean removal efficiency of nitrogen components. Since ammonia nitrogen in the denitrification-treated water increased slightly, the denitrification-treated water was not sent directly back to the water tank but sent back to the filtration tank (nitrification tank). In this denitrification tank, a removal efficiency of approximately 97% was achieved, and the target value of NO_3^--N less than 10 mg-N/L in the tank water was achieved.

2. Microflora analysis of denitrifying bacteria

On the 60th day after the start of the water flow test, DNA was extracted from sludge sampled from the bottom of the USB denitrification tank, including granules. A microflora analysis focusing on bacteria and identification of dominant species was then conducted.

An UltraClean Soil DNA Isolation Kit (Bio-Rad, Hercules, CA, USA) was used to extract DNA from the sludge sample, and EUB 338F-mix and UNIV 1500R primers specific to bacteria were used to amplify the 16S rRNA gene. A TOPO®TA Cloning kit (Invitrogen, Carlsbad, CA, USA) was used for cloning PCR products. Sequencing of each clone was performed by the Dragon Genomics Center of Takara Bio. Inc. Among the bacterial 16S sequences obtained, those with greater

Table 10.3 Mean efficiency of removal of nitrogen components

		Breeding water	Denitrification treatment water
NH_4^+-N	(mg-N/L)	0.1 ± 0.2	0.2 ± 0.3
NO_2^--N	(mg-N/L)	0.02 ± 0.02	0.06 ± 0.17
NO_3^--N	(mg-N/L)	4.2 ± 2.1	0.3 ± 0.7
NO_3^--N removal efficiency	(%)	96.8 ± 4.1	

Table 10.4 Results of genetic analysis of denitrifying bacteria

Genealogical classification	Lower-level classification	Ratio (%)
Proteobacteria	α-proteobacteria	1%
	Azospirillum	
	β-proteobacteria	19%
	Tauera	
	γ-proteobacteria	3%
	Pseudomonas	
	γ-proteobacteria	1%
	Marinobacter	
	δ-proteobacteria	1%
	Desulfobacter	
	δ-proteobacteria	3%
	Denitromonas	
	Others	35%
Others	–	35%

than 97% nucleotide identity obtained in a BLAST analysis against the Greengenes nucleotide database (http://greengenes.lbl.gov) were judged to be genetically identical. These sequences were then used to identify and categorize species sampled from the sludge.

Results of this analysis are shown in Table 10.4. Many of the sampled bacteria from the phylum *Proteobacteria* were found in the denitrification tank. Among the denitrifying bacteria of the phylum *Proteobacteria*, most belong either to *Thauera* (Macy et al. 1993; Song et al. 2000; Garrity 2005b), a β-proteobacteria-class genus that degrades acetic acid, or to *Marinobacter* (Gauthier et al. 1992; Yoon et al. 2004; Garrity 2005a; Takai et al. 2005), a marine γ-proteobacteria-class denitrifying genus. Species from *Thauera* were found to predominate among the bacterial populations in this denitrification tank.

10.2.2.4 Demonstration Using USB Denitrification Tank with Granules

1. Short-term demonstration test in a 1-m^3 tank

For this demonstration, the recirculating denitrification system developed as described in the previous chapter was evaluated in a 1-m^3 backyard aquarium

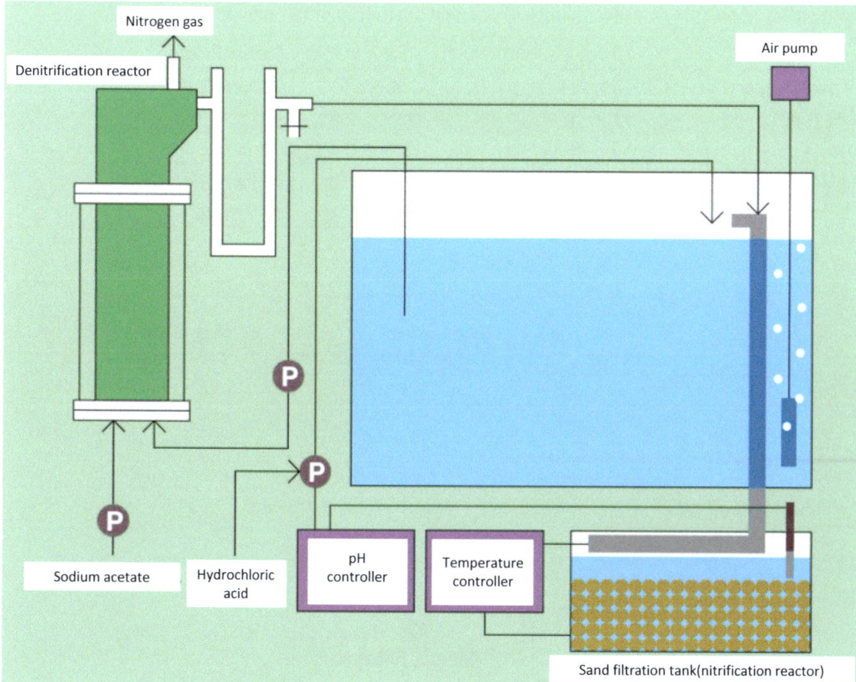

Fig. 10.7 Overview of test apparatus

tank. An overview of the system is shown in Fig. 10.7. For the demonstration, a granule (USB) type denitrification tank with an effective capacity of 10 L was used, combined with a sand filtration tank (nitrification tank) and a 1-m^3 water tank equipped with a temperature control device set to 25 °C located outdoors at the Shinagawa Aquarium, Tokyo, for 2 months. The test was begun after four tropical fish (double-lined fusilier, *Pterocaesio digramma*) with a mean weight of 245 g were placed into the water tank that had an accommodation density of 1 kg/m^3. Approximately 10 g of feed was provided per day (Kawamata et al. 2010).

In addition to supplying tank water directly to the denitrification tank, sodium acetate was provided as the electron donor needed for the denitrification reaction, with a C/N ratio = 3. The denitrification tank was operated with an HRT of 2.5 h (the ratio of the water treatment quantity to the total tank water quantity was 9.6% with a flow rate of 96 L/(m^3·day)). To ensure that the C/N ratio would not fluctuate greatly, an additional quantity of sodium acetate was calculated based on the nitrogen component concentration, which was determined by HACH analytical methods (HACH Company, Loveland, Colorado, USA).

The pH in the tank was maintained at 8.0 by the addition of a little hydrochloric acid, and the demonstration test was conducted for approximately 2 months. Table 10.5 shows the mean values for the parameters measured in the tank environment. These results showed that this system could provide the fish in the

Table 10.5 Analysis of water quality of the demonstration environment

	Target water quality	Tank water	Denitrification treatment water
Water temperature (°C)	25 ± 1	24.8 ± 0.3	24.2±0.2
DO (mg/L)	5–6	6.15	–
pH	7.8–8.3	8.0 ± 0.1	8.1 ± 0.2
ORP (mV)		130 ± 24	278 ± 44
Salinity (‰)		33.3 ± 0.3	33.0 ± 0.2

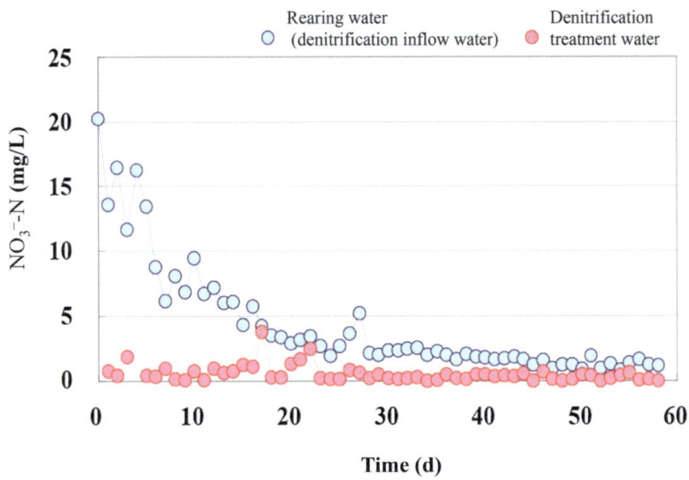

Fig. 10.8 Performance of the USB denitrification reactor with granules

demonstration tank with high-quality water and also maintain a stable water environment.

The concentration of NO_3^--N in the denitrification treatment water in the demonstration was roughly 0.1 mg-N/L or less. As shown in Fig. 10.8, almost all of the NO_3^--N could be removed from the tank water. As a result, the concentration of NO_3^--N in the water gradually began to decrease after the test was begun, and by approximately 20 d after the start of operation, the concentration of NO_3^--N in the water had approximated the theoretical convergence concentration (4.2 mg-N/L).

Subsequently, the NO_3^--N in the tank water actually decreased to 2.1 mg-N/L, or half of the theoretical convergence concentration, with a calculated denitrification efficiency of 95%. A mean NO_3^--N removal efficiency of 93.3% was achieved in the period during which the treatment performance had stabilized and the denitrification performance attained a maximal efficiency of approximately 97%.

2. Long-term demonstration in a 3-m^3 tank

A long-term demonstration was conducted in a 3-m^3 tank to which a 30-L denitrification reactor had been introduced (Figs. 10.9 and 10.10). Figure 10.11

Fig. 10.9 30 L denitrification reactor

Fig. 10.10 Long-term demonstration using a 3-m^3 saltwater tank

Fig. 10.11 Denitrifying bacteria granular carriers

shows the granular carriers (spherical, diameter approximately 2 mm) with which the denitrification tank was filled.

For this demonstration, a granule (USB) denitrification reactor with an effective capacity of 30 L was combined with a 3-m^3 water tank equipped with a sand filtration tank (nitrification reactor) and a temperature control device set to maintain a water temperature of 25 °C. The test was conducted with tropical fish and an accommodation density of approximately 1 kg/m^3 tank water.

The water from the sand filtration tank was introduced directly into the denitrification tank, and sodium acetate was provided as the electron donor for the denitrification reaction, with a C/N ratio = 3. The denitrification tank was operated with an HRT of 2.5 h. The pH in the tank was maintained at 8.0 using hydrochloric acid suitably, and the test was continued for more than 2 years without any water replacement.

3. Test results and conclusions

The short-term and long-term demonstrations showed that the denitrification system has high efficiency for NO_3^--N removal and also creates a stable problem-free water environment for the organisms in the tank. Furthermore, this denitrification system could also be installed separately with other recirculation and filtration systems for water treatment equipment in the same aquaria and can also be added to the water treatment scheme for existing aquaria. At present, the system described here has been introduced to the Sumida Aquarium (Fig. 10.12), where it has attained normal operation and good water quality. This denitrification system will also be an effective technology for wastewater recirculation and will be adopted for water treatment at inland aquaria that are constructed in the future. The system is expected to make a major contribution to the efficiency of a closed-circulation system.

Fig. 10.12 Denitrification system at the Sumida Aquarium, Tokyo

10.2.3 Development of a Saltwater Reuse System

Background The filters in the high-performance filtration system discussed in the previous section must be cleaned periodically to prevent internal clogging. For this purpose, previous systems have conducted backwashing of the filters in saltwater tanks with fresh saltwater. Aquaria constructed in coastal regions can obtain natural saltwater directly from the ocean, so the backwashing requirement presented no operational problems even with the conventional systems. In the case of inland aquaria, however, the constant use of natural saltwater for backwashing increases operating costs and requires higher capacity storage tanks. Accordingly, a saltwater reuse system was constructed to recycle used saltwater to use instead of natural saltwater for cleaning the filters by backwashing.

Objectives The objective was to recycle the used saltwater and reuse it for saltwater filter backwashing instead of using natural saltwater, overflow water from the exhibition tanks, or the wastewater from filter cleaning and thereby reduce the quantity of drain water discharged into the sewer system.

System Functions The saltwater reuse technology described here is used to purify the reused saltwater into water fit for filter backwashing.

The overflow water from the exhibition tanks and the wastewater from filter cleaning was treated by first removing the suspended proteins, then by sterilization, and finally by suspended solid treatment and biofiltration.

Saltwater Reuse In the saltwater reuse system, the overflow tank water and the filter cleaning wastewater are stored in the reuse tank. The saltwater stored in the reuse tank is constantly purified. This purified saltwater is sent to the backwashing tank and used again as backwashing water for cleaning the filter. During the course of repeated use, the repurified saltwater that has become increasingly polluted is purified once again and sent to the drain water tank, from which it is discharged as drain water (Fig. 10.13).

10.3 Conclusions

This chapter has described the closed-circulation systems that have been introduced for aquaria recently in Japan and covers the three major components in detail: an artificial saltwater manufacturing system, a high-performance filtration system, and a saltwater-reuse system. These systems were developed with the aim of addressing the issues that needed to be resolved for the operation of inland aquaria: reducing initial and operating costs and reducing environmental load. The artificial saltwater manufacturing system completely eliminated the need for natural saltwater as replenishment saltwater. The high-performance filtration system described here reduced the amount of water required for replenishment, while it maintained the quality of the tank water. The saltwater reuse system made it possible to purify and

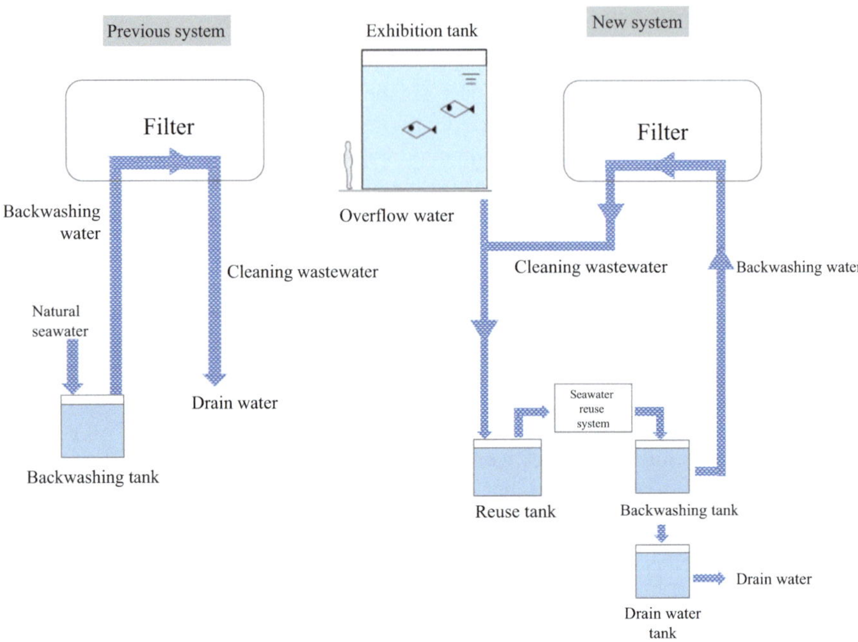

Fig. 10.13 Process of water treatment in saltwater reuse system

reuse wastewater for filter backwashing instead of using fresh saltwater. These systems have been introduced into the Kyoto Aquarium and the Sumida Aquarium that opened in March 2012 and May 2012, respectively.

However, these closed-circulation systems that have been used at aquaria have not yet been introduced to aquaculture facilities that use similar water treatment equipment. The reason is likely because, as in the aquarium sector, most aquaculture facilities have been situated in coastal areas, where they have access to an inexhaustible supply of new saltwater, and thus little need to develop technologies for reusing wastewater. Nevertheless, as in the case of aquaria, some aquaculture facilities have recently been built in inland regions quite far from the coast. Needless to say, the most pressing issues have been pointed out, such as the high cost of drawing and transporting saltwater, the environmental impact of CO_2 emissions resulting from long-distance transport, and the discharge of large quantities of wastewater containing high concentrations of nitrogen components into drains. This system represents one effort to resolve these issues, but further research and development is needed.

The water treatment technologies introduced in this section have been developed precisely because of the need for rigorous control of water quality at aquaria, where water transparency is required. In the future, these technologies will likely be deployed at aquaculture facilities.

References

Garrity GM (ed) (2005a) Bergey's manual of systematic bacteriology, vol 2 (The proteobacteria), part B (The Gammaproteobacteria), 2nd edn. Spring, New York

Garrity GM (ed) (2005b) Bergey's manual of systematic bacteriology, vol 2 (The proteobacteria), part C (The Alph-, Beta-, Delta-, and Epsilonproteobacteria), 2nd edn. Spring, New York

Gauthier MJ, Lafay B, Christen R et al (1992) *Marinobacter hydrocarbonoclasticus* gen. nov., sp. nov., a new, extremely halotolerant, hydrocarbon-degrading marine bacterium. Int J Syst Evol Microbiol 42:568–576

Hamaguchi T, Ono S, Kurabe M et al (2010a) Development of a recirculating nitrogen removal system for marine organism breeding water. JSCE research papers on environmental engineering 47:297–303

Hamaguchi T, Ono S, Takahashi M et al (2010b) Characteristics of wastewater treatment under high salinity conditions in a fixed-bed nitrification and denitrification process. In: Proceedings of the 44th annual meeting of Japan Society on Water Environment, Fukuoka, p 87

Hamaguchi T, Takahashi M, Kawakami S et al (2010c) Development of a system for biological denitrification in a high salt concentration environment. In: Proceedings of the 65th annual meeting of Japan Society of Civil Engineers, Hokkaido, pp 103–104

Honma A (1990) Encyclopedia of living fish. Fuji Technosystem Co. Ltd, Tokyo

Kawamata M, Mori M, Hamaguchi T, Yamaguchi T (2010) Demonstration test of denitrification system in saltwater fish breeding tanks. In: Proceedings of the 66th annual meeting of Japan Society of Civil Engineers, Ehime, pp 95–96

Kurabe M, Takahashi M, Yamaguchi T et al (2010) Continuous treatment test of wastewater containing high saline concentration in a denitrification reactor. In: Proceedings of the 44th annual meeting of Japan Society on Water Environment, Fukuoka, p 269

Macy JM, Rech S, Auling G et al (1993) *Thauera selenatis*, gen. nov., sp. nov., a member of the beta subclass of Proteobacteria with a novel type of anaerobic respiration. Int J Syst Evol Microbiol 43:135–142

Nishi G, Sawatari T (2007) Aquarium work. Tokai University Press, Kanagawa

Ono S, Hamaguchi T, Yamaguchi T et al (2010a) Development of technology for removal of biological nitrogen from wastewater with high salt concentration at marine aquaria. In: Proceedings of the 65th annual meeting of Japan Society of Civil Engineers, Hokkaido, pp 101–102

Ono S, Kawamata M et al (2010b) Removal of biological nitrogen from wastewater with high salt concentration. In: Proceedings of the 44th annual meeting of Japan Society on Water Environment, Fukuoka, p 268

Osaka T, Shirotani K, Yoshie S, Tsuneda S (2008) Effects of carbon source on denitrification efficiency and microbial community structure in a saline wastewater treatment process. Water Res 42:3709–3718

Song B, Palleroni NJ, Haggmlon MM (2000) Description of strain 3CB-1, a genomovar of *Thauera aromatica*, capable of degrading 3-chlorobenzonate coupled to nitrate reduction. Int J Syst Evol Microbiol 50:551–558

Suzuki K, Nishi G (2005) Aquarium science. Tokai University Press, Kanagawa

Takai K, Moyer CL, Miyazaki M et al (2005) *Marinobacter alkaliphilus* sp. nov., a novel alkaline serpentine mud from ocean drilling program site 1200 at South Chamono Seamount, Mariana forearc. Extremophiles 9:17–27

Yoon JH, Yeo SH, Kim IG, TK O (2004) *Marinobacter flavimaris* sp. nov. and *Marinobacter deapoensis* sp nov., slightly halophilic organisms isolated from seawater of the Yellow Sea Korea. Int J Syst Evol Microbiol 54:1799–1803

Chapter 11
Aquaponics

Toshio Takeuchi and Masato Endo

Abstract Freshwater aquaponics has been performed widely, especially in the United States, to culture channel catfish and tomato, tilapia and lettuce, or tilapia and basil/okra. However, many of these systems are used as learning tools in school science curricula, but not as industrial systems for harvesting aquaponics products. In Japan, because of marine fish aquaculture's dominance over freshwater fish aquaculture, and because the majority of offshore aquaculture use net cage culture, water pollution from offshore mariculture has become a serious issue. However, if mariculture can be conducted using a land-based recirculation system and the associated waste can be recycled for plant cultivation, an organic farming system to cultivate both fish and plants could be realized, and environmental conservation could be improved. In this section, we review previously reported approaches to aquaponics, present the results of our preliminary studies, and discuss the directional strategy of aquaponics in Japan.

Keywords Aquaculture • Aquaponics • Fish • Hydroponics • Plant • Recirculating fish culture system

11.1 Introduction

Aquaponics is a term coined to denote aquaculture, as in culture of fish, and hydroponics, as in hydroculture of plants. In Japan, which is currently in the middle of its third plant factory boom, a soil-free nutriculture hydroponics system is conventional. In this system, a nutrient medium containing chemical fertilizers solubilized in water is circulated in a cultivation bed, where plants can absorb nutrients necessary for growth via their roots. However, given the rising cost of chemical fertilizers required to manufacture hydroponic media due to the excessive

T. Takeuchi (✉) • M. Endo
Tokyo University of Marine Science and Technology, 4-5-7 Konan, Minato, Tokyo 108-8477, Japan
e-mail: take@kaiyodai.ac.jp

T. Takeuchi (ed.), *Application of Recirculating Aquaculture Systems in Japan*, Fisheries Science Series, https://doi.org/10.1007/978-4-431-56585-7_11

import, especially in Japan, of phosphorus, it is important to address the stable supply of phosphorus as well as the development of chemical fertilizer-independent cultivation techniques.

Riding the wave of Japan's plant factory boom, land-based recirculating aquaculture systems have also gained attention. The details regarding such land-based aquaculture in Japan are covered in Chaps. 1 and 2. Even though net cage culture is currently the mainstream aquaculture technique, the resultant eutrophication of nearby marine environments due to leftover feed and waste (containing nitrogen and phosphorus) from cultured fish creates serious problems, especially frequent red tides. Therefore, serious discussions have been conducted to develop measures for enhancing production levels of cultured fish while preventing environmental eutrophication. At present, there are two directional strategies. One is to place large-scale offshore aquaculture near fast-moving ocean currents to improve the quality of fish meat as well as to disperse the waste more broadly in the ocean. This strategy has just been incorporated into the net cage culture of yellowtail and bluefin tuna. The second directional strategy is the introduction of land-based recirculating aquaculture. Although this system effectively recycles uneaten feed and fish waste, it does not currently include an approach for effective subsequent processing. Because it overcomes these shortcomings of aquaculture, aquaponics has been gaining attention in recent years. This system can recycle fish waste for plant cultivation, thereby enabling the organic farming of both plants and fish. In many countries, not only are industrial aquaponics systems in place, but home and backyard aquaponics for freshwater fish are also common. In Japan, however, saltwater fish are much more popular than freshwater fish; hence it is more important to develop a saltwater aquaponics system. The establishment of a polyculture system for farming seaweeds, such as sea lettuce, adjacent to a net cage aquaculture system has been attempted; the details of this system are presented in Chap. 12. In the present section, we review the current situation regarding freshwater and saltwater aquaponics in Japan and explore future directions for their development.

11.2 Overview of Aquaponics

Aquaponics technology has been advanced by Rakocy and colleagues, who are well known for their combined tilapia and lettuce cultivation in the Virgin Islands (Rakocy 1989). As Table 11.1 shows, plants and freshwater fish are the most common aquaponics combination, particularly in the United States. Aquaponics cultivation of barramundi and tomato or lettuce was reportedly commercialized in Australia; however, in spite of the popular use of aquaponics at home or in school science classes, fewer than 20 commercial aquaponics facilities have been established in the United States (Aragon 2013). Aquaponics so far has been developed mainly in terms of recycling wastewater in a closed recirculation system to perform hydroponic culture of useful plants. However, because of the recent

Table 11.1 Past representative studies of aquaponics

Sutton and Lewis (1982)	Channel catfish and tomato
Watten and Busch (1984)	Tilapia and tomato
Seawright et al. (1998)	Tilapia and lettuce
Adler et al. (2000)	Rainbow trout and lettuce/sweet basil
Lennard and Leonard (2004)	Murray cod *Maccullochella peelii peelii* and lettuce
Tailor Made Fish Farms (2009)	Barramundi *Lates calcarifer* and tomato/lettuce (industrial)
Rakocy et al. (2004)	Tilapia and basil/okra
Jeong, G.S. (personal communication)	Catfish/carp and ornamental plants (orchid/miniature rose)

popularity of organic farming and promotion of local food production, companies with a large-scale hydroponics system have begun to co-establish closed recirculation systems for cultured fish in order to generate organic fertilizers.

In Table 11.1, we can see that all currently existing aquaponics facilities use freshwater instead of seawater and that the majority of the facilities are small scale, with only a few commercial facilities.

11.3 Preliminary Attempts at Tokyo University of Marine Science and Technology

In addition to the conventional breeding of carp and tilapia, our institution has been farming basil in co-culture with carp or tilapia, as well as white radish sprouts with tilapia. This section describes primarily the experimental approaches that have been performed in our laboratory.

11.3.1 Nitrogen and Other Minerals in Wastewater from Fish Culture

Table 11.2 shows nitrogen and mineral contents in wastewater from the respective closed recirculating aquaculture systems (CRASs) for culturing tilapia, Japanese flounder, tiger puffer, and kelp grouper. To more clearly show comparisons, data were converted into relative concentrations, using a relative nitrogen concentration of 100 mg/L. Large differences in the concentrations of phosphorus, potassium, and magnesium between freshwater and saltwater aquaponics are obvious. Potassium and magnesium are major elements in saltwater, and thus their high levels reflect saltwater-like composition. However, phosphorus is excreted by fish, and its concentration is relatively low in freshwater but high in saltwater. When pH decreases in freshwater, calcium is thought to be eluted from the coral sands contained in the

Table 11.2 Composition of nitrogen and other elements in rearing wastewater of respective CRAS for four kinds of freshwater and saltwater fish culture

		Tilapia				Japanese flounder	Tiger puffer	Kelp grouper
		Oreochromis niloticus				*Paralichthys olivaceus*	*Takifugu rubripes*	*Epinephelus bruneus*
		Okubo (2009)	Nishimura (2014)			Kusakari (2012)	Zhang (2013)	Takeuchi unpub.
		Freshwater	Freshwater	8‰		17‰	31‰	32‰
N	(mg/L)	100.0	100.0	100.0		100.0	100.0	100.0
P	(mg/L)	2.1	1.2	0.5		6.1	4.9	7.2
K	(mg/L)	16.4	17.4	44.0		51.1	159.3	106.1
Ca	(mg/L)	111.0	139.6	163.0		132.0	223.0	202.4
Mg	(mg/L)	6.7	9.7	90.2		126.3	325.6	220.4
Fe	(μg/L)	16.5	8.1	3.3		18.3	20.1	3.4
Mn	(μg/L)	43.4	5.0	61.5		10.9	27.8	1.4
Cu	(μg/L)	0.8	5.8	7.1		1.0	40.4	10.2
Zn	(μg/L)	0.5	12.7	13.2		9.0	25.9	6.8
Co	(μg/L)	NA	NA	NA		NA	22.2	2.0

NA not analyzed

Fig. 11.1 Photograph of hydroponic system of basil *Ocimum basilicum* on the water surface of fish tank of outdoor RAS for intensive culture of tilapia

experimental CRAS filtration system, reacting with phosphorus to form sediments (Endo et al. 2000). However, saltwater that contains various elements tends to prevent phosphorus from forming sediments. Because pH change greatly affects the dynamics between solubilization and sedimentation of iron and other trace elements (Sonneveld and Voogt 2009), it is important to take account of the total amount of elements in the CRAS, as well as the pH of the system, and optimize cultivation techniques to meet the needs of the cultivated plants and marine animals. In the case of halophytes, the salinity of rearing water needs to be optimized as well.

The research data shown in Table 11.2 can serve as basic and useful information to those who plan to incorporate aquaponics.

11.3.2 Hydroponics of Basil on Fish Tank

Figure 11.1 shows our simple hydroponics system of an outdoor water tank and self-made Styrofoam floating devices. This system differs from the conventional aquaponics system, which comprised a CRAS for culturing fish and a connected hydroponics device for cultivating plants. A special feature of our system is that plant roots are protected by pots placed on the floating device, and any roots

Table 11.3 Nitrogen and phosphorus budget between tilapia and white radish sprouts under the aquaponics in indoor system (%)

	Control (fish only)		Aquaponics	
	Nitrogen	Phosphorus	Nitrogen	Phosphorus
Accumulation in fish	25.6	39.4	25.8	37.4
Waste materials	2.1	6.7	1.9	9.9
Rearing water	67.9	49.3	33.9	20.6
Accumulation in radish	–	–	19.0	30.8
Unknown	4.9	4.6	19.5	1.3

outgrowing the pot will be eaten by the fish in the tank. Furthermore, as the water surface is partially covered by the floating devices, our system is thought to be less stressful to fish. As illustrated here, hydroponics can be performed by simply placing floating cultivation beds in a rearing fish tank.

11.3.3 Culture of Tilapia and White Radish Sprouts

We also developed an indoor rearing tank for tilapia associated with hydroponic cultivation beds for white radish sprouts. Tilapia with an average weight of 100 g and average body length of 17 cm were cultured in 24 °C for 8 days. As the results show (Table 11.3; Takeuchi 2014), no significant difference in the fractions of nitrogen and phosphorus accumulated in fish or in sediments was observed between the control section and the aquaponics section with the associated hydroponics devices. However, the fractions of nitrogen and phosphorus in the circulating rearing water in the aquaponics section were half that of the control section, the nitrogen and phosphorus appearing to have been accumulated in the white radish sprouts. It is unclear whether denitrification takes place during transpiration, but the fraction of missing nitrogen was higher in the aquaponics section. Although further study is needed to investigate this phenomenon, the present results clearly show that nitrogen and phosphorus excreted by fish quickly accumulate in plants in a relatively short period.

Although this result has been shown previously (Rakocy 2010), our study again showed the efficiency of aquaponics in which nitrogen and phosphorus from fish can be effectively absorbed by plants.

11.3.4 Saltwater Aquaponics

In Japan, assuming that fish in land-based aquaculture are being produced for commercial use, and assuming that such aquaculture will focus on saltwater fish

Fig. 11.2 Photograph of common ice plant *Mesembryanthemum crystallinum* culture

given their higher market price than freshwater fish, it will be necessary to use rearing water containing salts for farming. Edible halophytes that can be grown in hydroponic systems include New Zealand spinach, common ice plant (Fig. 11.2), *Suaeda japonica* Makino, common glasswort, and Swiss chard. Among these, common ice plant has been investigated by Saga University to evaluate its value as an edible vegetable and to establish a method for producing it (www.barafu.jp/). As indicated by sales of common ice plant in supermarkets and on the Internet, its value appears to be high.

Our laboratory aims to develop an aquaponics system for combined cultivation of a halotolerant or halophilic plant and a saltwater fish, such as Japanese flounder, tiger puffer, or kelp grouper. According to previous studies (Saitoh et al. 1995; Imai et al. 2010), even 8‰ was sufficient to farm Japanese flounder and tiger puffer and even had a minor growth-promoting effect. Because no previous information was available when we conducted this study, we farmed kelp grouper at 8, 16, 24, and 32‰ and found that their growth and survival were adversely affected at 8‰ (Matsumoto et al. 2014). For this reason, kelp grouper were farmed in 32‰

saltwater for an extended period of time, and the rearing water with a concentration of nitrate nitrogen of 500 ppm was then diluted to one-fourth for hydroponic cultivation of common ice plant at 20 °C. The initial rate of decline of nitrogen and phosphorus due to absorption by the common ice plant was 9.7 and 1.7 mg/kg plant fresh wt./day, respectively. Because common ice plant can gradually acclimate to saltwater, it is possible to farm it in 100% saltwater (Agarie et al. 2007) and to directly connect its cultivation system to an aquaculture system for a wide range of fish species, from stenohaline fish, which prefer saltwater, to euryhaline fish, which grow even in low-salinity water. However, as is the case with fish species, the relationship between salinity and optimal growth of halotolerant and halophilic plants varies by species. Therefore, it is important to evaluate and optimize cultivation methods by adjusting the salt content and diluting wastewater. In the future, we plan to improve the accuracy of aquaponics and to incorporate it with the polyculture of flatfishes and tiger puffer.

11.4 Conclusions

Although this section mainly introduces research efforts made by our institution, other research conducted in Japan includes land-based aquaculture of Japanese tiger prawn combined with plant cultivation (watercress and Chinese water spinach) (Nohara 2013), flexible manufacturing of tilapia and tomato (Nakamura 2013), aquaponics of goldfish and leafy vegetables (Japanese mustard spinach, red-leafed lettuce, green onion, and crown daisy) at Miyagi Prefectural Fisheries High School, and small-scale aquaponics of tilapia and soybean or tomato (Abe 2016), all of which use freshwater. For this reason, we feel that aquaponics in Japan, especially saltwater aquaponics, has only just begun. Before the study and development of aquaponics can make further progress, land-based aquaculture must be industrialized; nonetheless, studies exploring the early development of aquaponics should be continued.

One characteristic of common ice plant is that it accumulates high levels of heavy metals from the soil. However, we predict that the level of heavy metals excreted by fish will not cause increases to problematically high levels in common ice plant if fish are given a formula feed produced in accordance with the Law Concerning Safety Assurance and Quality Improvement of Feed. In other words, this new system enables the farming of safe-to-eat organic plants and fish. In the future, we believe it will be necessary to raise the general public's awareness of aquaponics by introducing and demonstrating aquaponics at schools and by developing a system to display aquaponics at restaurants and aquariums.

References

Abe Y (2016) The general training that adopted aquaponics in Miyagi Prefectural Fisheries High School. Rakusui No. 854, 2–5

Adler PR, Harper JK, Wade EM, Takeda F, Summerfelt ST (2000) Economic analysis of an aquaponic system for the integrated production of rainbow trout and plants. Intl J Recirculating Aquaculture 1(1):15–34

Agarie S, Shimoda T, Shimizu Y, Baumann K, Sunagawa H, Kondo A, Ueno O, Nakahara T, Nose A, Cushman JC (2007) Salt tolerance, salt accumulation, and ionic homeostasis in an epidermal bladder-cell-less mutant of the common ice plant *Mesembryanthemum crystallinum*. J Exp Bot 58:1957–1967

Aragon SC (2013) Action for aquaponics in foreign countries. RIKUJOUYOUSHOKU (Land Aquaculture), Jhohou Kikou, Tokyo, pp 261–268. (In Japanese)

Endo M, Takeuchi T, Yoshizaki G, Satoh S, Ohmori K, Oguchi M, Nakajima A (2000) Studies on the development of closed ecological recirculating aquaculture system (CERAS) VI. Phosphorus form and mineral budgets in the fish-rearing closed tank during a long-term feeding experiment with tilapia. CELSS J 13:19–26. (In Japanese with English abstract)

Imai T, Arai D, Morita T, Kogane T, Yamamoto Y, Chida N, Endo M, Takeuchi T (2010) Effect of low salinity on growth and survival of tiger puffer *Takifugu rubripes*, and the purification of rearing water in closed recirculating seed production. Aquaculture Sci 58:373–380. (In Japanese with English abstract)

Kusakari R (2012) Basic study on the aquaponics applied to marine recirculating aquaculture systems. Masters dissertation, Tokyo University of Marine Science and Technology

Lennard WA, Leonard BV (2004) A comparison of reciprocal flow verses constant flow in an integrated, gravel bed, aquaponic test system. Aquacult Int 12:539–553

Matsumoto N, Endo M, Takeuchi T (2014) The influence of low salinity water on the rearing of kelp grouper *Epinephelus bruneus* in closed recirculating fish culture systems. Abst Metg Eco-Engineering:43–44. (In Japanese with English abstract)

Nakamura K (2013) Action for aquaponics in Japan. RIKUJOUYOUSHOKU (Land Aquaculture), Jhohou Kikou, Tokyo, pp 254–260. (In Japanese)

Nishimura T (2014) Studies on Spirulina culture with waste water and solid wastes discharged from recirculating tilapia culture system. Masters dissertation, Tokyo University of Marine Science and Technology

Nohara S (2013) Aquaponics-Advanced experiments in oversea and situation in Japan. Proc Joint Symp SEE and RASJ, Tokyo, 08 Nov. 29–34. (In Japanese)

Okubo Y (2009) Studies on the establishment of closed ecological recirculating aquaculture system by material cycle between tilapia and Spirulina. Masters dissertation, Tokyo University of Marine Science and Technology

Rakocy JE (1989) Hydroponic lettuce production in a recirculating fish culture system. Virgin Islands agricultural experiment station, island. Perspectives 3:4–10

Rakocy JE (2010) Aquaponics: integrating fish and plant culture. In: Timmons MB, Ebeling JM (eds) Recirculating aquaculture, 2nd edn. NRAC Pub, New York, pp 807–864

Rakocy JE, Bailey DS, Shultz C, Thoman ES (2004) Update on tilapia and vegetable production in the UVI aquaponic system. In: Bolivar RB, Mair, GC, Fitzsimmons K (eds) Proceedings from the Sixth International Symposium on Tilapia in Aquaculture, Manila, Philippines. pp 676–690

Saitoh S, Sasaki M, Le K (1995) Effects of low salinity on the growth and metabolism of juvenile Japanese flounder (*Paralichthys olivaceus*). Sci Rep Hokkaido Fish Exp Stn 34:1–8. (In Japanese with English abstract)

Seawright DE, Stickney RR, Walker RB (1998) Nutrient dynamics in integrated aquaculture-hydroponics system. Aquaculture 160(3–4):215–237

Sonneveld C, Voogt W (2009) Plant nutrition of greenhouse crops. Springer, Dordrecht. 431p

Sutton RJ, Lewis WM (1982) Further observations on a fish production system that incorporates hydroponically grown plants. Progress Fish Cult 44:55–59

Tailor Made Fish Farms (2009) http://www.tailormadefishfarms.com.au/

Takeuchi T (2014) Current trend of recirculating aquaculture system. Environmental Solution Tech 13(2):29–35. (In Japanese)

Watten BJ, Busch RL (1984) Tropical production of tilapia (*Sarotherodon aurea*) and tomatoes (*Lycopersicon esculentum*) in a small-scale recirculating water system. Aquaculture 41 (3):271–283

Zhang X (2013) Studies on microalgal cultures with waste water, solid waste and fractionated foam waste discharged from recirculating aquaculture system with marine fish. Masters dissertation, Tokyo University of Marine Science and Technology

Chapter 12
Advantages of Environmentally Sound Poly-eco-aquaculture in Fish Farms

Shusaku Kadowaki and Yuuki Kitadai

Abstract Environmentally sound poly-eco-aquaculture enables the preservation of aquatic environments to be compatible with that of sustainable aquaculture. With this method, not only healthy fish can be cultured in purified water, but also the productivity will increase by recycling seaweed to feed the fish. The maximum nitrogen uptake rate of each seaweed per square meter of seaweed area was 2.9 mg N/m_s^2/day for "Konbu" *Laminaria japonica*, 3.1 mg N/m_s^2/day for "Wakame" *Undaria pinnatifida*, and 3.6 mg N/m_s^2/day for sea lettuce *Ulva pertusa*. The maximum phosphate uptake rate was 0.43 mg P/m^2/day, 0.54 mg P/m^2/day, and 0.19 mg P/m^2/day, respectively. The calculated values of nitrogen and phosphate uptake rates, obtained by integrating the nutrient concentrations, light intensity, and water temperatures, corresponded well with each observed value. The minimum seaweed cultural density necessary per unit area of yellowtail *Seriola quinqueradiata* farm was calculated using the values of the maximum nitrogen uptake rate. The maximal production rates were 0.75 mg O_2/g wet/h for *L. japonica*, 0.83 mg O_2/g wet/h for *Un. pinnatifida*, and 6.39 mg O_2/g wet/h for *Ul. pertusa*. The minimal weight of cultured seaweeds necessary to accommodate the oxygen consumption of an individual *S. quinqueradiata* was calculated as 1.17 kg wet/fish, 0.83 kg wet/a fish, and 0.21 kg wet/fish.

Keywords Poly-eco-aquaculture • Seaweed • N uptake • P uptake • O_2 production • Fish farm • Yellowtail

12.1 Introduction

The industry of marine aquaculture in the twenty-first century is expected to be practiced in harmony with the environment. It is our responsibility to hand down a blue and abundant sea for our future generations to inherit. Environmentally sound poly-eco-aquaculture is a technical innovation of aquaculture used to purify water

S. Kadowaki (✉)
Emeritus Professor, Kagoshima University, 20-14-901 Tenpozan, Kagoshima 890-0061, Japan
e-mail: kshu@384.jp

Y. Kitadai
"Hakase" Cramming School, 1-1 Minobayashi, Anan, Tokushima 774-0017, Japan

T. Takeuchi (ed.), *Application of Recirculating Aquaculture Systems in Japan*, Fisheries Science Series, https://doi.org/10.1007/978-4-431-56585-7_12

and promote a balanced ecosystem by breeding seaweed and shellfish around coastal fish farms. The seaweed is used to feed for fish and shellfish. "Eco" in the word "eco- aquaculture" means "a harmony of ecology for nature" and "economy for humanity."

In order to establish coastal fish farms which enable sustainable aquaculture, there is a recent requirement for concrete measures to be taken to improve water quality in the farms. Biological water purification is necessary for preventing eutrophication and for reducing oxygen deficiency in water (Kadowaki 2001, 2004). Seaweed cultivation is currently attracting much attention as a plausible measure in this plight.

When seaweed is cultivated in the eutrophic water of coastal fish farms, it uptakes dissolved inorganic nitrogen and phosphate while supplying dissolved oxygen, which is essential for the farms.

Aiming to improve the water quality of coastal yellowtail *Seriola quinqueradiata* farms in the warmwater zone over a year-long period, different seaweed species were cultivated in the farm during each season. "Wakame" *Undaria pinnatifida* was grown during the winter months, "Konbu" *Laminaria japonica* in the spring, and sea lettuce *Ulva pertusa* in the summer and autumn. Following this, the relationship between nitrogen and phosphate uptake rates, oxygen production rate by the different seaweeds, and the nutrient concentration, light intensity, and temperature of the water in the farm was estimated (Kitadai and Kadowaki 2003, 2004a, b; Kitadai 2005). Next, the improvement in nitrogen uptake in relation to nitrogen load by yellowtail aquaculture, as well as the cultivation scale of seaweeds necessary for oxygen production in relation to oxygen consumption per individual yellowtail and per cubic meter of the net cage, was estimated.

This study exemplifies the extent to which water quality of feeding fish farms was improved by cultivating seaweed, proposing specific measures.

12.2 Heavy Environmental Load by Mono-aquaculture

The mainstream in predominant fish cultivation has been that of mono-aquaculture breeding only one kind of fish. With this method, oxygen consumption by the cultured fish increases, and the load of carbon dioxide becomes heavier. In addition, nutrients, such as nitrogen and phosphates from feces or remaining fish feed, dissolve in the seawater, making it eutrophic. This causes red tides, fish pathologies, and oxygen deficiency in seawater, resulting in the mass mortality of fish, auto-pollution, etc. This was a significant problem in the Southwestern part of the Yatsushiro Sea, Kumamoto, Japan, in 2002, when the nitrogen load by fish aquaculture reached 700 to 2600 times that of the nitrogen uptake by seaweed breeding (Kadowaki and Kitadai 2005).

12.3 Environmentally Sound Poly-eco-aquaculture

In order to create a truly rich production by cultured fish, we would like to propose that in the part where the balance of the ecosystem has been broken, the balance is restored by the introduction of poly-eco-aquaculture which directly utilizes solar energy as shown in Fig. 12.1. The primary principle of poly-eco-aquaculture is breeding seaweeds, such as *Un. pinnatifida*, *L. japonica*, and *Ul. pertusa*, throughout the year to create an artificial sea forest around cultured fish cages. The seaweed will uptake nutrients, such as nitrogen and phosphate from fish feces and remaining feed. The seaweed also inhibits pathogenic bacteria (Nagahama and Hirata 1990) and red tide organisms (Hirata et al. 1986). Grown seaweeds will be fed to abalone *Haliotis discus hannai*, *H. discus discus*, *H. gigantea*, sea urchin *Stichopus japonica*, *Holothuria pervicax*, yellowtail, and red sea bream *Pagrus major*. Sea cucumber *Stichopus japonica* is grown in symbiosis with abalone in aquaculture net cages. Feces generated by the abalone are fed to sea cucumber. Scallop *Chlamys nobilis* can be cultured because they eat organic suspended substances, such as remaining feed and fish feces. Environmentally sound poly-eco-aquaculture enables the preservation of aquatic environments to be compatible with that of sustainable aquaculture. With this method, not only healthy fish can be cultured in purified water, but also the productivity will increase by recycling seaweed to feed fish.

When cultured abalone and sea cucumber are dried, they can be stored for long periods and can be shipped long distances at room temperature. It was also found that half pearls could be grown in cultured giant abalone *H. gigantea* in 5 months after a pearl nucleus was inserted into them. The shells can also be used for mother

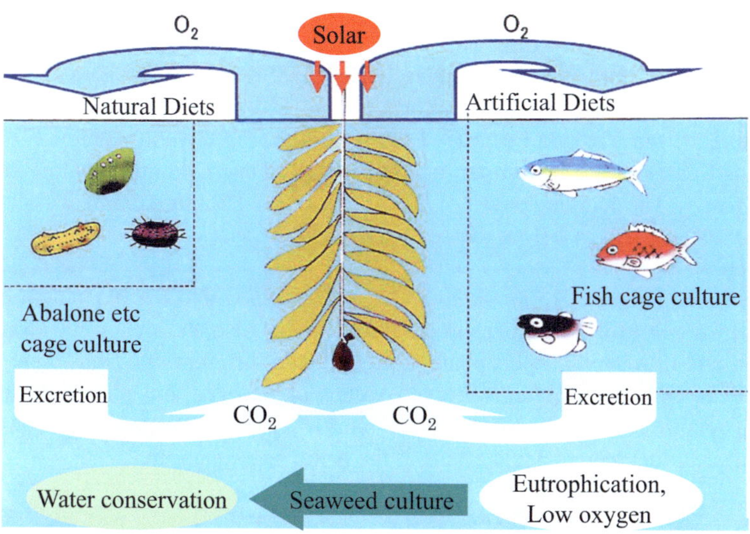

Fig. 12.1 Environmental conservation by eco-polyculture with fish, seaweed, shellfish, and sea cucumber in coastal fish farm

of pearl work. With poly-eco-aquaculture, there is a higher additional value, as well as expectation of increasing job opportunities.

The capability at which seaweeds can uptake nitrogen has been researched, and it was found that the purification of aquatic environments to allow a large amount of cultured fish became feasible when the area of seaweed breeding was larger than the area of fish aquaculture. Based on these research results, Azuma-cho Fisheries Co-operative Association decided in 2000 to employ seaweed breeding near marine aquaculture farms in an effort to increase the proportion of the area of seaweed to the area of fish aquaculture, aiming at safe, sustainable aquaculture. The fishermen themselves are practicing poly-eco-aquaculture.

12.4 The Cultural Density of Seaweed Necessary for Water Purification in Fish Farms

This study was conducted in fish farms producing *S. quinqueradiata*, *P. major*, and puffer fish *Takifugu rubripes* in Goshoura-cho, Kumamoto Prefecture, located in the southeastern part of the Yatsushiro Sea as shown in Fig. 12.2. During this research, the water temperature and oxygen concentration 3 m below the sea surface in fish farms A to E were measured every 3 h. The dissolved inorganic nitrogen (DIN) and dissolved inorganic phosphate (DIP) concentrations 2 m below the sea surface in fish farms were analyzed once a week (Strickland and Parsons 1972). The light intensity at 1.3 m and 2.0 m below the sea surface was measured automatically every hour using a photo sensor (190SA of LI-COR Biosciences).

L. japonica was cultivated between December of 2000 and July of 2001, *Ul. pertusa* was cultivated during August of 2002 and November of 2002, and *Un. pinnatifida* was cultivated from November of 2002 to May of 2003, respectively. Fish farms in Goshoura-cho, Yatsushiro, obtain seed yarns of *Un. pinnatifida* and *L. japonica* from Yoshida Fisheries Ltd. in Shimabara City and Aomori Fisheries Farming Center in late November (below 21 °C) and late December (below 20 °C), respectively. The hanging layer of *L. japonica* and *Un. pinnatifida* was located between 1 m and 4 m below the sea surface, while that of *Ul. pertusa* was only between 0.5 m.

The blade length of *L. japonica* and *Un. pinnatifida* and the blade area of *Ul. pertusa* were identified and measured twice a month. Also, the area, wet weight, dry weight, and nitrogen and phosphate content of the different seaweeds were measured every month to obtain observed values of nitrogen and phosphate uptake rates. The nitrogen and phosphate uptake rates ($P_{\mathrm{N,P}}$, mg N,P/m^2/day) in relation to the seaweed area were calculated using the following equation:

$$P_{\mathrm{N,P}} = (C_{\mathrm{N,P}t} - C_{\mathrm{N,P0}}) \cdot \alpha / t$$

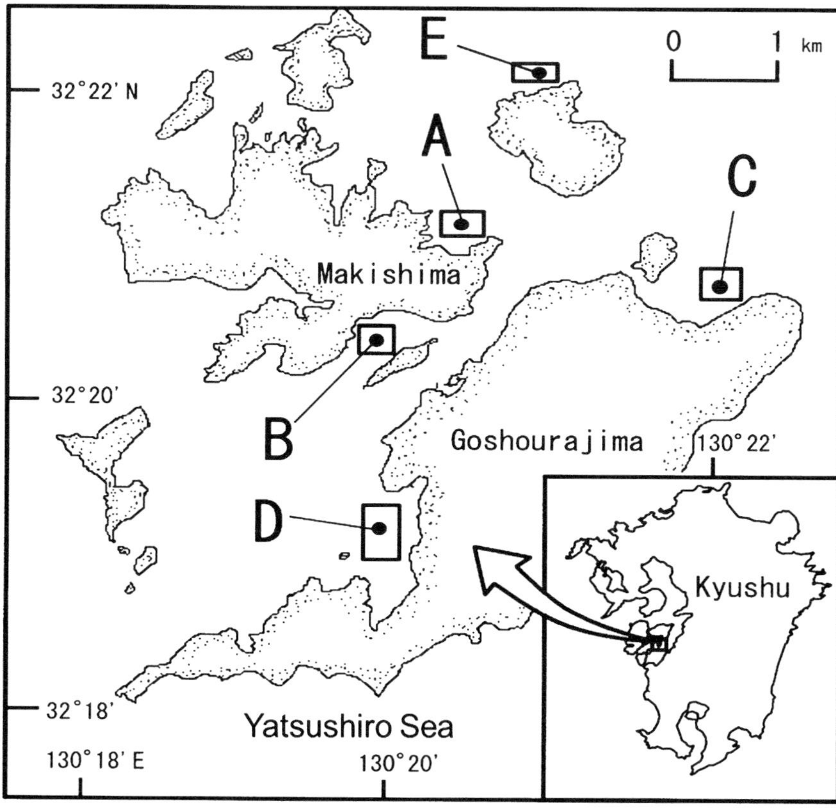

Fig. 12.2 Map showing the cultured sites of seaweeds at stations A–E of Goshoura coastal fish farms in the Yatsushiro Sea

where $C_{N,P0}$ represents the nitrogen and phosphate content (mg N,P/g dry) on the initial day of the experiment, while $C_{N,Pt}$ represents the nitrogen and phosphate content (mg N,P/g dry) t days after the experiment started. α represents the dry weight of the seaweed per square meter of seaweed area (g dry/m^2), while t represents the number of cultivation days.

The oxygen production and consumption rates were measured over 4 h from 10:00 to 14:00 during fine weather conditions using light and dark oxygen bottles, whose value per unit of chlorophyll-a was shown. The hanging level of the oxygen bottles for *L. japonica* and *Un. pinnatifida* was 2 m below the sea surface, while that for *Ul. pertusa* was only 0.5 m. Oxygen concentration was titrated using Winkler's method. The relationship between the nitrogen and phosphate uptake rates ($P_{N,P,}$ mg N,P/m^2/day), as well as the oxygen production rate ($P'c_{N,P}$, mg O_2/mg chl.a/h) of the seaweed area and DIN and DIP concentrations in the fish farm, was analyzed using the Michaelis-Menten formula (Dudale 1967). The $Pm_{N,P}$, $P'cm$, and K were calculated by the following formula:

$$P_{N,P} = Pm_{N,P} \cdot S_{N,P}/(K_{N,P} + S_{N,P})$$

where $Pm_{N,P}$ represents the maximum nitrogen and phosphate uptake rates of seaweed area (mg N,P/m_s^2/day) and $S_{N,P}$ represents DIN and DIP concentrations (μg N,P/L). $K_{N,P}$ represents the minimum DIN and DIP concentrations for growth of seaweed as Michaelis-Menten constants (μg N,P/L):

$$P'c_{N,P} = P'cm_{N,P} \cdot S_{N,P}/(K_{N,P} + S_{N,P})$$

where $P'cm_{N,P}$ represents the maximum oxygen production rates of the seaweed's chlorophyll-*a* (mg O_2/mg chl.a/h).

The minimum nitrogen and phosphate concentrations necessary to obtain the maximum nitrogen and phosphate uptake rates and the maximum oxygen production rate were calculated. The relationship between the nitrogen and phosphate uptake rates of the seaweeds and light intensity was analyzed using the Steel formula (Steel 1962) in order to obtain the optimum light intensity for the maximum nitrogen and phosphate uptake rates. The saturation irradiance *Im* (μmol/m_f^2/s) to $Pm_{N,P}$ was calculated by the following formula:

$$P_{N,P} = Pm_{N,P} \cdot (I/\mathrm{Im}) \cdot \exp(1 - I/\mathrm{Im})$$

where *I* represents the downward irradiance per fish cage area (μmol/m_f^2/s) and *Im* represents the optimum light intensity to $Pm_{N,P.}$ (μmol/m_f^2/s).

The relationship between the nitrogen and phosphate uptake rates of the seaweeds and water temperature was analyzed by the allometric formula (Kadowaki and Tanaka 1994) in order to obtain water temperature coefficients, $Q_{01N,P}$. The $Q_{01N,P}$ was calculated by the following formula:

$$P_{\theta N,P} = P_{TN,P} \cdot Q_{01N,P^{(\theta - T)}}$$

where $P_{\theta\ N,P}$ represents the $P_{N,P}$ at θ°C (mg N,P/m_s^2/day), $P_{T\ N,P}$ is the $P_{N,P}$ at *T*°C (mg N,P/m_s^2/day), $Q_{01N,P}$ represents water temperature coefficients, θ represents water temperature (°C), and *T* represents water temperature 20 °C for *L. japonica*, 16 °C for *Un. pinnatifida*, and 25 °C for *Ul. pertusa*.

In addition, seaweed cultural density (kg wet/m_f^2) in relation to the area of a fish farm was calculated, which is necessary for determining the uptake of nitrogen load in fish aquaculture. Furthermore, the seaweed cultural weight necessary for the oxygen consumption of individual *S. quinqueradiata* (g wet/a fish) and the seaweed cultural density necessary for oxygen consumption per cubic meter of the net cage (kg wet/m^3) were calculated.

12.4.1 Environment of the Seaweed Cultivation

The water temperature ranged between 12 °C and 28 °C, and the oxygen concentration was in the range of 5.7 mg/L and 10.7 mg/L in fish farms during the seaweed cultivation period. The nitrogen concentration hovered between 31 μg N/L and 150 μg N/L, while the phosphate concentration was between 7.0 μg P/L and 27 μg P/L. The ratio of nitrogen to phosphate was in the range of 3.1 and 8.4. The mean for downward irradiance of the layer 2 m below the sea surface (±standard deviation) was 650 ± 74 μmol/m^2/s.

12.4.2 Growth of Seaweeds

The blade length of *L. japonica* and *Un. pinnatifida* grew up to 250 cm and 182 cm, respectively, in the layer 2 m below the sea surface. The maximum daily growth rate of *L. japonica* and *Un. pinnatifida* was 3.0 cm/day and 4.2 cm/day, respectively. The blade area of *Ul. pertusa* grew up to 640 cm^2 in the layer 0.5 m below the sea surface, and the maximum growth rate was 7.6 cm^2/day (Table 12.1).

12.4.3 Nitrogen and Phosphate Uptake Rates of Seaweed Species

The maximum nitrogen uptake rate of each seaweed species per square meter of seaweed area was 2.9 mg N/m$_s^2$/day for *L. japonica*, 3.1 mg N/m$_s^2$/day for *Un. pinnatifida*, and 3.6 mg N/m$_s^2$/day for *Ul. pertusa*, respectively. The nitrogen uptake rate of *Ul. pertusa* was the highest of all. The maximum phosphate uptake rate was 0.43 mg P/m^2/day, 0.54 mg P/m^2/day, and 0.19 mg P/m^2/day, respectively. In addition, the minimum nitrogen concentration necessary for the growth of *L. japonica*, *Un. pinnatifida*, and *Ul. pertusa* was 29 μg/L, 17 μg/L, and 26 μg/L, while the minimum phosphate concentration necessary for growth was 8.7 μg/L, 6.2 μg/L, and 8.0 μg/L, respectively.

Table 12.1 Maximum growth and growth rate of blade length (BL) of *L. japonica* and *Un. pinnatifida* and thallus area (TA) of *Ul. pertusa*

Items	Unit	*L. japonica*	*Un. pinnatifida*	*Ul. pertusa*
Layer	(m)	2.0	2.0	0.5
Blade length	(cm)	250	182	–
Thallus area	(cm^2)	–	–	640
Growth rate of BL	(cm/day)	3.0	4.2	–
Growth rate of TA	(cm^2/day)	–	–	7.6

Table 12.2 Maximum N and P uptake rates ($Pm_{N,P}$), the maximum irradiance to $Pm_{N,P}$ (Im), Michaelis-Menten constants (K), and the water temperature coefficients (Q_{01}) of *L. japonica*, *Un. pinnatifida*, and *Ul. pertusa*

Items	Unit	*L. japonica*		*Un. pinnatifida*		*Ul. pertusa*	
		N	P	N	P	N	P
Water temp.	(°C)	16–23		12–20		18–28	
Pm	(mg/m_s^2/day)	2.9	0.43	3.1	0.54	3.6	0.19
Im	($\mu mol/m_f^2/s$)	720		670		730	
K	(μg/L)	29	8.7	17	6.2	26	8.0
Q_{01}		1.071	1.062	1.090	1.081	1.076	1.084

The optimum light intensity for nutrient uptake of *L. japonica*, *Un. pinnatifida*, and *Ul. pertusa* was calculated as 720 μmol/m^2/s, 670 μmol/m^2/s, and 730 μmol/m^2/s, respectively. Additionally, the water temperature coefficient (Q_{01}) in relation to the nitrogen uptake rate of each seaweed was 1.071, 1.090, and 1.076, respectively, while the Q_{01} in relation to the phosphate uptake rate was 1.062, 1.081, and 1.084, respectively (Table 12.2). The calculated values of nitrogen and phosphate uptake rates, obtained by integrating the nutrient concentrations, light intensity, and water temperatures, corresponded well with each observed value.

12.4.4 Production and Consumption of Oxygen by the Seaweeds

Oxygen production and consumption rates of *L. japonica*, *Un. pinnatifida*, and *Ul. pertusa* were maximized when water temperature was 23 °C, 20 °C, and 28 °C, respectively. The maximum oxygen production rate of each seaweed was 2.6, 2.7, and 2.8 (mg O_2/mg chl.a/h), while that for the oxygen consumption rate was 0.29, 0.24, and 0.35 (mg O_2/mg chl.a/h), respectively.

With these values, the maximum oxygen production rate of the seaweed in reference to the oxygen consumption rate of individual fish was calculated as 8.9 for *L. japonica*, 11.2 for *Un. pinnatifida*, and 8.0 for *Ul. pertusa* (Table 12.3). This means that the oxygen production rate of the seaweed during the daytime under fine weather conditions is eight to eleven times as high as the oxygen consumption rate, indicating that seaweed cultivation would be effective for supplying oxygen to water in fish farms.

12.4.5 Seaweed Cultural Density in Relation to Nitrogen Load in Fish Farm Area

It has been reported that the nitrogen load rate per square meter of area of a yellowtail *S. quinqueradiata* farm during the seaweed cultivation period is

Table 12.3 Maximum O_2 production rates (*P′cm*) and O_2 consumption rates (*R′c*) of *L. japonica*, *Un. pinnatifida*, and *Ul. pertusa*

Items	Unit	*L. japonica*	*Un. pinnatifida*	*Ul. pertusa*
Water temp.	(°C)	23	20	28
P′cm	(mg O_2/mg chl.a/h)	2.6	2.7	2.8
R′c	(mg O_2/mg chl.a/h)	0.29	0.24	0.35
P′cm/*R′c*		8.9	11.2	8.0

290 mg N/m^2/day for *L. japonica*, 115 mg N/m^2/day for *Un. pinnatifida*, and 520 mg N/m^2/day for *Ul. pertusa* (Kouchi Fisheries Experimental Station 1989). The minimum seaweed cultural density necessary per unit area of a *S. quinqueradiata* farm was calculated using the values of the maximum nitrogen uptake rate obtained above Sect. 12.4.3. With the maximum nitrogen uptake rates of the seaweeds mentioned above, the minimum seaweed cultural density necessary for the area of this particular *S. quinqueradiata* farm was obtained. The cultural density of *L. japonica* was 105 kg wet/m^2, 2 kg wet/m^2 for *Un. pinnatifida*, and 7.6 kg wet/m^2 for *Ul. pertusa* (Table 12.4).

It is considered that the effective cultural density of *Ul. Pertusa* in a fish farm is 3.0 kg wet/m^2 (Maesako et al. 1985). For this reason, the nitrogen uptake rate of *Ul. pertusa* in relation to the nitrogen load in a fish farm is calculated up to approximately 40% (3.0/7.6) when the area of cultured seaweed is the same as the area of the farm. In other words, the cultured area of *Ul. pertusa* necessary to purify the nitrogen load in an inner-bay fish farm is 2.5 times (7.6/3.0) as large as the fish farm.

12.4.6 Seaweed Cultural Density to Oxygen Consumption by Cultured Fish

The oxygen consumption rate of an individual *S. quinqueradiata* while *L. japonica*, *Un. pinnatifida*, and *Ul. pertusa* were cultured was calculated as 879 mg O_2/a fish/h, 695 mg O_2/a fish/h, and 1392 mg O_2/a fish/h, respectively (Kadowaki 1990, 1994). The required mass of the seaweed necessary to accommodate oxygen consumption by an individual *S. quinqueradiata* was calculated using the seaweeds' maximal oxygen production rate per unit weight of each seaweed species. The production rates were 0.75 mg O_2/g wet/h for *L. japonica*, 0.83 mg O_2/g wet/h for *Un. pinnatifida*, and 6.39 mg O_2/g wet/h for *Ul. pertusa*.

In addition, the minimal weight of cultured seaweeds necessary to accommodate the oxygen consumption of an individual *S. quinqueradiata* was calculated as 1.17 kg wet/a fish, 0.83 kg wet/a fish, and 0.21 kg wet/a fish, respectively. The minimum seaweed cultural density necessary to accommodate the oxygen consumption per cubic meter of the net cage was calculated as 5.6 kg wet/m^3 for *L. japonica*, 4.0 kg wet/m^3 for *Un. pinnatifida*, and 1.3 kg wet/m^3 for *Ul. pertusa* (Table 12.5).

Table 12.4 Comparison of minimum density of seaweeds cultured per fish farm area for nitrogen load of *S. quinqueradiata* in *L. japonica*, *Un. pinnatifida*, and *Ul. pertusa*

Items	Formula	Unit	*L. japonica*	*Un. pinnatifida*	*Ul. pertusa*
N load rate of *S. quinqueradiata*[a]	A	(mg N/m_f^2/day)	290	115	520
Pm_N	B	(mg N/m_f^2/day)	2.9	3.1	3.6
Weight of seaweed	C	(kg wet/indiv.)	0.116	0.192	–
Area of seaweed	D	(m_s^2/indiv.)	0.11	0.26	–
Area per weight of seaweed	D/C = E	(m_s^2/kg wet)	0.95	1.35	19
Minimum density of seaweed per fish farm area	A/(B· E)	(kg wet/m_f^2)	105	27	7.6

[a]Kouchi Fisheries Experimental Station (1989)

Table 12.5 Comparison of minimum amount of seaweeds cultured per a fish and minimum density of seaweed cultured per fish cage volume for O_2 consumption of *S. quinqueradiata* in *L. japonica*, *Un. Pinnatifida*, and *Ul. pertusa*

Items	Formula	Unit	*L. japonica*	*Un. pinnatifida*	*Ul. pertusa*
Water temperature		°C	23	20	28
Body weight of *S. quinqueradiata*[a]		kg	2.0	1.8	2.6
Density of *S. quinqueradiata* per cage volume[a]	A	fish/m^3	4.8	4.8	6.3
O_2 consumption rate of *S. quinqueradiata*[b]	B	mg O_2/fish/h	879	695	1392
Maximum O_2 production rate of seaweed	C	mg O_2/g wet/h	0.75	0.83	6.39
Minimum amount of seaweed per a fish	B/C	kg wet/fish	1.17	0.83	0.21
Minimum density of seaweed per fish cage volume	A·B/C	kg wet/m^3	5.6	4.0	1.3

[a]Kadowaki (1990); [b]Kadowaki (1994)

12.5 Conclusions

From the results, it was obvious that all of the seaweed species had the capacity to take in nitrogen and phosphate loads and that they fulfilled the role as oxygen producers. However, it may be difficult for seaweed to completely take in nitrogen and phosphate loads alone. Even with *Ul. pertusa* which uptakes nitrogen and

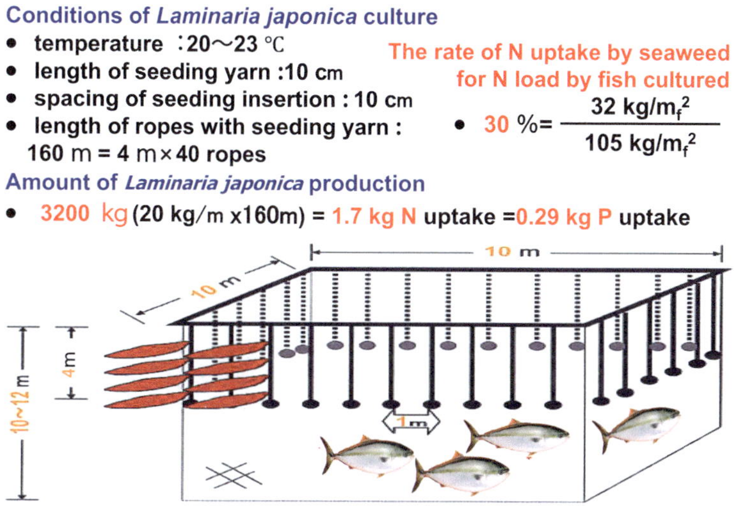

Fig. 12.3 An example of seaweed cultured around the fish culture cage

phosphate loads most effectively, it would take an area two and a half times that of a fish farm in order to take in the loads completely. Still, it is considered important to cultivate effective seaweed for the eutrophication of each fish farm and improve the water quality.

It was also shown that seaweed worked effectively in supplying oxygen. This indicates that it is both possible to reduce the environmental load and supply the oxygen necessary for feeding fish, in addition to managing the water environment, using seaweed.

Kadowaki (2006) has proposed a specific model on how to cultivate seaweeds as shown in Fig. 12.3. It assumes that *L. japonica* and *Un. pinnatifida* are cultivated around cultured fish cages. The cultivation area is 4 m below the sea surface in consideration of the growth of seaweed and water exchange. Stem ropes, in which 10 cm length of seeding yarns are inserted at 10 cm spacings, are hung down at 1 m intervals around the cage (Ohyama et al. 2005). It is expected that 3.2 mt of *L. japonica* can be grown, and about 30% of the nitrogen load of the feeding fish will be reduced with the grown seaweed.

The time to fully implement poly-eco-aquaculture is now! In order to reproduce marine aquaculture farms, concrete measures need to be taken to promote year-round seaweed breeding inside and outside of fish farms, artificial formation of seaweed beds, and reuse and circulation of output biomass of seaweed. Environmentally sound poly-eco-aquaculture answers the needs of both environment and industry, because it will enable environmental conservation through water purification, compatible with sustainable aquaculture that would culture healthy fish. When various groups of living things support the sea, the productivity of fish farms will be developed, and a rich sea having sustainable productivity might be realized. It is our hope that those in the aquaculture industry will try this eco-friendly approach to promote sustainable aquaculture.

References

Dudale RC (1967) Nutrients limitation in the sea. Limnology. Oceanography 12:685–695

Hirata H, Suehara H, Kawaguchi T (1986) Some experiments on prevention of breeding and extermination of red-tide organisms. Aquiculture 34:61–68. (In Japanese)

Kadowaki S (1990) Production ecological studies on allowable capacity of aquaculture in yellowtail, *Seriola quinqueradiata*. Doctoral thesis of Hokkaido University, 123pp. (In Japanese)

Kadowaki S (1994) Energy budget for a yellowtail, *Seriola quinqueradiata* in pen culture. Environmental management in Aquaculture, Bulletin of National Research Institute of Aquaculture, Supplement 1, pp 45–59

Kadowaki S (2001) Creations of sea, fishes and human in 21th century for environmental conservation of mariculture. The Society of Naval Architects of Japan, 16th Ocean Engineering Symposium, pp 1–16. (In Japanese)

Kadowaki S (2004) Advantages of environmentally sound poly-eco-culture in coastal fish farms for the way of recycling of resources. Jpn Fish Resour Conserv Assoc Mon 407:7–9. (In Japanese)

Kadowaki S (2006) The minimum density of seaweed cultured for water purification in coastal fish farms. Yousyoku Mon 541:14–17. (In Japanese)

Kadowaki S, Kitadai Y (2005) Fish and seaweed aquaculture in the Yatsushiro Sea. Kaiyo Mon 415:69–73. (In Japanese)

Kadowaki S, Tanaka Y (1994) Oxygen consumption rates by phytoplankton and bacteria in sea water. Suisanzoshoku 42:267–272. (In Japanese with English abstract)

Kitadai Y (2005) Studies on water purification in coastal fish farm by seaweeds. Doctoral thesis of the United Graduate School of Agricultural Sciences Kagoshima University, 132 pp. (In Japanese with English abstract)

Kitadai Y, Kadowaki S (2003) The growth process and N, P uptake rates of *Laminaria japonica* cultured in coastal fish farms. Suisanzoshoku 51:15–23. (In Japanese with English abstract)

Kitadai Y, Kadowaki S (2004a) The growth and N, P uptake rates of *Ulva pertusa* cultured in coastal fish farms. Suisanzoshoku 52:65–72. (In Japanese with English abstract)

Kitadai Y, Kadowaki S (2004b) The growth process and N, P uptake rates of *Undaria pinnatifida* cultured in coastal fish farms. Suisanzoshoku 52:365–374. (In Japanese with English abstract)

Kouchi Fisheries Experimental Station (1989) Load volume of nitrogen. Bull Kochi Prefecture Fish Exp Station 85:34–41. (In Japanese)

Maesako N, .Fujii A, Yotui T (1985) Production of *Ulva* sp in land-based tank. Bull Nagasaki Prefecture Fish Exp Station 11: 21–23. (In Japanese with English abstract)

Nagahama T, Hirata H (1990) Nitrogen uptake by the sterile *Ulva pertusa*, Ohmura strain, and their interaction to the bacterial communities. Suisanzoshoku 38:285–290. (In Japanese with English abstract)

Ohyama H, Kitadai Y, Kadowaki S (2005) Production of cultured *Laminaria japonica* and its nitrogen and phosphate uptake in a coastal fish farm. Mem Fac Kagoshima Univ 54:29–34. (In Japanese with English abstract)

Steel JH (1962) Environmental control of photosynthesis in the sea. *Limnology*. Oceanography 7:137–150

Strickland JDH, Parsons TR (1972) A practical handbook of seawater analysis. Bull Fish Res Board Can Ott 177:45–90

Chapter 13
Closed Ecological Recirculating Aquaculture Systems

Toshio Takeuchi

Abstract Closed habitation experiments are essential for global environment improvement and maintenance of self-sustaining ecosystems. Hydro-ecosystems, which utilize aquatic organisms, are a promising prospect in habitation experiments. The utilization of aquatic organisms in closed systems can essentially be divided into three parts: (1) cultivation of phytoplankton for oxygen supply, (2) rearing of fish as a protein source, and (3) utilization of fish and shellfish for recycling of organic waste. Utilization of aquatic organisms (e.g., phytoplankton, zooplankton, and fish) to realize CO_2 and O_2 circulation and the construction of food chains are essential for the promotion of closed ecological systems. Recently, fundamental data were collected using the Closed Ecological Recirculating Aquaculture System (CERAS). Essential biomass was estimated from feeding trials between phytoplankton and fish, between phytoplankton and zooplankton, and between zooplankton and fish, and the CO_2 and O_2 budgets between phytoplankton and fish were investigated during development of the experimental equipment. In this chapter, we summarize simple CERAS food chains in freshwater and saltwater environments and give the results of photoperiod experiments.

Keywords Ecological aquaculture • Closed system • Algae • Zooplankton • Fish • Fish culture waste • Food chain • Photoperiod

13.1 Introduction

If humans are to establish permanent space colonies, the production of food, purification of water, regeneration of oxygen from carbon dioxide, and recycling of waste into usable resources through various physicochemical and biological methods will become vitally important. Controlled ecological life support systems (CELSS) are life-support systems that are capable of performing these regenerative

T. Takeuchi (✉)
Tokyo University of Marine Science and Technology, 4-5-7 Konan, Minato, Tokyo 108-8477, Japan
e-mail: take@kaiyodai.ac.jp

T. Takeuchi (ed.), *Application of Recirculating Aquaculture Systems in Japan*, Fisheries Science Series, https://doi.org/10.1007/978-4-431-56585-7_13

functions (Matsumoto and Hamazaki 1992). Many studies are currently being carried out with the aim of constructing atmospheric, geospheric, and hydrospheric CELSS subsystems (Nitta et al. 1996).

As one such CELSS prototype, Closed Ecological Recirculating Aquaculture Systems (CERAS) (Takeuchi et al. 1997) have recently been developed for hydrospheric use with aquatic organisms such as phytoplankton, zooplankton, and fish (Fig. 13.1: Takeuchi 2011). In CERAS, fish are spawned and hatched, and the larvae, juveniles, and adults are reared on plankton. The use of aquatic organisms (e.g., phytoplankton, zooplankton, and fish) to realize the circulation of CO_2 and O_2 and the construction of food chains are essential for the promotion of these closed ecological systems. CERAS is a relatively large project requiring a large amount of energy for food production. However, for use in space stations or during development of lunar bases, a small low-energy system is required. Accordingly, to establish a compact fish-culturing system, removal of the zooplankton stage from the CERAS food chain has been suggested (Takeuchi and Omori 2005).

Recently, recirculating system technology for aquaculture has been advancing rapidly due to growing concern over environmental conservation and food safety. For application of CERAS, aquaculture factories are based on saltwater fish. This system aims to provide efficient aquaculture within an artificial environment that

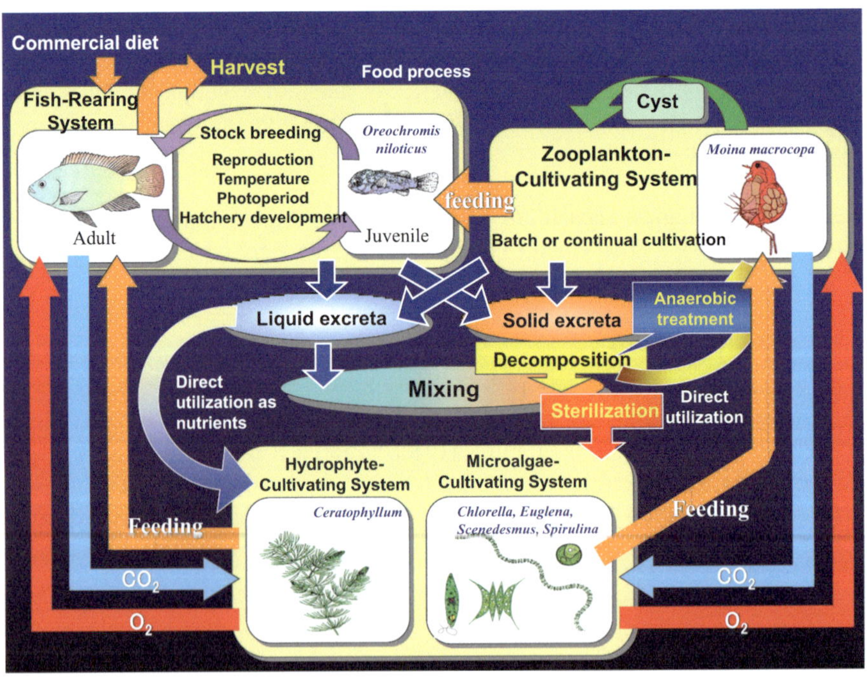

Fig. 13.1 Material cycles in the Closed Ecological Recirculating Aquaculture System (CERAS) (Modified from Takeuchi 2011)

closely resembles the natural environment. The following are therefore necessary: (1) culture of fish in an environment completely free of medicines as well as harmful biological, chemical, and physical substances, (2) an established tracking system, (3) established methods for the prevention of environmental pollution (N, P), and (4) aquaponics, which is the combination of aquaculture (fish) and hydroponics (plant) in a recirculating system.

13.2 The Closed Recirculating Fish-Rearing System

13.2.1 Freshwater

The freshwater fish Nile tilapia *Oreochromis niloticus* has many attributes suitable for culture on Earth as well as in space. For example, it has an excellent growth rate and is tolerant of a wide range of environmental conditions including dissolved oxygen levels as low as 0.7 mg/L, temperatures as high as 42 °C, and NH_4^+-N concentrations as high as 80 ppm (Yada and Miyashita 1988). Tilapia also shows low susceptibility to disease and is amenable to handling and captivity. Moreover, it has a short generation time and is capable of breeding in captivity (Takashima 1997). Most importantly, it is widely accepted as a food fish because of its high palatability and history of use in aquaculture. With all these advantages, tilapia shows particular potential for the use in aquaculture.

We developed a closed recirculating fish-rearing system (Fig. 13.2) consisting of an airtight tank with an O_2/CO_2 exchange unit and subsequently established a long-term feeding experiment with tilapia. In this experiment, with the exception of

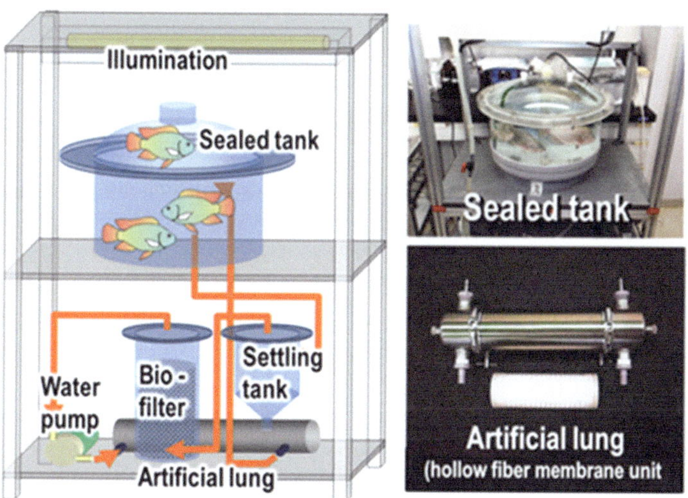

Fig. 13.2 A sealed fish-rearing tank with an artificial lung (Cited from Endo and Takeuchi 2013)

evaporation/condensation to and from the air and water analysis, the water remained unchanged. The experiment was conducted for 189 days at 28 °C with two additional open tank systems under different experimental conditions for comparison. In this closed fish-rearing system, tilapia showed normal growth and high feeding efficiency over the entire 189-day period compared with the recirculating open tank under a high concentration of NO_3-N. After completion, nitrogen (N) and phosphorus (P) budgets were calculated: 38.2% and 50.7% were retained as fish growth; 3.4% and 38.5% were removed as solid waste; and 49.3% and 1.4% accumulated in the rearing water, respectively (Endo et al. 1999). Other minerals, such as calcium (Ca), magnesium (Mg), potassium (K), iron (Fe), manganese (Mn), zinc (Zn), and copper (Cu), were examined as well. Among the elements excreted by tilapia, the main constituents Mg and K accumulated in the rearing water, while others, such as P, Ca, Fe, Mn, Zn, and Cu, were deposited and removed as solid waste. Over 80% of the P in these solids was bound to Ca (apatitic P), with the remaining forms (non-apatitic inorganic P and organic P) negligible when compared with fish diet (Takeuchi and Omori 2005).

13.2.2 Saltwater

Recirculating aquaculture in saltwater is described in prior chapters. Our group chose the tiger puffer *Takifugu rubripes*, Japanese flounder *Paralichthys olivaceus*, and longtooth grouper *Epinephelus bruneus* for the closed recirculating aquaculture system (CRAS). However, these experiments did not include a closed recirculating fish-rearing system (Fig. 13.2). Tiger puffer waste in CRAS was therefore collected for the use in algae culture. These experiments are described in detail later (see Sect. 13.5.4).

13.3 Microalgae–Tilapia Culture

Construction of simplified production links such as the food chain between phytoplankton and fish is necessary to minimize the size and save energy and labor. In its natural habitat, tilapia changes its diet and feeding mode from carnivorous to omnivorous at a total body length of 2–3 cm and then again to phytoplanktivorous at about 6–7 cm (Yada 1982; Getachew 1987; McDonald 1987). Adult tilapia are filter-feeding herbivores with blue-green algae being common components of their diet (Fry and Iles 1972; Trewavas 1983; Yada 1982; McDonald 1987). To investigate the feasibility of the food chain between phytoplankton and tilapia, we examined the growth and reproduction of tilapia fed solely on raw *Spirulina platensis* from the onset of exogenous feeding. *Spirulina* is generally regarded as a rich source of protein (60–70%), vitamins, essential amino acids, minerals, essential fatty acid [e.g., γ-linolenic acid (GLA)], and antioxidant pigments such as carotenoids (Cohen 1997). Its thin and soft cell wall (about 50 nm), made up of 80% pectin and 20% cellulose (Hedenskog and Hofsten 1970), is thinner than that of *Chlorella* (about 50–200 nm, with >90% cellulose) (Matsumoto and Hamazaki 1992) and can be readily digested. Moreover, *Spirulina* has unique characteristics, such as the presence of glycogen instead of starch as the photosynthetic stock nutrient (Cohen 1997). In addition to its nutritional value, it is also effective in immune modulation and radiation protection (Belay et al. 1996) and is reported to be an effective exchanger of O_2/CO_2 (Oguchi et al. 1987). Mass production is possible under conditions of high salinity and alkalinity, ensuring a stable supply of algae in harsh environments where water availability is limited (Takeuchi and Omori 2005).

13.3.1 Tilapia Fed a Microalgae Diet

In the first experiment, we confirmed that juvenile tilapia at the feed transition stage grew normally when fed solely on raw *Spirulina* (Takeuchi et al. 2002). Next, the effect of *Spirulina* on larval tilapia was compared with that of two other species of microalgae (*Euglena gracilis* and *Chlorella vulgaris*). Larval tilapia ingested significantly more *Spirulina* than *Euglena* and *Chlorella*, and *Spirulina* was more readily assimilated (61.4–80%) from the onset of exogenous feeding. Comparisons of assimilated levels further showed that *Spirulina* appeared superior to *Euglena* and *Chlorella* in terms of available carbon for accumulation in the larval tilapia body. These results suggest that *Spirulina* is more acceptable than *Euglena* and *Chlorella* as a sole food source for larval tilapia from the onset of exogenous feeding (Omori et al. 2000; Lu et al. 2002; Takeuchi and Omori 2005).

13.3.2 Tilapia Egg Quality

Studies of spawning and egg quality in tilapia fed solely on raw *Spirulina* throughout three generations were subsequently conducted to investigate the effect on maturation and reproduction (Lu and Takeuchi 2004). There were no significant differences in spawning performance (including first spawning, spawning periodicity, and fecundity) or egg and larval quality (egg size, hatchability of fertilized eggs, the survival time of starved hatchlings, and the rate of hatchling abnormalities) between the three generations of broodstock tilapia fed solely on raw *Spirulina* and commercial diets, respectively (Fig. 13.3: Lu and Takeuchi 2004). These findings suggested that the investment in reproductive effort remains remarkably consistent throughout sexual differentiation and growth. That is, tilapia fed solely on raw *Spirulina* are able to maintain normal reproduction throughout three generations (Takeuchi and Omori 2005).

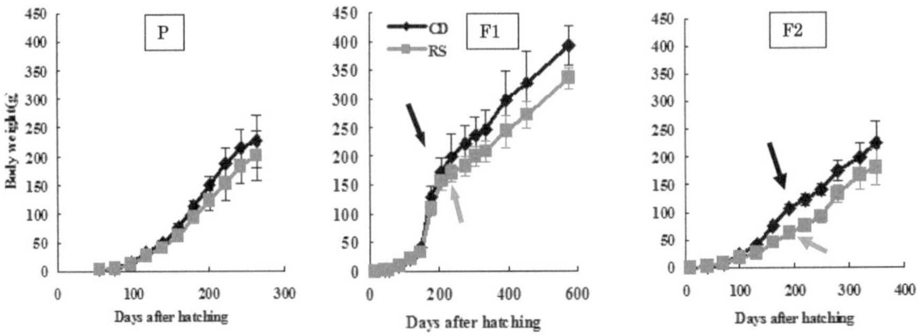

CD, commercial diet; RS, raw *Spirulina*. Arrows indicate first spawning day.

Fig. 13.3 Growth of tilapia (P, F_1, F_2) fed raw *Spirulina* and a commercial diet and their first spawning (Cited from Lu and Takeuchi 2004)

13.3.3 Tilapia Taste

This set of experiments focused on the ultimate objective: the taste and flesh quality of tilapia fed solely on raw *Spirulina* compared with those fed commercial diets. One worry was that the tilapia might acquire off-flavors as a result of feeding solely on raw *Spirulina*, because the most prevalent off-flavors in aquaculture are related to blue-green algae (Boyd and Tucker 1999). The flesh of raw *Spirulina*-fed fish contained a higher amount of proteins and a lower amount of lipids than those fed the control diet. Moreover, the color, odor, texture, and fatness evaluations of the raw *Spirulina*-fed tilapia were slightly superior. Furthermore, the rheological parameters of the muscle were found to coincide with the results of sensory assessment (Lu and Takeuchi 2002; Lu et al. 2003). These findings demonstrate that tilapia fed solely on raw *Spirulina* have high flesh quality, to the extent that they are suitable for use as sashimi (Takeuchi and Omori 2005).

13.3.4 Summary

This study verified the feasibility of rearing tilapia solely on raw *Spirulina*, clearly confirming the feasibility of constructing a food chain between phytoplankton and tilapia in CERAS. Simplified CERAS with a food chain between *Spirulina* and tilapia is therefore a promising approach, with potential ecological engineering applications in space exploration and exploitation. Furthermore, construction of the *Spirulina*–tilapia food chain opens up the possibility of aquaculture practice using recirculating aquaculture systems in harsh environments such as deserts, where regular aquaculture in ponds or flow-through tanks is not possible because of the scarcity or absence of suitable resources.

13.4 Transgenic Nile Tilapia

To strengthen the safety and security of aquaculture, we propose closed recirculating aquaculture technology that allows for full control of pollution and contamination both to and from the surrounding environment. The aim is to thereby provide safe and secure aquaculture products through selection of optimal cultivating species suitable for closed recirculating aquaculture systems (CRASs). Such species should possess characteristics that help reduce the environmental load from CRAS by decreasing waste discharge. Moreover, since CRAS tends to be energy intensive, with large running costs and high initial investment, the productivity of the cultivated species should also be high. Growth hormone (GH)-transgenic fish are currently the first choice, because they showed markedly enhanced growth.

Our research group developed "all fish," GH-transgenic tilapia (GHTi), that overexpress the GH gene throughout their bodies. They carry a tilapia GH complementary DNA fragment gene spliced to a medaka β-actin gene promoter construct integrated into the tilapia genome (Kobayashi et al. 2007). This GHTi has become the first choice for CRAS culture since it exhibits highly efficient productivity. Compared with pedigree common domestic strain tilapia (CTi) reared to the same edible size under visual satiation feeding, rearing of GHTi was achieved 69 days faster at a 1.4-fold higher specific growth rate and 37.3% higher feeding efficiency when fed 38.1% less (Lu et al. 2009). Moreover, the total N and P discharge loads from GHTi were lower: 50.1 and 61.1% of the control fish, respectively (Fig. 13.4). Thus, GHTi is suitable for CRAS because it possesses characteristics that reduce discharge load, thereby minimizing the overall cost of the water recycling treatment (Lu et al. 2009).

However, these fish have undergone morphological changes, resulting in a short body that is less acceptable for consumers. Early adult GHTi showed incomplete bone mineralization, while late adult GHTi had skeletal abnormalities. We therefore focused attention on the association between these morphological changes and the altered mineral requirements. In CTi, body ash, Ca, and P contents in the whole body increased with growth; however, they decreased in GHTi. Furthermore, the Ca/P ratio increased from 1.70 to 1.88 in CTi but decreased from 1.66 to 1.57 in GHTi. It was therefore speculated that when fed conventional commercial diets to satiety, GHTi with a body weight (BW) of more than approximately 230 g were unable to retain sufficient Ca, while those with a BW of more than approximately 429 g could not retain sufficient P (Lu et al. 2013). This significantly lower total Ca retention may be attributable to the synergistic effect of the 38.1% lower total food intake resulting from the shortened rearing period, as well as the significantly lower Ca retention rate (42.6%) in GHTi compared to CTi (50.9%) when both were fed the same commercial diet. The significantly lower total P retention, on the other hand, can only be attributed to the lower total P intake since the P retention rates of GHTi and CTi were similar (38.0% and 37.3%, respectively). These findings may be physiologically associated with changes in Ca and P metabolism and homeostasis in GHTi (Lu et al. 2009).

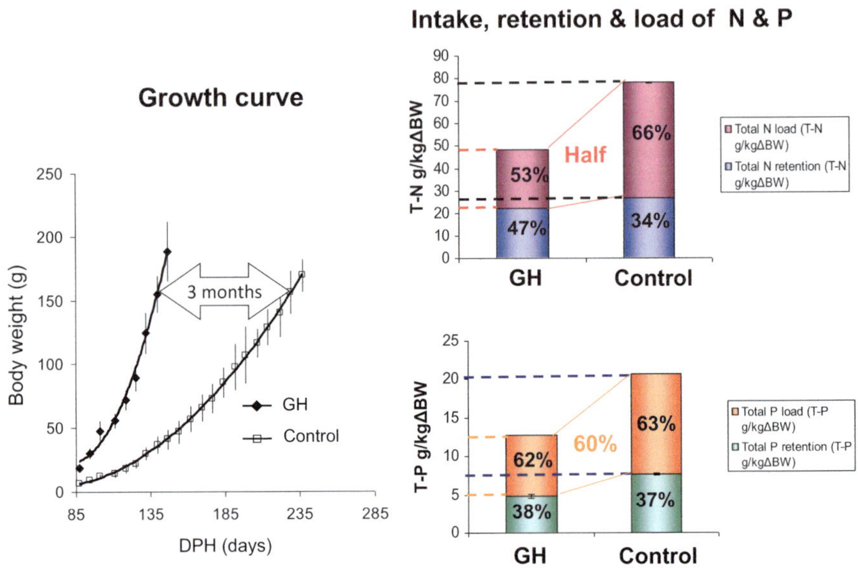

Fig. 13.4 Efficient productivity and the lowered nitrogen and phosphorus discharge load of growth hormone (GH)-transgenic tilapia under visual satiation feeding (Modified from Lu et al. 2009)

After determining the association with changes in nutritional status, we then attempted to reduce the prevalence of morphological deformities in GHTi via dietary compound regulation by supplying a high amount of dietary Ca [6.82%, Ca supplementation diet (Ca-S)] or P [4.75%, P supplementation diet (P-S)]. Both the Ca-S and P-S improved mineral retention, and fish fed the Ca-S showed a lower prevalence of deformities, close to size-matched CTi. These results suggest that the requirements of dietary Ca and P were significantly higher in GHTi than CTi and, consequently, that conventional levels of Ca and P in commercial diets cannot meet the mineral requirements of GHTi (Lu et al. 2013). We therefore propose development of an optimal diet that meets these altered nutritional requirements of GH-transgenic tilapia for optimal growth promotion. This may become a prerequisite for rearing the GHTi line in CRASs, thereby allowing application of these new "domesticated fish" in CRAS and CERAS.

13.5 Fish Waste–Microalgae Culture in Freshwater

To determine whether the waste generated from fish culture in closed culture systems could promote algal growth and efficiently substitute algal culture medium, the waste composition was compared with the elemental composition of algal culture media used for *Chlorella*, *Spirulina*, and *Scenedesmus*, respectively. The

Cu content of the rearing water was sufficient for culture of all three microalgae, as were the contents of P, Fe, Mn, and Zn in the removal solids. However, the latter, being insoluble precipitate, cannot be used directly for algal culture because it requires ionization prior to utilization. The K and Mg content were insufficient for algal culture. These results show that several elements in the waste require ionization and/or supplementation if they are to serve as nutrients for algal growth in a closed culture system (Endo et al. 2000) and that sedimentation of these minerals is regulated by the chemical equilibria in the system (Takeuchi and Omori 2005).

13.5.1 *Scenedesmus*

In order to determine algal growth in fish culture wastewater and the effect of adding digested fish culture solid waste, 7-day culture experiments with the green alga *Scenedesmus quadricauda* were conducted at a constant temperature of 15 °C under continuous light and aeration. Media treatments were derived from diluted wastewater (WW) and solid waste (SW) digested with concentrated sulfuric acid and hydrogen peroxide at 440 °C. Treatments included diluted wastewater, diluted wastewater with solid waste after digestion (WW + SW), and diluted wastewater with sulfuric acid and hydrogen peroxide (WW + SA + HP) as a control. Algal growth occurred in all media treatments, although significantly lower growth and N removal were recorded in WW and WW + SA + HP. Final P readings in WW and WW + SA + HP could not be detected. These results indicate that the two treatments without solid waste had insufficient P and were comparable to algal growth observed in synthetic medium. The final readings in WW/SW for N, P, K, Fe, Mn, Zn, and Cu were reduced by more than 70% compared to the initial readings. Overall, it was clear from these findings that the addition of digested solid waste resulted in superior N removal from fish culture wastewater by *S. quadricauda* (Endo and Takeuchi 2004; Takeuchi and Omori 2005).

13.5.2 *Spirulina*

Using culture waste of tilapia fed solely on *Spirulina* under closed conditions, we subsequently cultured *Spirulina* to elucidate the tilapia–*Spirulina* matter cycle with the aim of evaluating the possible production of *Spirulina* biomass (Endo et al. 2009). Results revealed that WW alone was insufficient, whereas *Spirulina* grew well in the presence of a WW and SW blend. However, growth in this medium was poorer than that in *Spirulina*–Ogawa–Terui (SOT) medium. Moreover, the addition of P and Zn enhanced growth in the WW + SW medium to the same level as that in SOT. These results were consistent with previous reports whereby P in particular was effective for culture of *Spirulina* (Endo et al.2009).

WW and SW generated under the optimal condition for tilapia growth (salinity, 8 psu) were also used to prepare media for *Spirulina* culture (Nishimura et al. 2012). The resulting biomass was 0.95 g/L, 1.18 g/L, and 1.21 g/L when used for SOT medium, WW + SW medium at a salinity of 4 psu, and WW + SW medium at a salinity of 8 psu, respectively. This indicates that more biomass was produced in the salt-containing media than SOT, thereby suggesting that the appropriate salinity for culture of *Spirulina* is 4 or 8 psu.

13.5.3 Chlorella–Moina *Feedback Culture*

To develop a total production system involving the above culture steps, we established an experimental system using fish culture waste products of tilapia to culture *C. vulgaris* and *Moina*, which were then used to support further tilapia aquaculture (Mori et al. 2006). Following this, we constructed a prototype of the *Moina*-cultivating system, to provide effluent from the *Moina* cultivation supply for continuous consumption and growth of *C. vulgaris* (Fig. 13.5) (Endo and Takeuchi 2013). Tilapia culture waste was absorbed at a twofold higher efficiency by this *Moina* culture system than by *Moina* fed solely on *C. vulgaris*. The results reveal that the use of fish culture waste from tilapia enables effective production of feed, in this case *Moina*, via *C. vulgaris* cultivation.

Fig. 13.5 A nutrient feedback rearing system for water fleas. The system was constructed using an algal cultivation flask (left side) (Cited from Endo and Takeuchi 2013)

13.5.4 Fish Waste–Microalgae Culture in Salt Water

Chaetoceros gracilis contains a high level of eicosapentaenoic acid (EPA), while *Tetraselmis tetrathele* contains high levels of protein and linolenic acid; therefore, both can be effective feed for prawn larvae and short-neck clams. Zhang (2013) evaluated the composition of medium containing fish culture waste products generated by tiger puffer, for the cultivation of *C. gracilis* and *T. tetrathele*. One component of a saltwater recirculating aquaculture system is a foam fractionator, which also discharges WW and SW from tiger puffer culture. Previous studies suggest that the WW + SW medium obtained from tiger puffer aquaculture waste lacks certain ingredients essential for algae culture, namely, iron and manganese. Here, WW + SW medium supplemented with Fe and Mn proved an effective medium for growth of both algal species. The time to reach maximum density was longer when *C. gracilis* was cultured in medium supplemented with WW + SW compared to control medium (F medium), while the crude lipid content and nonessential amino acid content were 1.4-fold and 2.2-fold higher, respectively, in the WW + SW supplemented medium than F medium. The yields and compositions (crude protein content, crude lipid content, amino acid composition, and fatty acid composition) of the cultured *T. tetrathele* were similar between the WW + SW supplemented and F medium.

These findings suggest that the waste generated during aquaculture of tilapia and tiger puffer, which are freshwater and saltwater fish species, respectively, can be used for algae cultivation. Supplementation with minerals may be necessary, depending on the algae species under cultivation. These results further suggest the feasibility of constructing a CERAS, an aquaculture system that performs both feed production and water purification.

13.6 Gas Exchange between *Chlorella* and Tilapia

To more precisely elucidate gas circulation and the behavior of oxygen and carbon dioxide, Endo and Takeuchi (2013) developed an experimental aquaculture system that enables gas exchange between tilapia and microalgae chambers (Fig. 13.6). A 2-week experiment examining gas exchange between *C. vulgaris* and tilapia showed that levels of oxygen and carbon dioxide gradually stabilized with growth of *C. vulgaris*, subsequently allowing a sufficient amount of oxygen for tilapia culture and, as a result, successful fish survival and maintenance of water quality.

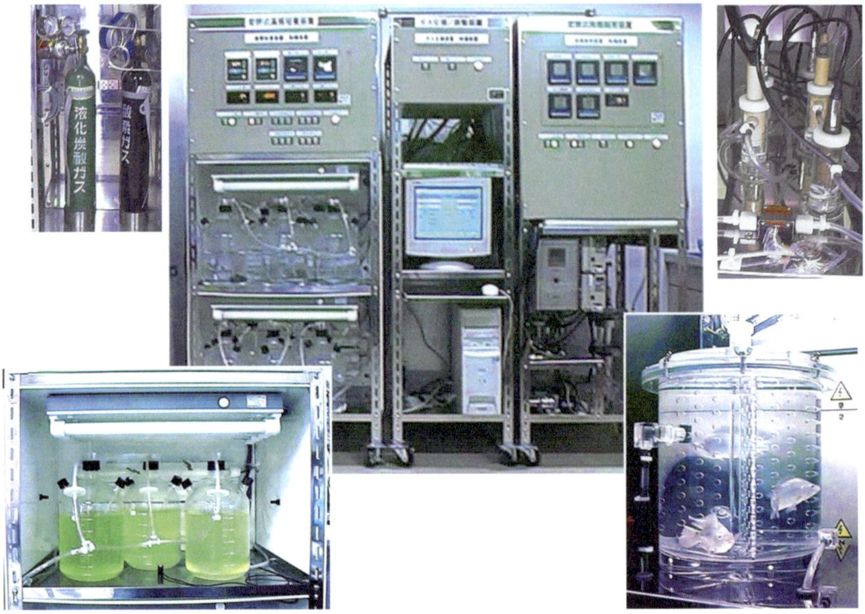

Fig. 13.6 Experimental unit for gas exchange between microalgae and fish

13.7 Estimation of the Mass Balance

In the introduction of this chapter, we mentioned that the final goal in the development of CERAS is smooth loops for transfer of substrates that function with a higher overall bioconversion rate. The initial loop of the food chain starts with the waste processor, where water-soluble nutrients are cycled to the microalgal component. These nutrients are utilized for synthesis of *C. vulgaris* biomass, which are then transferred to the next component. The *Moina macrocopa* and tilapia biomass are then returned to the waste processor and recycled into nutrients for algal growth. Gases flow, namely, oxygen and carbon dioxide, and the resulting concentrations are controlled by a storage device.

Various experiments were conducted to estimate N and P flow and the biomass conversion rate of the artificial CERAS food chain. First, we determined the percentages of N and P removal from the tilapia cultural medium with *C. vulgaris* (75.2% and 71.9%, respectively). Next, the percentages of biomass conversion, and N and P retention between *M. macrocopa* and *C. vulgaris*, were obtained. Values were 20.4%, 24.9%, and 17.0%, respectively. As a third step, a feeding trial with tilapia fed solely on *M. macrocopa* was conducted. The biomass conversion and the N and P retention rates were subsequently calculated as 25.4%, 27.3%, and 45.3%, respectively. Using these values, a schematic flow diagram using estimated biomass values was obtained (Fig. 13.7). Details of the calculation

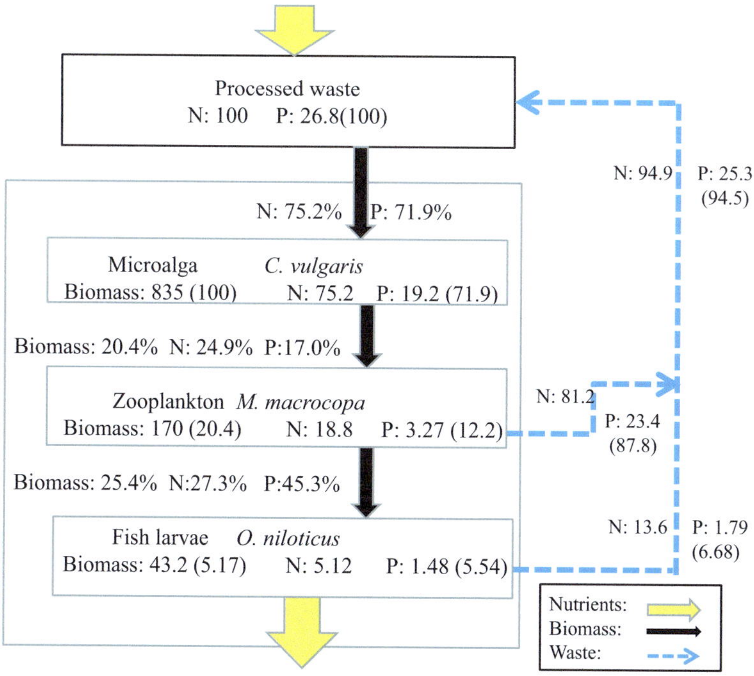

Fig. 13.7 Schematic flow diagram showing the estimated values of biomass, nitrogen (N) and phosphorus (P) conversions, and their retention in each aquatic organism based on the nitrogen-containing processed waste (Modified from Endo and Takeuchi 2005)

formula were described in a previous paper (Endo and Takeuchi 2005). Values in the figure are expressed by the amount of N contained in the processed waste converted to a value of 100. Values of P retention and biomass conversion in parentheses are represented by the converted P contained in the waste and the biomass of microalga as a value of 100, respectively.

In this study, the overall mass flow of the artificial food chain, that is, aquacultural waste to *C. vulgaris* to *M. macrocopa* to tilapia larvae, and oxygen balance in the system were calculated. Values were obtained according to the assumption that fish were reared from larval tilapia (body weight, 0.01 g) fed on *M. macrocopa* acquired through the food chain. The estimated fish biomass provided by the food chain was 5.17% of the *C. vulgaris* biomass, while N and P retention in the fish body from aquaculture waste were 5.12% and 5.54%, respectively. The amount of oxygen regenerated by *C. vulgaris* was estimated as more than twice that consumed by tilapia and *M. macrocopa*. These results suggest that the oxygen supply is sufficient for normal operation of the system (Endo and Takeuchi 2005).

13.8 Photoperiod Studies

It is well known that the photoperiod can influence growth, reproduction, and many other physiological functions in various animal species. However, in fish, photoperiod experiments have been carried out only under 24-h light/dark cycles. A 24-h light/dark cycle is one of the most pervasive epigenetic influences on organisms, from single-celled organisms to humans. Any discrepancy from this circadian rhythm has been shown to influence physiological performances. It is further thought that extreme circadian cycles exert differing influences on physiological functions, subsequently reflected in differential physiological performance. Thus, it is possible that growth may occur at different speeds if an organism lives through more cycles than another within the same time span. This possibility has been evidenced in some mammals where it was concluded that the rate of body weight increase is not determined by real time but rather by subjective time represented by the number of light/dark cycles (Madrid et al. 1992; Vilaplana et al. 1996). This reinforces the necessity of investigating this phenomenon in fish.

13.8.1 Tilapia

The overall goal here was to determine the optimal photoperiod regime for improved growth rate and to determine the feasibility of using photoperiod manipulation to control excessive reproductive activity, which can lead to problems of overcrowding in CERAS tanks. Our previous studies (Biswas and Takeuchi 2002; Biswas et al. 2002) showed that oxygen consumption in both young and adult tilapia is influenced by different artificial photoperiod regimes (e.g., 3 L:3D, 6 L:6D, 12 L:12D, and 24 L:24D). The higher postprandial increase in metabolic rate or energy loss does not always act to reduce the scope of activity. Therefore, the effects of different photoperiod regimes and feeding interval on food intake and growth of tilapia were investigated (Biswas and Takeuchi 2003). Fish exposed to a 6 L: 6D had a significantly higher growth rate than those exposed to other photoperiods. Moreover, when fed to satiation, significantly higher food consumption and feeding efficiency as well as lower adiposity were observed in fish exposed to 6 L:6D compared to 12 L:12D. These results indicate that a 6 L:6D regimen results in improved growth through stimulated food intake; however, it would be premature to suggest this as the optimal photoperiod for rearing fish without careful analysis of how photoperiod affects stress levels. Therefore, further experiments were carried out to determine the levels of stress indicators in fish exposed to the different photoperiod regimes. Physiological responses (cortisol, glucose, Cl^-, hematocrit, lymphocyte, and neutrophil counts) were used as stress indicators. The results demonstrated that an artificial photoperiod regime does not cause a significant acute or chronic stress response, although a slightly higher cortisol level was observed in fish exposed to 6 L:6D. These elevated levels of cortisol may play an important role in positively controlling the physiological functions of fish exposed to this regime (Biswas et al. 2004). One interesting observation was that fish exposed to 6 L:6D showed higher blood lymphocyte concentrations than those exposed to 12 L:12D. This may indicate a positive effect of 6 L:6D, because lymphocytes possess all the necessary components required for an independent extra-neuronal cholinergic system for the regulation of immune function. A major problem in tilapia aquaculture is excessive reproduction of female fish, which leads to increased competition for food and stunted somatic growth. The feasibility of using photoperiod manipulation to arrest reproductive performance was therefore examined. Fish exposed to 12 L:12D and 14 L:10D spawned successfully throughout the experiments. Although fish exposed to 6 L:6D spawned successfully immediately after initiation of the photoperiod regime, spawning was arrested after two to four spawning cycles. This was further paralleled by a significant decrease in plasma levels of estradiol-17β (Fig. 13.8: Biswas et al. 2005a). In contrast, there were no major differences in testosterone levels among treatments. These findings suggest that photoperiod manipulation can be used to arrest spawning in tilapia and, moreover, that growth and reproduction could be controlled by manipulation of photoperiod and feeding interval (Takeuchi and Omori 2005).

13.8.2 Saltwater Fish

The convincing findings in tilapia motivated scientists in Japan to carry out studies on the effect of photoperiod manipulation on the growth performance of various commercially important marine species. Red sea bream *Pagrus major* is considered one of the most commercially important marine species in Japan, as it has a range of

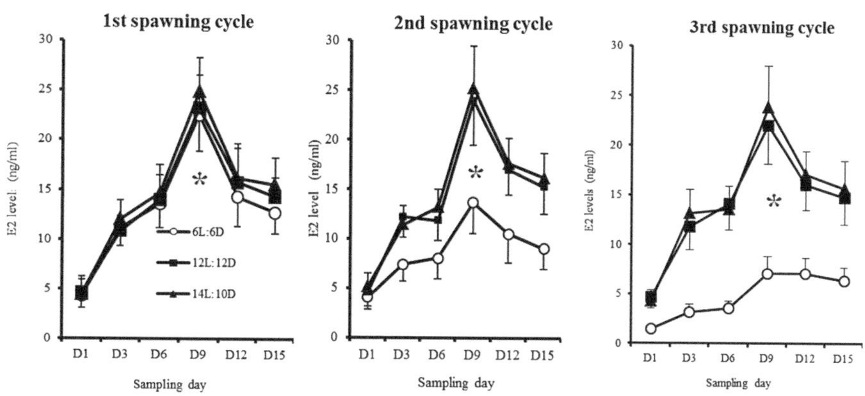

Fig. 13.8 Changes in plasma levels of estradiol-17β (E2) in the reproductive cycle of tilapia maintained under different photoperiod regimes as a function of days post-spawning (Modified from Biswas et al. 2005a). * Means ± SD ($n = 3$ fish/treatment)

uses and is traditionally consumed at festive occasions due to it symbolizing good fortune. A series of experiments were therefore carried out to determine whether photoperiod manipulation could be used to stimulate growth performance at different growth stages of this species. Four photoperiods (6 L:6D, 12 L:12D, 16 L:8D, and 24 L:0D) were designed to investigate the effect at the following growth stages: 1–30 g, 20–100 g, 200–400 g, and 1.0–2.0 kg. In all cases, both continuous (24 L:0D) and long (16 L:8D) photoperiods resulted in a 20–50% increase in weight gain compared to a 12 L:12D photoperiod (Biswas et al. 2005b, 2006b, 2008a, b, 2009, 2010a). The higher growth, especially that at 24 L:0D, was attributed to higher food intake, digestibility of nutrients and energy, and feed conversion efficiency (Biswas et al. 2005b, 2006b, 2008a, b, 2009, 2010a). The higher food intake under a continuous photoperiod is believed to have been the result of increased activity of these diurnal fishes under this regime as well as increased foraging activity when food was delivered or to have been related to the positive effect of growth hormone on appetite (Johnsson and Björnsson 1994). In a different attempt using a self-feeder with different photoperiods, the 24 L:0D photoperiod similarly resulted in increased growth performance compared to the control (12 L:12D) (Biswas et al. 2011). Here, the longer access time to the self-feeder under a continuous photoperiod compared to the 12L12D regime might have allowed slower and more efficient digestion, thus improving conversion efficiency. Because stress is known to reduce growth and disease resistance in fish, investigation of the physiological responses to artificial photoperiods is also necessary. Therefore, a number of stress indicators were investigated under each approach. The results demonstrated that levels of the different stress indicators in fish exposed to the manipulated photoperiods were far lower than the stress-induced levels observed in this species (Biswas et al. 2006a, b, 2008a, 2010a, 2011).

A number of experiments have been carried out in striped knifejaw *Oplegnathus fasciatus* which has attracted great interest from Japanese fish farmers due to its high market value and consumer demand. Different photoperiods (6 L:6D, 12 L:12D, 16 L:8D, and 24 L:0D) were designed to investigate the effect at 130–290 g and 280–350 g, respectively. Although there were no significant differences between 6 L:6D, 16 L:8D, and 24 L:0D, significantly higher growth performance was observed compared to fish under the 12 L:12D regime at both growth stages (Biswas et al. 2008c, d). Similar to the case in red sea bream, the higher feed intake under the manipulated photoperiods may have been due to the feeding strategy of these fish, which was closely reflected in the times of maximum appetite (Azzaydi et al. 1999; Biswas et al. 2006b). Investigation of feeding interval under 16 L:8D and 24 L:0D at 10–70 g revealed no significant difference in growth performance among 6-, 9-, and 12-h intervals under both photoperiods, indicating that striped knifejaw can adjust its feeding activity between 6- and 12-h intervals to maximize growth (Biswas et al. 2010b). In addition, the levels of different stress indicators under the manipulated photoperiods were shown to be similar to those of unstressed levels, suggesting that the artificial photoperiods caused no physiological disturbance in striped knifejaw (Biswas et al. 2008c, 2010b).

In both species, improved appetite, greater food intake and higher feed conversion efficiency, higher digestibility, and superior retention efficiency were factors reportedly responsible for the faster growth rate under the manipulated photoperiods. These findings show that photoperiod manipulation is therefore a major factor in the promotion of production in both species.

13.9 Conclusions

So far, we have been investigating the construction of CERAS, with application of the experimental results (Fig. 13.9) currently being explored. Recently, recirculating system technology for aquaculture has been rapidly advancing due to the growing concern over environmental conservation and food safety. In order to establish successful CERAS technology, the following are important: (1) complete control of the commercial diet, with establishment and full analysis of the permissible values of hazards and methods for removal of these hazards from the diet; (2) complete control of the water environment, with monitoring of the aquaculture ground and culturing of organisms in water free of hazards; and (3) establishment of "visible" aquaculture methods, the hazard analysis critical control point. Development of computer technology to support the above is also necessary.

So far, basic CERAS experiments have been concerned with the following: achievement of highly efficient fish production through artificial control of

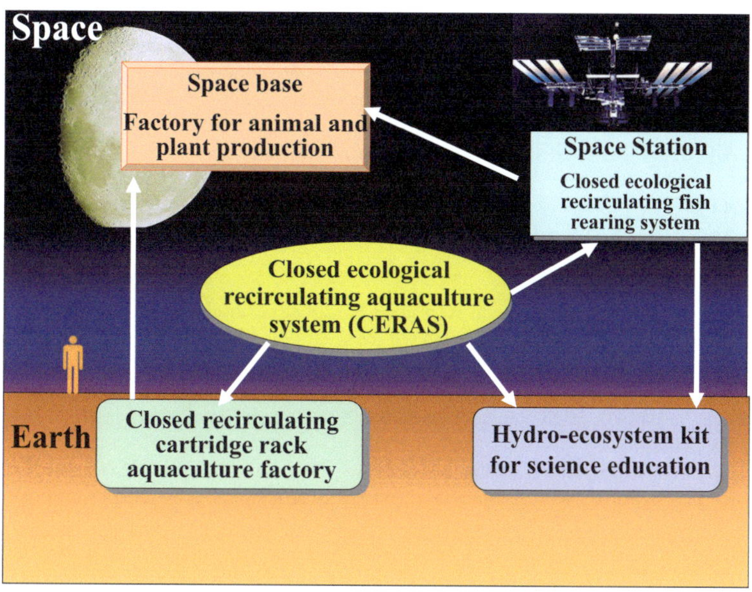

Fig. 13.9 Scenario for the strategic application of CERAS

illumination, utilization of low-priced and unused resources in the development of new fish feeds, and development of a cartridge rack fish culture system (Omori et al. 2006; Yamada et al. 2009). In the future, various results are expected as development of CERAS progresses, for example, improvements in food quality, safety, and security, stabilization of distribution, promotion of the aquaculture industry, employment of elderly individuals, creation of regional industry, reduction in environmental loading and pollution, and the creation of urban and desert aquaculture.

References

Azzaydi M, Martinez FJ, Zamora S, Sánchez-Vázquez FJ, Madrid JA (1999) Effect of meal size modulation on growth performance and feeding rhythms in European sea bass (*Dicentrarchus labrax,* L.) Aquaculture 170:253–266

Belay A, Kato T, Ota Y (1996) *Spirulina (Arthrospira)*: potential application as an animal feed supplement. J Appl Phycol 8:303–311

Biswas AK, Takeuchi T (2002) Effect of different photoperiod cycles on metabolic rate and energy loss of both fed and unfed adult tilapia *Oreochromis niloticus*: Part II. Fish Sci 68:543–553

Biswas AK, Takeuchi T (2003) Effects of photoperiod and feeding interval on food intake and growth rate of Nile tilapia *Oreochromis niloticus*. Fish Sci 69:1010–1016

Biswas AK, Endo M, Takeuchi T (2002) Effect of different photoperiod cycles on metabolic rate and energy loss of both fed and unfed young tilapia *Oreochromis niloticus*: Part I. Fish Sci 68:465–477

Biswas AK, Maita M, Yoshizaki G, Takeuchi T (2004) Physiological responses in Nile tilapia exposed to different photoperiod regimes. J Fish Biol 65:811–821

Biswas AK, Morita T, Yoshizaki G, Maita M, Takeuchi T (2005a) Control of reproduction in Nile tilapia *Oreochromis niloticus* (L.) by photoperiod manipulation. Aquaculture 243:229–239

Biswas AK, Seoka M, Inoue Y, Takii K, Kumai H (2005b) Photoperiod influences the growth, food intake, feed efficiency and digestibility of red sea bream (*Pagrus major*). Aquaculture 250:666–673

Biswas AK, Seoka M, Takii K, Maita M, Kumai H (2006a) Stress response of red sea bream *Pagrus major* to acute handling and chronic photoperiod manipulation. Aquaculture 252:566–572

Biswas AK, Seoka M, Takii K, Kumai H (2006b) Effect of photoperiod manipulation on the growth performance and stress response in juvenile red sea bream, *Pagrus major*. Aquaculture 258:350–356

Biswas AK, Seoka M, Ueno K, Takii K, Kumai H (2008a) Stimulation of growth performance without causing stress response in young red sea bream, *Pagrus major* (Temminck & Schlegel) by photoperiod manipulation. Aquac Res 39:457–463

Biswas AK, Seoka M, Tanaka Y, Takii K, Kumai H (2008b) Use of photoperiod manipulation to stimulate the growth performance of juvenile red sea bream (*Pagrus major*). World Aquaculture 39(2):12–15

Biswas AK, Seoka M, Ueno K, Yong ASK, Biswas BK, Kim YS, Takii K, Kumai H (2008c) Growth performance and physiological responses in striped knifejaw, *Oplegnathus fasciatus*, held under different photoperiod. Aquaculture 279:42–46

Biswas AK, Seoka M, Inoue K, Takii K, Kumai H (2008d) Photoperiod manipulation stimulated growth performance in young striped knifejaw, *Oplegnathus fasciatus*. Bull Fish Lab Kinki Univ 11:157–164

Biswas AK, Seoka M, Tanaka Y, Ueno K, Takii K, Kumai H (2009) Continuous photoperiod can be used to get higher growth performance in juvenile red sea bream (*Pagrus major*). World. Aquaculture 40(1):4–6

Biswas A, Seoka M, Inagaki H, Takii K (2010a) Reproduction, growth and stress response in adult red sea bream (*Pagrus major*) exposed to different photoperiods at spawning season. Aquac Res 41:519–527

Biswas A, Inoue K, Takii K (2010b) Feeding interval and photoperiod influence the growth performance of striped knifejaw, *Oplegnathus fasciatus*. Aquac Res 41:e517–e523

Biswas A, Inoue K, Takii K (2011) Effect of self-feeding and photoperiod on the growth performance and stress response of red sea bream, *Pagrus major*. J Appl Icht 27:1104–1107

Boyd CE, Tucker CS (1999) Off-flavors and harmful algae. In: Boyd CE, Tucker CS (eds) Pond aquaculture water quality management. Kluwer Academic Publishers, Norwell, pp 439–461

Cohen Z (1997) The chemicals of *Spirulina*. In: Vonshak A (ed) Spirulina Platensis-Arthrospira: physiology, cell biology and biotechnology. Taylor & Francis Ltd, London, pp 176–179

Endo M, Takeuchi T (2004) Nutrient removal from aquacultural waste using the green alga *Scenedesmus quadricauda*. Eco-Engineering 16:195–201. (In Japanese with English abstract)

Endo M, Takeuchi T (2005) Estimation of the mass balance in a closed ecological recirculating aquaculture system, CERAS. Proc Int Symp Closed Habit Expt Mater Circ Tech:458–464

Endo M, Takeuchi T (2013) Effect of production on fish under the closed systems and microgravity. Int J Microgravity Sci Appl 30(2):111–119. (In Japanese with English abstract)

Endo M, Takeuchi T, Yoshizaki G, Toyobe M, Kanki R, Omori (Suzuki) K, Oguchi M, Kibe S (1999) Studies on the development of closed ecological recirculating aquaculture system (CERAS) IV. A long term feeding experiment with *Oreochromis niloticus* in the fish-rearing closed tank. CELSS J 11:17–24. (In Japanese with English abstract)

Endo M, Takeuchi T, Yoshizaki G, Satoh S, Omori K, Oguchi M, Nakajima A (2000) Studies on the development of closed ecological recirculating aquaculture system (CERAS) VI. Phosphorus form and mineral budgets in the fish-rearing closed tank during a long-term feeding experiment with tilapia. CELSS J 13:19–26. (In Japanese with English abstract)

Endo M, Okubo Y, Takeuchi T (2009) Production of *Spirulina platensis* using aquacultural wastes discharged from recirculating fish culture system with tilapia. Abst Metg Eco-Engineering:95–96. (In Japanese with English abstract)

Fry G, Iles TD (1972) The cichlid fishes of the great lakes of Africa: their biology and evolution. Oliver & Boyd, Edinburgh

Getachew T (1987) A study of an herbivorus fish, *Oreochromis niloticus* L., diet and its quality in two Ethiopian Rift Valley lakes, Awassa and Zwai. J Fish Biol 30:439–449

Hedenskog G, Hofsten AV (1970) The ultrastructure of *Spirulina platensis*. A new source of microbial protein. Physilogia Plantarum 23:209–216

Johnsson JI, Björnsson BT (1994) Growth hormone increases growth rate, appetite and dominance in juvenile rainbow trout, *Oncorhynchus mykiss*. Anim Behav 48:177–186

Kobayashi S, Alimuddin MT, Miwa M, Lu J, Endo M, Takeuchi T, Yoshizaki G (2007) Transgenic Nile tilapia (*Oreochromis niloticus*) over-expressing growth hormone show reduced ammonia excretion. Aquaculture 270:427–435

Lu J, Takeuchi T (2002) Taste of tilapia *Oreochromis niloticus* fed solely on raw *Spirulina*. Fish Sci 68:987–988

Lu J, Takeuchi T (2004) Spawning and egg quality of the tilapia *Oreochromis niloticus* fed solely on raw *Spirulina* throughout three generations. Aquaculture 234:625–640

Lu J, Yoshizaki G, Sakai K, Takeuchi T (2002) Acceptability of raw *Spirulina platensis* by larval tilapia *Oreochromis niloticus*. Fish Sci 68:51–58

Lu J, Takeuchi T, Ogawa H (2003) Flesh quality of tilapia *Oreochromis niloticus* and fed solely on raw *Spirulina*. Fish Sci 69:529–534

Lu J, Li J, Furuya Y, Yoshizaki G, Sun H, Endo M, Haga Y, Satoh S, Takeuchi T (2009) Efficient productivity and lowered nitrogen and phosphorus discharge load from GH-transgenic tilapia (*Oreochromis niloticus*) under visual satiation feeding. Aquaculture 293:241–247

Lu J, Yoshizaki G, Endo M, Takeuchi T (2013) Effect of dietary high amount of calcium and phosphorus on reducing the prevalence of morphological deformities in GH-transgenic Nile tilapia. Fish Sci 79:647–658

Madrid JA, Lax P, Matas P, Navarro I, Matin-Cuena E (1992) Modifications of growth rate and feeding behavior in rats during exposition to circadian diurnal LD cycles. J Interdiscipl Cycle Res 23:209–210

Matsumoto H, Hamazaki A (1992) Utilization in space. In: Yamaguchi K (ed) Utilization of microalgae. Kouseisha-kouseikaku, Tokyo, pp 116–117. (In Japanese)

McDonald ME (1987) Interaction between a phytoplanktivorous fish, *Oreochromis niloticus* and two unialgal forage populations. Env Biol Fish 18:229–234

Mori Y, Endo M, Takeuchi T (2006) Live food production for larval and juvenile fish by waste utilization in a closed recirculating fish culture system. Proc 2006 SEE Conf:153–157. (In Japanese with English abstract)

Nishimura T, Endo M, Takeuchi T (2012) Productivity of *Spirulina platensis* and its ability of nutrient removal from aquacultural wastes of recirculating fish culture system with tilapia. Abst Metg Eco-Engineering:15–16. (In Japanese with English abstract)

Nitta K, Ashida A, Otsubo K (1996) Closed ecological experiment facilities (CEEF) construction planning and present status. Life Support Biosphere Sci 3:101–115

Oguchi M, Otsubo K, Nitta K, Hatayama S (1987) Food production and gas exchange system using blue-green alga (*Spirulina*) for CELSS. Adv Space Res 7:7–10

Omori K, Ogai K, Lu J, Yoshizaki G, Takeuchi T, Oguchi M, Nakajima A (2000) Research on removal of zooplanktons from the culturing system of tilapia, *Oreochromis niloticus*. CELSS J 13:13–17. (In Japanese with English abstract)

Omori K, Endo M, Yoshizaki G, Takeuchi T (2006) Studies on technology for fish culture by using cartridge type tanks -trials of individual rearing of tilapia *Oreochromis niloticus* and examinations of reasonable tank width. Eco-Engineering 18:9–14. (In Japanese with English abstract)

Takashima F (1997) Suizokuikuseiron. Seizandou, Tokyo. 236pp. (In Japanese)

Takeuchi T (2011) Development of fish feed and aquaculture in space age. Biophilia 7(1):6–10. (In Japanese)

Takeuchi T, Omori K (2005) Utilization of aquatic organisms in closed systems. Proc Int Symp Closed Habit Expt Mater Circul Tech:240–250

Takeuchi T, Noto S, Yoshizaki G, Toyobe M, Kanki R, Oguchi M, Kibe S, Tanaka H (1997) Study on the development of closed ecological recirculating aquaculture system (CERAS) I: development of fish-rearing closed tank. CELSS J 10:1–4. (In Japanese with English abstract)

Takeuchi T, Lu J, Yoshizaki G, Satoh S (2002) Effect on the growth and body composition of juvenile tilapia *Oreochromis niloticus* fed raw *Spirulina*. Fish Sci 68:34–40

Trewavas E (1983) Tilapiine fishes of the genera Sarotherodon, Oreochromis and Danakilia. Br Mus Nat Hist, London

Vilaplana J, Madrid JA, Sanchez-Vazquez J, Campuzano A, Cambras T, Diez-Noguers A (1996) Influence of period length of light/dark cycles on the body weight and food intake of young rats. Physiol Behav 58:9–13

Yada Y (1982) Study on the feeding habit of *Oreochromis nilotica* I. The change on feeding habit of *Oreochromis nilotica* in feeding and no feeding pond. Suisanzoshoku 29:229–233. (In Japanese with English abstract)

Yada Y, Miyashita Y (1988) Tilapia. Midorishobou, Tokyo. (In Japanese)

Yamada N, Endo M, Yoshizaki G, Takeuchi T, Omori K (2009) Studies on technology for fish culture by using cartridge type tanks—effect of tank size on the growth performance and fin damage for individual rearing of tilapia *Oreochromis niloticus* for a long period. Eco-Engineering 21:161–169. (In Japanese with English abstract)

Zhang XC (2013) Ph.D. thesis, Tokyo University of Marine Science and Technology, Tokyo, Japan. (In Japanese)

Chapter 14
Local Survey and Consideration of Land-Based Factory for Closed Recirculating Aquaculture Using Waste Heat Discharged from Biomass Power Plants

Masato Endo, Kunihiko Mouri, and Toshio Takeuchi

Abstract To assess the potential for realization of biomass power plants and for utilizing their waste heat in water temperature control of closed recirculating aquaculture system (CRAS), here we examined project feasibility of closed recirculating land-based aquaculture factory (CRLAF) using waste heat generated from biomass power plants in the northern Tohoku region of Japan, namely, Kuji City, Hirono Town, and Noda Village located in Iwate Prefecture. Three types of biomass were selected from survey data of biomass abundance, and the processing capabilities, construction costs, and energetic budgets of three relevant power plants—combustion of poultry manure, methane fermentation of swine manure, and combustion of wood chips—were examined to uncover the feasibility of implementation for each system. A model system for a CRAS was also established for tiger puffer culture, and the required heat energy and production costs were calculated. The results indicated that the biomass power plant utilizing poultry manure combustion is feasible, due to the ability to easily purchase manure for processing and generate additional income by selling processed fertilizer. Processing capacity and biomass utilization policy were examined for other power plants as well. The cost of tiger puffer production was calculated to be 2650 yen/kg, and 250 yen/kg of this cost in energy could be reduced by waste heat utilization. The waste heat discharged from a power plant processing 50 tons/day of poultry manure was estimated to be capable of providing the heat power for nine units of a 100-ton CRAS.

M. Endo (✉) • T. Takeuchi
Tokyo University of Marine Science and Technology, 4-5-7 Konan, Minato, Tokyo 108-8477, Japan
e-mail: asteroid@kaiyodai.ac.jp

K. Mouri
eL Power Technology Ltd, 2-26-6 Misakidai, Hachinohe, Aomori 031-0842, Japan

T. Takeuchi (ed.), *Application of Recirculating Aquaculture Systems in Japan*, Fisheries Science Series, https://doi.org/10.1007/978-4-431-56585-7_14

Keywords Closed recirculating aquaculture system • Waste heat utilization • Biomass power plant • Tiger puffer • Wood chips • Poultry manure • Swine manure • Methane fermentation

14.1 Introduction

The closed recirculating aquaculture system (CRAS) is currently seeing active research and development to secure the sustainable supply of aquatic resources as well as the safety of food, and its technical maturity is advancing remarkably. To this end, various studies have been conducted, especially in fundamental research of the elemental technologies used in culture systems, such as residue management, efficient ways of feed supply, and developments of effective feed and seedling fish. Studies have also focused on fundamental technologies for industrialization, comprising the application of engineering technologies for the exploitation of animals' biological traits (Endo 2008, 2014). However, because initial facility expenses and running costs are high relative to the cost of the products, economic efficiency has been considered low. In particular, the running cost is a source of this inefficiency, as an extreme amount of energy is consumed for the temperature regulation. The development of a 100% closed recirculating land-based aquaculture factory (CRLAF) with a stable and inexpensive heat supply would allow for the establishment of a new culture industry without restrictions on location near a saltwater supply and with no need to obtain fishery rights. The development of this technology is anticipated to become the backup of conventional fishery industry within Japan, as well as contribute greatly to exports to the global fish market.

The Tohoku region of Japan is rich in biomass resources, but few investigations have emphasized livestock biomass as an untapped resource. Swine manure, poultry manure, wood (from forest thinning), and recirculating land-based aquaculture residue can be transformed into energy sources adapted to the specific regional characteristics, which can be used efficiently to provide a stable heat source for recirculating aquaculture systems. By simultaneously utilizing the minerals (ash), and methane fermentation digested liquid produced as by-products of fertilizing grains for animal consumption, it can be expected that a recycling society in which the fishery, forestry, and livestock industries cooperate will be established. However, because each region has its own distinct biomass resources, fishery catch, and processing methods, proposal of model plans that suit each region's individual characteristics and needs is crucial. Surveys and studies were therefore conducted as a part of promoting the establishment of CRAS that make use of exhaust heat generated by the operation of power plants, with a focus on biomass. Specifically, as a starting point for solving the abovementioned problems, a model plan was assembled for CRLAF that makes use of clean energy to establish novel industries in the Tohoku region. This chapter describes these results.

14.2 Research Outline

Provision of a stable supply of secure and safe products throughout the year is one of the advantages of CRLAF. However, for the establishment of this project, various simulations were necessary to understand the specific regional characteristics for the selection of fish species, feed price, quantity of production, electricity/ fuel costs, labor costs, construction, operation, and maintenance fees. Whereas ordinary aquaculture utilizing natural bodies of water must face business constraints such as acquisition of fishery rights, CRASs do not have such restrictions. This leads to the expectations of reducing costs through scale enlargement and system optimization.

Therefore, in this study, the Kuji Area (comprising Kuji City, Noda Village, and Hirono Town in Iwate Prefecture) was selected as the model construction site, and simulations and plans for the industrialization and implementation of CRLAF powered by waste heat from biomass power generation were conducted. Currently, this area is attempting to push forward with a post-earthquake revival plan based on the improvement of employment opportunities and activation of the regional economy while reducing its carbon dioxide emissions.

Accordingly, this project is defined as a necessary condition for the establishment of an enterprise that utilizes the regionally abundant supplies of poultry manure, swine manure, and wood chips and gives an internal rate of return (IRR) of 5–10%. In the current plan, the optimization of systems such as the transformation of aquacultural residues into energy by gasification and incineration will be promoted in view of energy usage, rearing apparatus, and methods in fish. The energy supply plan by biomass energy type is shown in Fig. 14.1.

Specific evaluation of industrialization such as fishery supply chains on site was performed for the Kuji Area. In addition, investigations and research were performed on the selections of fish species with high feasibility and estimation of ecological energy source and the possibilities for the construction of smart community to select the appropriate construction fee and effective business model. For the evaluation of industrialization, the feasibility of a CRAS that uses poultry manure power generation as a heat and power source, as well as one that considers the recycling of its residues, was evaluated. Also, model cases that utilize swine manure and wood biomass were investigated.

In addition, a graphic representation of the enterprise was constructed for the visual materialization of an integrated recirculating food production plant called “eco-farm” that utilizes renewable, natural energy.

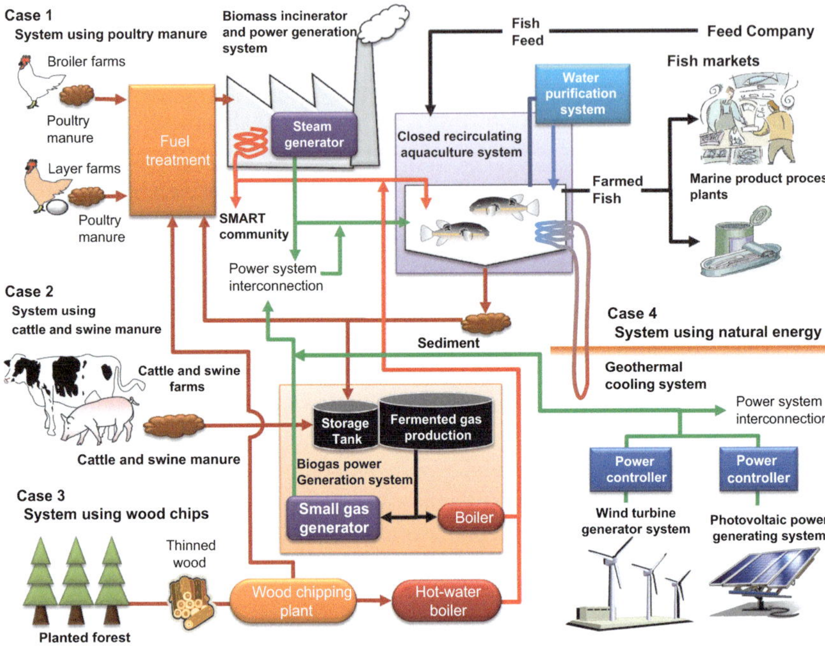

Fig. 14.1 Conceptual scheme of the relationship between biomass energy supplies and the closed recirculating aquaculture systems

14.3 Survey of Biomass Abundances and Needs Assessment on Fisheries Products in the Kuji Area

14.3.1 Biomass Abundances in the Kuji Area

Using the Kuji Area as a model, fundamental research and analyses were performed in preparation of discussion and decisions for the industrialization plan of a land-based aquaculture that utilizes biomass energy and takes characteristics and needs of the area into consideration, as well as supports the use of renewable energy and production of marine produces. This research was performed in February and March of 2012.

First, examinations on the amount of available renewable energy in the region and composition analyses of the biomass resources were performed. The total amount of livestock biomass as farm animal manure and forestry biomass as logging residue and construction waste products contained across the entire Kuji Area is estimated in the “New Energy Vision” (Iwate Prefecture 1998), which was drafted by the government of Iwate Prefecture together with each municipality. However, in the specific investigation of feasibility, the available amount of biomass decreases due to the necessity for individual business owners and farmers

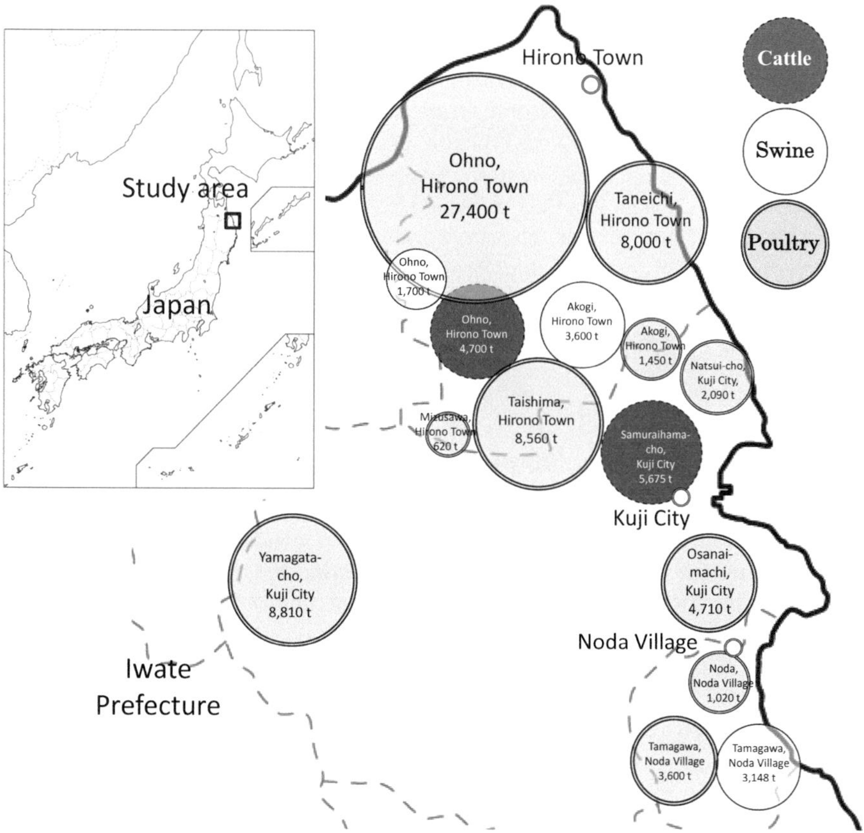

Fig. 14.2 Annual livestock manure production by region in Kuji City, Hirono Town, and Noda Village, Iwate Prefecture, Japan

to make the business judgment to support the project, and agreements must be reached regarding the price as well. Accordingly, to clarify the actual available amount of biomass, the structure of "biomass energy utilizing CRLAF" was introduced to the business owners in the subject area. A biomass availability study was conducted by questionnaire on the cattle, swine, and poultry farmers and wood chip-producing lumber mills in Hirono Town, Kuji City, and Noda Village, which were located based on the listings at a web telephone directory (i Town Pages; http://itp.ne.jp/). Of the sources of livestock biomass, three beef/dairy cattle farmers, four swine breeders, and 62 poultry farmers collaborated, and of the wood biomass source, two business owners answered, for a total response rate of 64.5%. Composition analysis of the regional biomass was performed based on the responses.

Figure 14.2 shows the regional annual discharge quantity of livestock excrements reported by business owners. The amount of discharge is represented per category of livestock (cattle, swine, poultry), together with the name of the relevant administrative unit and village, town, or city. The size of the circle corresponds to the quantity of discharge. The individual data are not shown; however, there is little variation in the cattle and swine manure by month. For poultry manure, at the individual farm level, there are some farms with a large, nearly fourfold fluctuation over a 3-month period due to the breeding cycles, but for the industry as a whole, these fluctuations are leveled due to the peaks occurring at different times for each farm. Poultry excrement can thus be considered to have a near-constant amount of discharge throughout the year. The survey on wood chips revealed that one lumber mill produces 550 m^3 of wood chips per year from conifers. From the above results, an investigation was conducted on biomass power generation that best utilizes these distinct characteristics of the Kuji Area.

14.3.2 Needs Assessment on Fisheries Products

Furthermore, an opinion poll among fisheries was conducted in the form of a questionnaire, for the reference purpose on selecting the fish species upon construction of the CRAS. Thirty-seven responses were obtained from business owners in the aquatic products processing industry and aquatic products retail businesses, for a response rate of 34%. There were high percentages of responses stating that salmonids, mackerel, squid, sea urchin, and abalone were current local species for which augmented production is desired. However, 70% of respondents did not respond to the question concerning the circulation of nonlocal aquatic products. A high percentage named tunas, salmonids, flatfishes, broadbanded thornyhead *Sebastolobus macrochir*, sea urchins, and abalones as personally preferred products, and a high percentage named tunas, broadbanded thornyhead, tiger puffer *Takifugu rubripes*, abalones, and sea urchins as preferred luxury products. These results clearly indicated desire for increased production of salmonids, sea urchin, and abalone. Recognition of a CRAS was estimated at 31% at the most.

The result suggests that a high degree of support of or even participation in CRAS could be expected, as "expansion of fish farming" was the most numerous answers to the question of what is necessary for the future revival of Tohoku's fisheries area; together with the response "expansion of aquaculture," they accounted for over half of the total response.

From these results, tiger puffer, flatfish, and abalone became apparent as feasible fish species for aquaculture productions in this region, taking the profitability of CRLAF into considerations.

14.4 Consideration of a Land-Based Facility for Recirculating Aquaculture Using Waste Heat Discharged from Biomass Power Plants

Based on the results of survey and analyses described in the previous section on the biomass available as energy source around the Kuji Area, the construction and operation of a biomass power generation facility that utilizes poultry manure, swine manure, and wood chips was investigated, using the unitized CRLAF system as a model.

14.4.1 CRAS

A closed recirculating aquaculture system was designed based on a unit consisting of two 45-kL rearing tanks and one 10-kL seedling tank, and estimations were made for the construction cost, production capacity, and energy requirement. Figure 14.3 shows the schematic flow diagram. The apparatuses are connected sequentially from rearing tanks, liquid cyclone separators (Itoh et al. 2010, 2011), foam fractionator, biological filtration tank, and ultraviolet sterilizer. Water temperature of each tank is controlled by the temperature monitoring and heat exchange with hot water produced by waste heat of biomass power plants. The circulation of rearing water to the heat exchanger is regulated by a control panel.

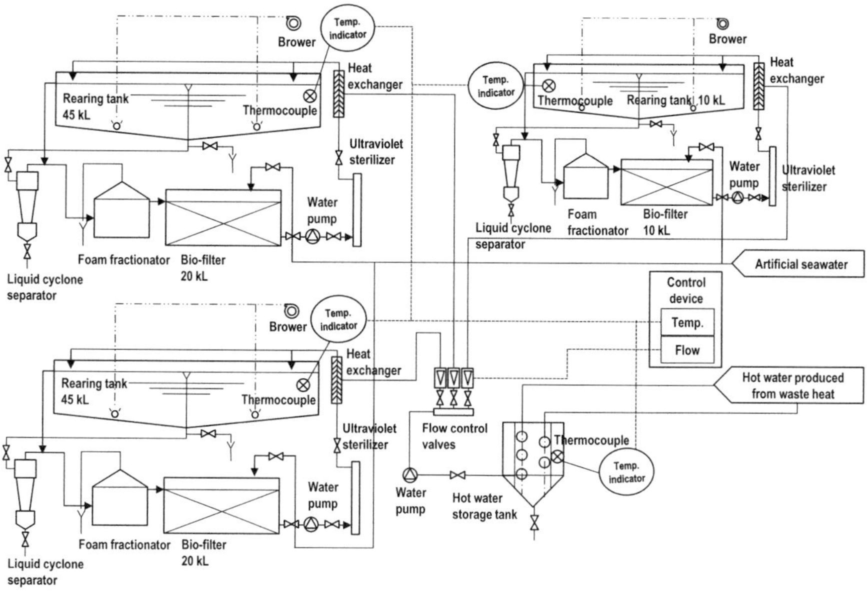

Fig. 14.3 Schematic flow diagram of one unit of the recirculating aquaculture system for tiger puffer *Takifugu rubripes* production

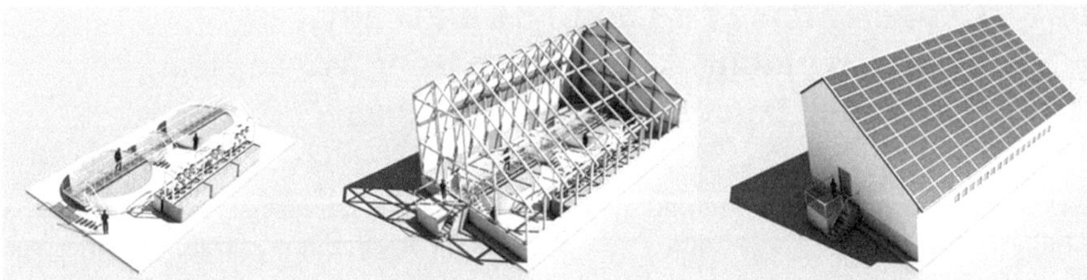

Fig. 14.4 Graphic of one unit of the recirculating aquaculture system. Images display the rearing system (left), the constructed roof frame made from thinned lumber (center), and solar battery panels (right)

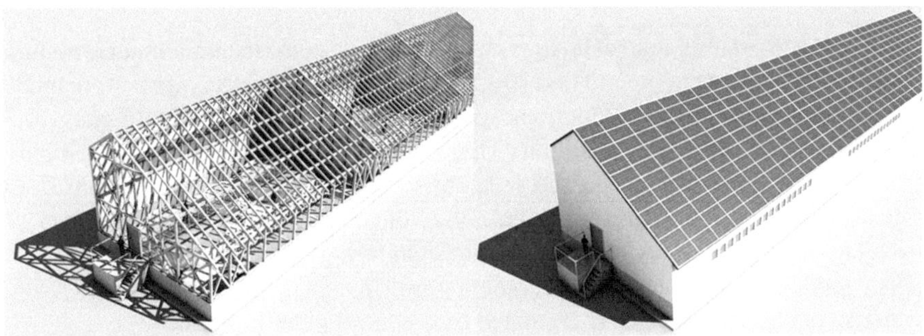

Fig. 14.5 Graphic of multiple units of the recirculating aquaculture system. Images display the rearing system (left), the constructed roof frame made from thinned lumber (center), and solar battery panels (right)

A graphic model of the building construction is given in Fig. 14.4. The necessary floor space of a building containing a single culture system, which is the basic unit, is 12.6 m × 21.6 m. The layout is such that a nursery tank is placed between two rearing tanks, and each tank has independent water treatment equipment placed in its vicinity. The building was designed as a wooden building with 100-mm-thick urethane insulation, in consideration of the effective use of thinned wood provided by the forest products industry. In addition, snow-melting solar battery panels were placed on the roof, as electricity is necessary to power circulation pumps and other machinery required for the effective use of thermal energy produced from biomasses. Figure 14.5 depicts joined basic units. Each unit can be accessed via the work platform placed over the aquaria. Shipments can be executed through the rear part, which is closest to the rearing tank. Rearing tanks will be made from FRP panels and vinyl sheets, and the final construction cost for one unit, excluding the solar panels, was estimated to be 75 million yen.

Figure 14.6 represents an aquaculture unit in which small-scale combined marine species aquacultures would be performed. The required lot size for this unit is 12.6 m × 14.4 m × 2 units. Based on the results of the questionnaire, one of

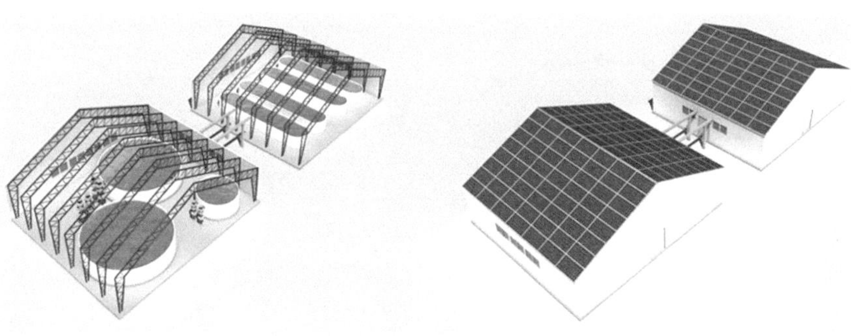

Fig. 14.6 Graphic of the recirculating aquaculture system for production of multiple species (fish, sea urchins, sea cucumbers, etc.). Images display the rearing system with constructed roof frame (left) and with solar battery panels (right)

the two units is designed for rearing of fish such as tiger puffer and flatfish and the other for culturing shellfish such as sea urchin and abalone. The closed recirculating aquaculture system unit for fish rearing consists of the abovementioned basic unit with the rearing tanks placed to minimize required floor space. The system for shellfish cultures employs recirculating raceway-type tanks. Water treatment equipment is placed three dimensionally in a similar arrangement to the basic unit.

Requirement of heat for management water temperature of CRAS was determined by the calculation of the overall heat transmission about the CRAS building under the condition that temperature of rearing water for tiger puffer is maintained at 22 °C. The surface area of building excepting floor area was 416 m^2, and the coefficient of overall heat transmission was set to 0.329 $W/m^2{\cdot}K$. The floor area was 182 m^2 and that was 0.465 W/m^2. The estimation of the requirement of heat was moved forward with the condition that specific latent heat of water at 22 °C is 649 kW/kg and the evaporation rate of rearing water is 10 kg/h·unit.

Figure 14.7 represents the thermal energy required for the basic unit, based on the monthly temperature at Kuji City. The estimated thermal energy requirement was calculated on the basis of the monthly minimal temperature, which shifts from −12.3 to 14.7 °C. The required heat quantity is the highest in winter and the lowest in summer, shifting from an estimated 9.0–20.2 kW/unit. Power other than that from the thermostat is assumed to be supplied as electricity, at an estimated amount of 11.1 kW/unit. Electric power consumption of CRAS was estimated based on the power dissipation, load factor, and rete of operation of apparatus equipped to CRAS. The actual consumption of air blowers, water pumps, foam fractionators, ultraviolet sterilizer for rearing water, and luminaire of fluorescent tubes was 1.7, 2.9, 1.6, 4.3, and 0.6 kW, respectively.

However, these are trial calculations, and in the future it would be necessary to correct these trial values in light of actual measurements in which energy, outside temperature, and rearing water temperature are monitored.

The aquarium usage plan of the culture unit for tiger puffer farming (Takii 2005; Kikuchi 2006) is shown in Fig. 14.8. The usage plan for one seedling tank and two

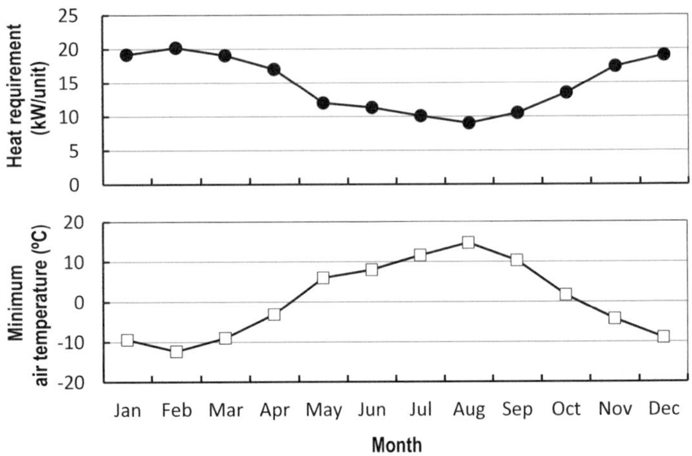

Fig. 14.7 Heat requirement of one unit of the closed recirculating aquaculture system based on the monthly minimum air temperature at Kuji City, Iwate, Japan, in 2012

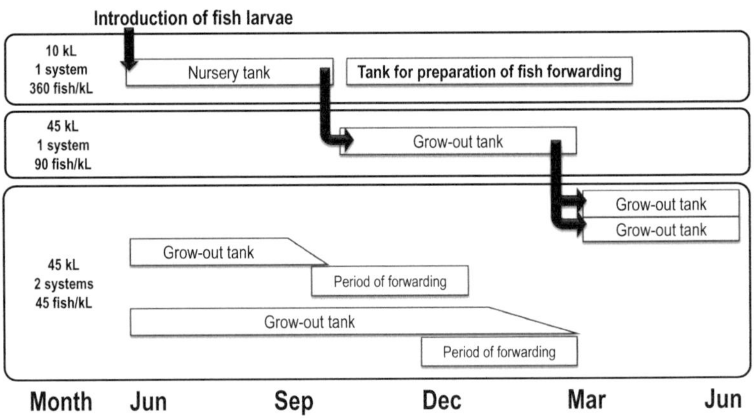

Fig. 14.8 Sample of fish growth, stock density, and feeding ration for tiger puffer *Takifugu rubripes* production

rearing tanks is represented, with shipping set at around 65 weeks after the introduction of seedling fish. The rearing of fish from introduction of seedling fish up to week 15 will be done in the juvenile tank, after which the fish will be transferred to rearing tank 1 and the teeth trimmed. After 20 additional weeks, teeth trimming and size sorting are performed, and some fish are sorted into rearing tank 2. The tiger puffers in one tank are eventually shipped 25–30 weeks after sorting, and those in the other tank are shipped 30–35 weeks after sorting. Between the transfer of juvenile fish and the introduction of new juvenile fish in the following year, the juvenile tank is used for preparing the rearing water and as a work tank during teeth trimming and shipping.

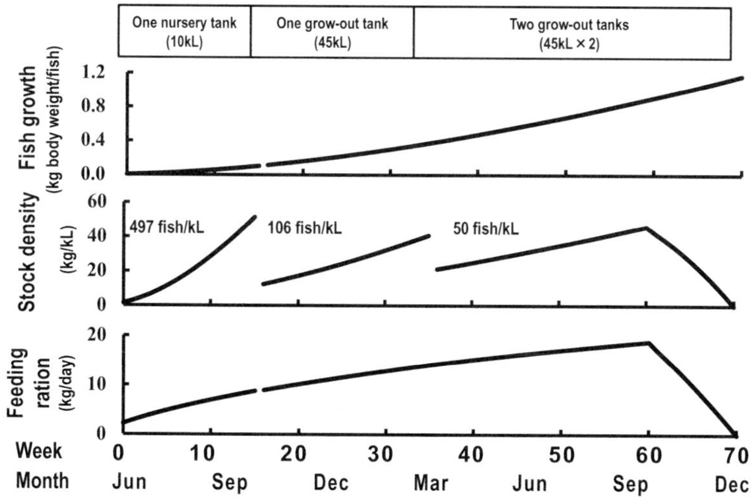

Fig. 14.9 Usage plan of tanks in the recirculating aquaculture systems for tiger puffer *Takifugu rubripes* production

Trial calculations for tiger puffer growth and required feed rations are represented in Fig. 14.9. Growth was calculated assuming that the fish are fed at the feeding rate by Mishiro (1997) and supposing a feed efficiency of 85%. Maximum population density was estimated at approximately 5% of rearing water volume, which was 5.1% in the juvenile tank and 4.6% in rearing tanks. Overall rations were estimated as 2.2 kg/day at the start of rearing and 18.8 kg/day at the maximum. When farming tiger puffers with the current conditions of shipping in 10 weeks, it was deemed possible to produce 4721 kg/year/unit with the production costs (material costs: seedling fish, rearing water, cost of feed) of 717 yen/kg. The production cost is estimated to be lower for heterosomata such as Japanese flounder *Paralichthys olivaceus* and brown sole *Pseudopleuronectes herzensteini* due to their higher feed efficiency than tiger puffers. From these results, the required amount of thermal energy and electricity to be provided by each biomass power plant was calculated at 20.2 kW/unit of thermal energy and 11.1 kW of electricity.

14.4.2 Biomass Plants

Construction costs and operation costs for energy plants were calculated for the following three locations, upon consideration of the biomass characteristics and other responses for each location extracted from the survey:

- Hirono Town (combustion): Poultry manure biomass, 50 tons/day
- Noda Village (methane fermentation): Swine manure biomass, 10 tons/day
- Kuji City (combustion): Wood chip biomass, 100 tons/day

Basic information was obtained from a well-established biomass power plant construction company, and construction fee and expenses for each power plant were calculated independently based on the obtained information. In investigating the current project, wood chips and poultry manure were considered as combustible energy sources (fuels). However, swine manure instead required that a methane fermentation system be employed, as the water content of the manure was too high for direct combustion. Methane fermentation has been widely introduced in small- and medium-scale livestock farms, but it is characterized by a relatively high construction cost per kW produced due to costs associated with operation and the management of gas. Wood chips and poultry manure combustion is a more ideal technology for large-scale implementation. Since 2003, a poultry manure power plant has been operating in south Kyushu, processing more than 300 tons/day (Takuma Corporation 2011). Measurement of the effects of economy of scale was set based on the data for 50 tons/day and 100 tons/day, which are attainable values, and the construction cost for 10 tons/day was calculated by substitution. Typically, projected construction costs are given in an official estimation upon determination of detailed specifications, and exact inlet and outlet conditions of biomass resources are exhibited. In the early stage of investigation for industrialization and other processes, it is possible to respond quickly to the drafting of project plan and evaluation of industrialization and present a sure direction by understanding the approximate numbers.

Therefore, the measure of the economy of scale for construction cost was set to 0.56, given the established data for 100 tons/day and 50 tons/day. It is common for the calculation of electronics to use 0.6–0.75; however, 0.56 was considered to be an appropriate factor as it lies close to the commonly used coefficient and also because supplies at private sector levels were taken into account.

The equation used is shown below:

$$\text{Construction cost of scale A} = (\text{Construction cost of scale C}) \times (\text{scale C}/\text{scale A})\ \alpha.$$

When the measure of the economy of scale $\alpha = 0.56$.

Measure of the economy of scale 0.56 was calculated by the following calculation on the ground that the construction cost of the design for 100 tons/day, excluding maintenance cost, is 1.084 billion yen and the actual construction cost for 50 tons/day is 733 million yen.

$$\alpha = \{\log(1084/733)\}/\log(100/50) = 0.56$$

If the data are accurate at the reference point of investigation, it would become possible to estimate the approximate construction cost and variable costs of a given scale using this coefficient. It would become possible to broadly understand the effects of scale without having to repeat drafts for each project. The combined construction costs of actual biomass plants and construction costs of those currently under investigation estimated by this method are as summarized in Table 14.1.

Table 14.1 Estimation of costs of biomass power plants

	(million yen)		
	Capacities of biomass processing		
Type of biomass power plant	10 t/day	50 t/day	100 t/day
Poultry manure-fueled power plant	326	803	1221
Methane biogas power plant	297	733	1084
Wood-burning power plant	1499	3691	5171

Cost data of power plants processing 50 t/day and 100 t/day biomass were estimated by plant company, and the data for processing 10 t/day were calculated using a scale factor

When construction costs of auxiliary facilities are taken into consideration with the costs shown, the total construction cost in the case of Hirono Town handling 50 tons/day of poultry manure is 923 million yen, including auxiliary facilities costing 120 million yen; the total construction cost for Noda Village handling 10 tons/day of swine manure methane fermentation is 319 million yen, including auxiliary facilities costing 22 million yen; and the total cost for Kuji City handling 100 tons/day of wood biomass is 5863 million yen, including auxiliary facilities costing 693 million yen.

For each biomass plant facility, the specifications of well-established power plants were used. The material flow of the power generation system for power generation from combustion of poultry manure is represented in Fig. 14.10. Poultry manure is transported from a loading hopper to a chopping device and then into the boiler via a conveyor belt. In this case, the type of conveyor belt would be determined by the moisture content of the poultry manure; in the current trial calculation, 30% moisture content was assumed. The moisture content of poultry manure is about 40% in broiler chickens and 75% in layers. In this investigation, the moisture was reduced to 30% using a ventilation desiccating technique. For boiler chicken manure, the stoker method was employed, which is known to be resistant to high calorific power when used in garbage incineration. Currently, an electricity-/heat-generating plant is operating in south Kyushu and handles 300 tons/day of poultry manure by fluidized bed combustion (Takuma Corporation 2011). Investigation of the use of fluidized bed boiler is necessary for possible future applications. As boilers must be supplied with water, water management is an essential factor, and the water will therefore be softened with chemical solutions. Because there will be heavy water usage, an apparatus for water retrieval may have to be set up in case industrial water is hard to obtain. For the current investigation, specifications are based on the assumption that there is sufficient water; thus, the construction cost is kept low. For the generation of power, the design employs a screw-type power generator, making it possible to use both electricity and heat, at least part of which will be recovered. Multi-cyclone, high-performance bag filter with a filtering efficiency of over 99% will be installed for dust collection to clean up the exhaust gas, which must be cooled to below 200 °C, to avoid damaging the filter cloth. Although this recovered heat can be used for recirculating aquaculture systems, this option was not adopted this time.

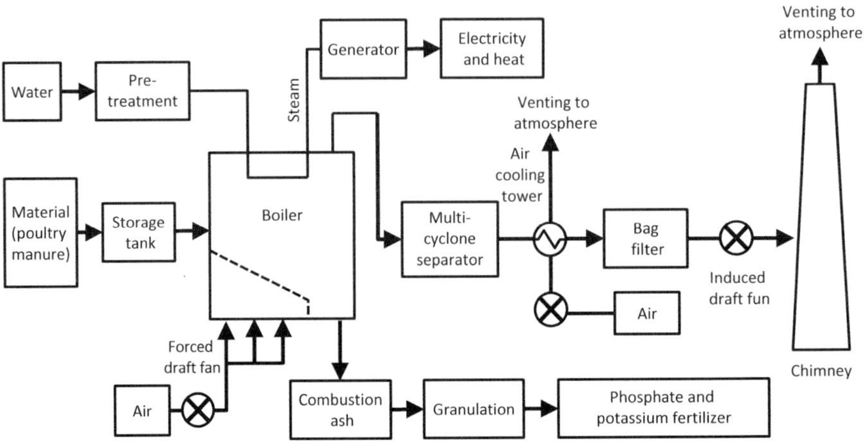

Fig. 14.10 Material flow diagram of electric power plant with combustion of poultry manure

The flow of the power generation system for methane fermentation of swine manure is shown in Fig. 14.11. Swine excrement is placed in the reception tank by suction using a vacuum truck, due to the manure's high moisture content. Swine manure is pumped from the reception tank to the methane fermentation tank, where biogas will be produced by mesophilic fermentation at around 50 °C. Biogas is stored in the gas holder after passing through the microbial desulfurization tank by air injection and is then used in the gas engine generator and recovery of exhaust heat. The liquid phase is transferred from the methane fermentation tank to the equalizing tank via plumbing and then stored for about 3 days before solid-liquid separation using coagulants. The sludge is eventually released to a composting facility and the liquid digest discharged after biological treatment. Raw swine manure, the separated solid phase, and sludge are composted and could be subsequently sold as fertilizer.

The flow of the power generation system for combustion of wood chips is presented in Fig. 14.12. First, the wood chips are transported from the loading hopper to a batch dryer by a chopping device. The wood chips are dried by the dryer because their moisture content varies depending on region, storage method, and season; the moisture content has been estimated as 20% in this investigation. The exhaust gas from engines and other machineries is used as the heat source for this drying process. Subsequently, the dried chips are heated in a pyrolyzer, and a reduced zone is formed to raise the temperature of pyrolysis gas for separating char and gas. Combustion is promoted in part by blowing air into the upper part of the reforming furnace, elevating the temperature of pyrolysis gas. Char and tar are gasified by steam. Unreacted char and ash that exit the reforming furnace are passed through a sieve to separate large char particles from small ash particles, after which char is recycled and reused. The produced gas is cooled after it leaves the reforming furnace and particulate matter is scavenged by a dust collector while the gas is sent to the engine. Electricity is produced by turning the generator with the gas engine.

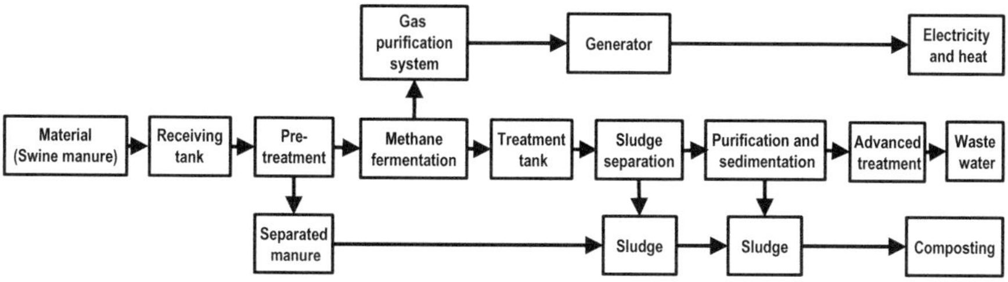

Fig. 14.11 Material flow diagram of electric power plant with methane fermentation of swine manure

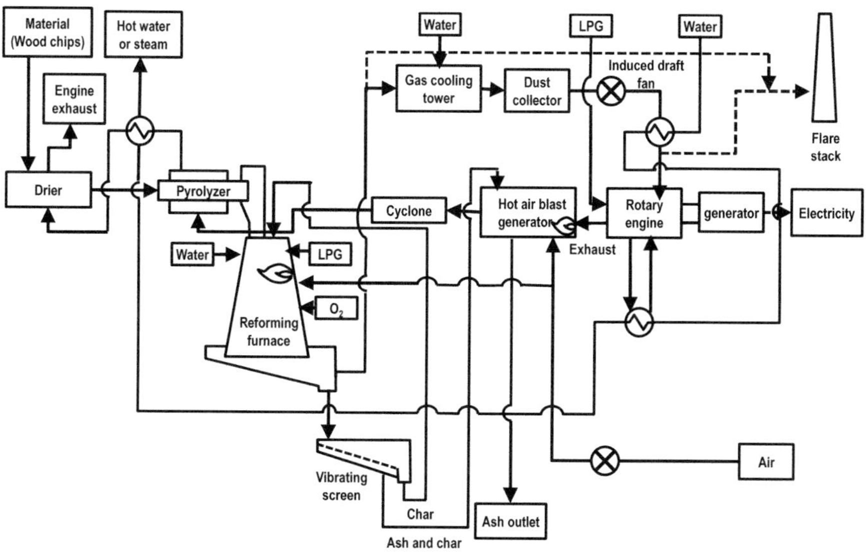

Fig. 14.12 Material flow diagram of electric power plant with combustion of wood chips

The exhaust gas of the gas engine is sent to the hot blast generator, where the ash and char from the reforming furnace are combusted to raise the temperature of the exhaust gas. The exhaust gas that leaves the outer case of the pyrolyzer is cooled to approximately 150 °C before being sent to the dryer to be used for drying the wood chips. Purified gas and exhaust gas are cooled by heat exchangers that perform temperature adjustment at each stage, and the cooling water is recovered as warm water and reused. In the event that purified gas from the reforming furnace is not used by the gas engine for power generation, it is released into the atmosphere after the combustible gas is burnt off in a flare stack.

For the calculation of energy balance by region, an energy balance model at the time of power generation was constructed, and each biomass power generation method was investigated accordingly. Figure 14.13 represents the constructed model. In this case, 13% of the heat input will be ultimately utilized for electricity,

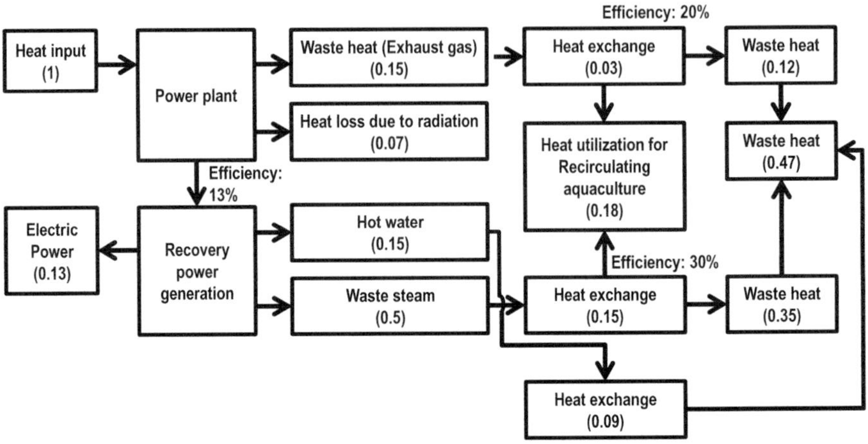

Fig. 14.13 Energy flow diagram of biomass power plant with heat utilization for closed recirculating aquaculture systems

and 18% will be exploited as thermal energy in the CRAS. Table 14.2 shows the result of adapting this model to the poultry manure, swine manure, and wood biomass power plants.

According to the questionnaire, it is possible to utilize poultry manure for biomass energy in Hirono Town; therefore, the survey results were used to investigate the potential power generation and usable energy balance. The annual biomass availability of poultry manure in Hirono Town is 18,250 tons. The total conversion equivalent in energy was calculated based on the net calorific value of standard poultry manure, which is 2300 kcal/kg. Assuming that approximately 13% of this total calorific value could be harnessed for power generation, annual energy production would amount to 266 MWh. Heat balance was estimated considering that 18% of heat input is heat with ability to do work (high-exergy heat), such as the heat discharged during energy production and exhaust gas heat, which can be used for the temperature adjustment of the CRAS. The remaining heat is discharged via heat dispersion and drainage, which can then be used for household functions such as floor heating, despite its low exergy. The energy balance investigated based on the outcome of the questionnaire of Noda Village and in the trends in thinking expressed for Hirono Town resulted in 3148 tons annually of swine manure in Noda Village, which converts to a total energy value of 600 kcal/kg based on the net calorific value of standard swine manure. Assuming that approximately 13% of this total calorific value could be used for power generation, annual energy production will amount to 11.8 MWh. Similar investigation on the energy balance on Kuji City based on the survey outcome, applying the same process as for Hirono Town and Noda Village, revealed that it is possible to collect an annual total of 36,500 tons of wood. Assuming that approximately 13% of this total calorific value could be used for power generation, annual energy production would be 623 MWh.

Table 14.2 Estimation of electric power production and available heat for recirculating aquaculture produced by biomass power plants

Type of biomass power plant	Assumed installation site	Input material (t/year)	Net calorific value*2 (kcal/kg w. b.)	Produced electric power (MWh)	Heat utilization for recirculating aquaculture*1 (MWh)
Poultry manure-fueled power plant (50 t/day)	Hirono Town	18,250	2300	266	367
Methane biogas power plant (10 t/day)	Noda Village	3148	600	11.8	16.3
Wood-burning power plant (100 t/day)	Kuji City	36,500	2700	623	863

Compared to thermal power generation, which is used in the general power industry (maximum efficiency of 55%, based on gross calorific value), biomass power generation is 13%. This is below 1/3 that of thermal power, as biomass is an inefficient system that produces redundant heat that is wasted. The aim of this project is to make use of this redundant heat source to realize a CRAS, which has a stable demand for heat. The feasibility of the project could be elevated through the combination with CRAS, even if there is no large revenue to the power generation industry. The use of heat cascades already has a proven track record as a cogeneration technology used in energy-saving strategies for buildings. This project plan proposes the integration of fisheries and industry by combining aquaculture and forestry/livestock biomass power generation.

14.4.3 Energy Matching

In matching the energy requirement of the CRLAF and the energy production by the biomass power generating system, calculations were performed on the possible scales of CRAS that could be operated using the amount of heat and electricity produced as shown by the results of trial calculations in Sect. 4.1 CRAS. In investigating the Hirono Town poultry manure power plant, trial calculation was based on the estimation that the energy demand will not greatly exceed that provided by the 50 tons/day scale energy plant. The thermal output of this facility is 484 kW and electricity production is 90 kW, which led to the calculation that 24 units of aquaculture facilities could be run thermal energy-wise and 8 units electricity-wise. For the swine manure power plant at Noda Village, the thermal output of the facility is 44 kW, and electricity is 24 kW which is just below the heat requirement for two units of CRAS that meets the electricity requirement for two units. For Kuji City's wood biomass power generation, forestry in the region is cooperative and the availability of biomass is high. Therefore, the scale of

100 tons/day was selected as the range in which the current balance of supply and demand will not be greatly influenced. The thermal output of this facility is 3354 kW and electricity production is 1820 kW, which led to the calculation that thermal energy was sufficient for 166 units of aquaculture facilities and the electricity was sufficient for 164 units.

It has become clear from the trial calculations of the three biomass power plants that the energy provision from those biomass power generation systems to the CRAS is sufficient to provide the heat source required for rearing water temperature adjustment in CRLAF and that there is a possibility for substantial energy saving. However, the direct use of electricity produced by biomass power generation is not suitable, because using part of the electricity as energy source for the water streaming pump and other machinery will lead to a shortage for operating the facility. It is desirable to introduce the electricity required for CRLAF from the main power system, because the biomass power generation facility itself utilizes part of its generated electricity as internal consumption.

Moreover, the Kuji Area has plentiful sunlight as well as biomass resources, and there are many suitable locations for solar panels. It is also a suitable area for installing offshore wind power systems; it will therefore become necessary to investigate a hybridized model of CRAS that includes renewable energy.

14.5 Evaluation of Projects on Heat Utilization of Biomass Power Plant for Closed Recirculating Aquaculture

Because effective use of exhaust heat requires that the CRAS is in the vicinity of biomass power plant, and because the fishery industry will not likely accept waste disposal in coastal areas, we considered it most appropriate to construct the power generation facility adjacent to swine, cattle, or poultry farms, or wood thinning disposal facility, and construct the aquaculture facility in the vicinity. This location is ideal for the additional reason that livestock excrement is associated with the problems of smell and high transportation cost. The specific candidate location for which the present hearing was conducted was situated on a hill, far from the coast. From this, we decided to perform the conceptual planning and trial calculations for a level area on a hill. From the result of hearings conducted with individuals with ties to fisheries and fishing households, it became apparent that the problem of transportation of livestock excrement would occur when the system would be introduced in the vicinity of supply chains for the marine products processing industry. The project was deemed to advance more smoothly by situating the energy plant and CRAS in the vicinity of livestock biomass resource rather than at a coastal location, due to past withdrawals of livestock farmers from the project.

14.5.1 CRAS

The evaluation of feasibility was carried out separately for CRAS and each of the energy production plants, to understand their respective characteristics clearly. For CRLAF, trial calculations of economic potential were conducted for the case of tiger puffer farming based on the aquaculture unit assumed in Sect. 14.4.1. The result is shown in Fig. 14.14. Trial calculations were performed for the costs of production per kilogram. The cost directly concerned with fish rearing, which comprises the cost of feed, fish seedlings, creation of rearing water, and labor, was 929 yen/kg production in total. The facility cost, which is the initial cost, was calculated as 1059 yen/kg production, assuming amortization in 15 years. Energy cost was calculated as rate of electricity conversion in this trial calculation. The costs of water quality management and illumination were calculated as the power cost, which was estimated to be 412 yen/kg production. Concerning this point, it is considered necessary to develop a water circulation system that aims for further energy consumption reduction. It was clear that the cost of 250 yen/kg production is required for the temperature adjustment of rearing water under the condition that a heat pump (coefficient of performance (COP) 3.0) is used, which can be covered by the exhaust heat from biomass power generation. The costs for temperature control were kept lower than expected in this trial calculation, as it was assumed that the aquaculture facility would be insulated using 100-mm-thick urethane foam. Conversely, the facility expenses were higher than anticipated due to the additional insulation cost. Concerning this point, it is desirable to vary the thickness of

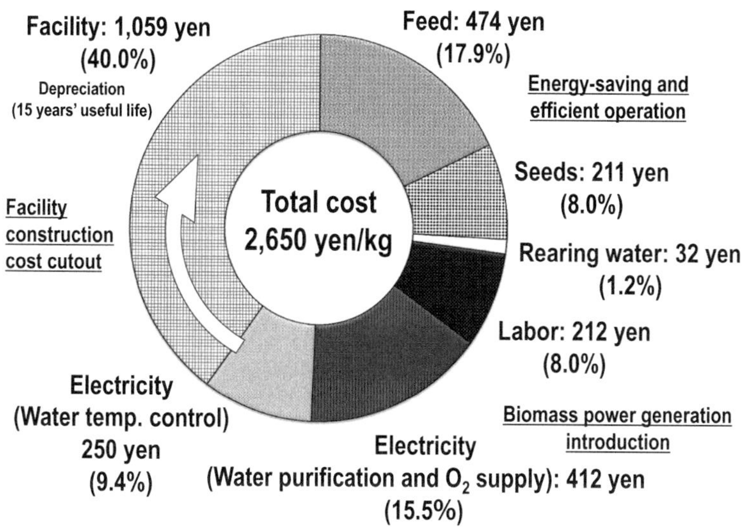

Fig. 14.14 Breakdown of costs for tiger puffer *Takifugu rubripes* production. Percentages represent the costs per 1 kg fish production. Energy usages were calculated as electric power costs (20 yen/kWh)

insulation according to the quality and quantity of exhaust heat obtained during power generation. For instance, it would be possible to bring down the initial cost by making the facility insulation thinner if there was abundant exhaust heat.

14.5.2 Biomass Plants

In the evaluation of biomass energy power plant, the present disposal of livestock waste is regarded as profit, as it will be disposed of by the biomass plant. In considering a long-term operation, a market value of 20 yen/kWh was applied to biomass power generation. Furthermore, the by-products after energy exchange were counted as profit in the trial calculation, as they have value as fertilizer. The amortization period of the energy plant was set to 15 years, and trial calculation was executed by evaluating labor, utility, and maintenance cost. Property and corporate taxes were evaluated as tax-free. Trial calculation was performed on the premise that the heat energy used for the CRAS would be used for increasing the feasibility of the aquaculture business instead of being sold.

It is necessary to clarify the actual disposal costs of excrement for power plants utilizing livestock biomass. However, disposal systems differ for swine, poultry, and cattle and their costs vary. According to the survey, the most common disposal method is fertilizer production via composting. Thus, investigation was pursued considering disposal via composting. The value employed for the cost of this method was 6000 yen/ton, which was taken from a survey response. However, as other business owners reported having spent 10,000 yen/ton for excrement disposal by composting, it was unclear what exactly was understood to be included in the disposal cost. This needs to be closely inspected for the specific execution of the project. This current expense of excrement disposal by the livestock farmers is considered to be reduced to “0” by the introduction of power plant, and thus this deferred disposal cost can be considered as profit by the power plant.

First, the result of feasibility evaluation for the 50 tons/day poultry manure power plant is shown in Table 14.3. Calculation of the internal rate of return (IRR) for the 50 tons/day scale was performed based on the biomass survey on the potentially available poultry manure. Total annual business income of biomass electric power plant with poultry manure combustion was estimated to be 283,167 thousand yen/year. The breakdown of the income is processing cost equivalent, electric power, and solid fertilizer selling. The processing cost equivalent of poultry manure was regarded 6000 yen/ton, and the processing of 16,425 tons/year can be performed in the power plant under 0.9 of the operation rate. The processing cost was expected to be 98,550 thousand yen/year. The revenue from sales of electric power was calculated to be 102,492 thousand yen/year. The ash of the portly manure can be sold as a mineral fertilizer, and the income was estimated to be 82,125 thousand yen/year. While the ordinary expenditure consists of plant depreciation, labor cost, utilities cost, industrial water, and maintenance cost, these were estimated to be 61,533, 35,000, 30,747, 5913, and

Table 14.3 Project evaluation of biomass electric power plant with poultry manure combustion at 50 t/day capacity

Item	Amount (thousand yen)	Bases	Remarks
(A) Business income			
(1) Processing cost equivalent	98,550	Processing cost 6000 yen/t, annual throughput 16,425 t	Low-cost subscription
(2) Electric power (all-quantity buyback)	102,492	Annual energy production = 350 kWh × 24 h × 365 days × 0.9 = 2759 MWh (20 yen/kWh)	Electric power for transmission
(3) Fertilizer	82,125	Unit price 20,000 yen/t × 16,425 t × 0.25	
Sum	283,167		
(B) Ordinary expenditures			
(1) Plant depreciation	61,533	923,000,000 yen/15 years	No subsidy
(2) Labor cost	35,000	7 employees: chief electrical engineer, etc.	
(3) Utilities cost	30,747	Internal consumption of electricity (2050 MWh × 15 yen/kWh)	
(4) Industrial water	5913	Unit price 100 yen/m^3 × 7.5 m^3 × 24 h × 365 days × 0.9	
(5) Analysis cost	1000	Exhaust gas, water quality and ash, etc. (5000 yen/item × 200 items)	
(6) Maintenance cost	27,690	3% of total construction costs (machinery cost: 923,000,000 yen)	
(7) Internal electricity cost	0	Exclude	
(8) Insurance premium	2769	0.3% of machinery cost (923,000,000 yen)	
(9) Taxes and public dues	12,922	1.4% of machinery cost (923,000,000 yen)	Real estate tax

(continued)

Table 14.3 (continued)

Item	Amount (thousand yen)	Bases	Remarks
Sum	177,574		
(C) Project evaluation			
Pretax profit	105,593	40.87% of pretax profit	
Corporation tax	43,564		
After-tax profit	62,029		
Cash flows from operating activities	123,562		
IRR	18%		

27,690 thousand yen/year, respectively. The total expenditure was 177,574 thousand yen/year including analysis cost, insurance premium, and taxes and public dues. The internal electricity cost was excepted from this calculation. The resulting IRR was 18%, which is a high score for feasibility. This calculation took into account the high calorific value of poultry manure (four times more than other livestock excrements), the large power output, the full amount tariff fixed at 20 yen/kWh (twice the current tariff of 10 yen/kWh, including the environmental value set by power company in south Kyushu), and the poultry manure disposal cost and profit of fly ash fertilizer sales. Another factor is that combustion-based poultry manure power plants have a simpler system and cheaper construction cost compared to plants for fermentation-based poultry manure power generation. Poultry manure power generation is considered to be a biomass resource technology with high feasibility, as it has a proven track record in south Kyushu, but whether it is possible to consider the annual disposal cost as profit is the key to the accelerated introduction. Additionally, investigation into the possibility for poultry farmers to distribute the costs of poultry manure disposal as part of the costs of power generation is highly relevant for the overall evaluation of livestock biomass plants.

Next, the feasibility of the 10 tons/day swine manure methane fermentation plant is shown in Table 14.4. The calculations were done similarly as for Hirono Town, based on the potential availability of swine manure reported in the biomass survey. Total annual business income of biomass electric power plant with biogas from the methane fermentation of swine manure was calculated to be 25,142 thousand yen/year. The breakdown of the income is processing cost equivalent, electric power, and liquid fertilizer selling. The processing cost of swine manure same as the cases of processing of poultry manure was assumed 6000 yen/ton, and the processing of 3148 tons/year can be performed in the power plant under 0.9 of the operation rate. The processing cost was expected to be 18,888 thousand yen/year. The revenue from sales of electric power was calculated to be 3626 thousand yen/year. The digested liquid which is the coproduct of methane fermentation of swine manure can be sold as a liquid fertilizer, and the income was estimated to be 2628 thousand yen/year. On the other hand, the ordinary expenditure mainly consists of plant depreciation, utilities cost, maintenance cost and the values. The total expenditure was 20,559 thousand yen/year including analysis cost, internal electricity cost, and insurance premium. The labor cost, cost of water usage, taxes, and public dues were excepted from this calculation. The labor cost is coordinated as that for swine farm. IRR was calculated for the scale of 10 tons/day, which resulted in a negative value and the conclusion that commercialization would be difficult with an annual availability of 3148 tons. The factors behind this is the relatively high construction cost that causes high water content and low energy potential of swim manure, which lowers the profit from power generation. To improve its feasibility, it is essential to bring down the construction cost of the methane fermenter, as well as increasing the scale of swine farming to increase power output. Trial calculation has not taken any particular subsidies for facilities and construction cost into account. The future condition of business feasibility

Table 14.4 Project evaluation of biomass electric power plant with methane fermentation of swine manure at 10 t/day capacity

Item	Bases	Remarks	
(A) Business income			
(1) Processing cost equivalent	18,888	Processing cost 6000 yen/t, annual throughput 3148 t	
(2) Electric power (all-quantity buyback)	3626	Annual energy production = 23 kWh × 24 h × 365 days × 0.9 = 181 MWh (20 yen/kWh)	Electric power for transmission
(3) Fertilizer	2628	Unit price 8000 yen/t × 1 t production/day × 365 days × 0.9	
Sum	25,142		
(B) Ordinary expenditures			
(1) Plant depreciation	10,633	319,000,000 yen/15 years	Subsidy: 50%
(2) Labor cost	0	Labor cost is coordinated as that for swine farm	
(3) Utilities cost	591	Internal consumption of electricity (39.4 MWh × 15 yen/kWh)	Approximation
(4) Industrial water	0	Well water	Approximation
(5) Analysis cost	1000	Exhaust gas, water quality and ash, etc. (5000 yen/item × 200 items)	
(6) Maintenance cost	7500	3% of total construction costs (machinery cost: 249,000,000 yen)	Approximation
(7) Internal electricity cost	35	Electricity under plant inspection 80 yen × 438	
(8) Insurance premium	800	0.3% of machinery cost (249,000,000 yen)	
(9) Taxes and public dues	0		Tax benefits
Sum	20,559		
(C) Project evaluation			
Pretax profit	4583		
Corporation tax	0		Tax benefits
After-tax profit	4583		
Cash flows from operating activities	15,216		
IRR	−4%		

would include the desired subsidization policy of more than 50% as well as an appropriate setting of electricity price based on feed-in tariff.

Finally, feasibility evaluation for 100 tons/day wood chip power plant was conducted. The results are shown in Table 14.5. Total annual business income of biomass electric power plant with wood chip combustion was estimated to be 422,582 thousand yen/year. The breakdown of the income is the selling of electric power and solid fertilizer made of an ash after wood combustion. The revenue from selling electric power and solid fertilizer was calculated to be 383,162 and 39,420 thousand yen/year, respectively. While the ordinary expenditure consists of plant depreciation, labor cost, utilities cost, wood chip cost, and maintenance cost, these were estimated to be 390,933, 35,000, 72,138, 197,100, and 175,000 thousand yen/year, respectively. The total expenditure was 888,671 thousand yen/year including other costs. The procurement of the wood chips is mostly needed for the management of the power plant. IRR was calculated on the potentially available wood biomass indicated by the biomass survey based on the scale of 100 tons/day. It was not possible to calculate IRR, as the profit was negative, leading to the conclusion that feasibility is poor. The main reason for this is that wood chip acquisition is purchase based and was thus accounted as expenditure, unlike livestock excrement, where the disposal costs could be considered as profit. The supply of wood chips must be substantially increased to scale up power generation and improve feasibility. There are current examples of commercialized wood biomass power generation, but this is thought to be due to the economies of scale of large-quantity cheap imports of wood biomass and the scale of the facility exceeding 1000 tons/day, which ameliorates the cost of construction and purchase value.

As there are no detailed statistical records of biomass utilization in Japan, and as there are many varied types of biomass energy, the characteristics of the country and region in question must be considered. Although Japan is rich in forestry resources, there has not been much development in large-scale power generation from wood as in Europe and the USA, due to the geographical difficulties in collecting the wood biomass. For livestock biomass, there is decentralization of farms to avoid infectious outbreaks of diseases such as influenza. However, recently there has been movement in the poultry industry toward the secondary sector, and some business owners are keeping approximately 2 million birds per farm. Therefore, large-scale power generation utilizing the immense volume of poultry manure can be expected.

The developments of biomass energy from unutilized sources such as sludge disposal, food debris, and garbage combustion are issues worth noting in the future. There should be a special focus on the price of whole tariffs for biomass power generation, after which the current feasibility evaluation will need to be revised. In addition, as the technical issues of energy plants could be addressed using existing technologies, there is a need to evaluate cost reduction and the effective use of energy from the consumer's point of view by attempting a comprehensive energy accommodation via a decentralized energy network, which aims to accelerate the introduction of renewable energy. The balance between the energy demand of the

Table 14.5 Project evaluation of biomass electric power plant with wood chip combustion at 100 t/day capacity

Item	Bases	Remarks	
(A) Business income			
(1) Electric power (all-quantity buyback)	383,162	Annual energy production = 2430 kWh × 24 h × 365 days × 0.9 = 19,158 MWh (20 yen/kWh)	Electric power for transmission
(2) Fertilizer	39,420	Unit price 20,000 yen/t × 1971 t	Ash content: 15%
Sum	422,582		
(B) Ordinary expenditures			
(1) Plant depreciation	390,933	5,864,000,000 yen/15 years	No subsidy
(2) Labor cost	35,000	7 employees: chief electrical engineer, etc.	
(3) Utilities cost	72,138	Internal consumption of electricity (4809 MWh × 15 yen/kWh)	Approximation
(4) Wood chips cost	197,100	Unit price 6000 yen/t, throughput: 32,850 t/year	Utilization rate: 90%
(5) Industrial water	0	Well water	Approximation
(6) Analysis cost	1000	Exhaust gas, water quality and ash, etc. (5000 yen/item × 200 items)	
(7) Maintenance cost	175,000	3% of total construction costs (machinery cost: 5,834,000,000 yen)	Approximation
(8) Internal electricity cost	0	Exclude	
(9) Insurance premium	17,500	0.3% of machinery cost (5,834,000,000 yen)	
(10) Taxes and public dues	0		Tax benefits
Sum	888,671		
(C) Project evaluation			
Pretax profit	−466,089		
Corporation tax	0		Tax benefits
After-tax profit	−466,089		
Cash flows from operating activities	−75,156		
IRR			

CRAS and the supply of biomass energy must be considered as well. Energy plants produce excessive heat, but this is not always utilized effectively. It is essential to set the scale of CRAS to be applicable to the range of exhaust heat.

14.5.3 Eco-farm

To clarify the abovementioned trial calculations and future prospects, as well as to deepen the understanding of CRLAF hybrid model that makes use of natural and renewable energy, including that from biomass, an integrated food production system (eco-farm) that utilizes renewable and natural energy was designed and modeled (Fig. 14.15). This image represents a relatively flat location in the Kuji Area with a circular road encircling an area of about 10 ha, within which a large-scale CRAS complex consisting of multiple aquaculture facilities is formed. One unit of such a facility is comprised of two 45-ton recirculating culture tanks and one 10-ton seed fish tank. The buildings for aquaculture and related facilities are equipped with solar-powered 1-MW batteries, and energy plants for production of electricity from swine manure, wood chips, and poultry manure are situated on the outside edge of the circular road. Also, wind power stations are placed in the vicinity of a transmission line system, and an upper reservoir of a small-scale pumped-storage hydroelectricity system was placed for nighttime leveling of surplus electric power. The image reflects an efficient use of heat, with the residence of the employees working at the plant and culture facilities placed inside the circular road, and vegetable, mushroom, and processing facilities are placed on the downstream side of energy plant in the view of "winter farm" that utilize excessive heat energy from the energy plant. A senior citizen-friendly urban development is expected for the further effective use of heat, where the facilities are integrated on the outside of the circle.

Investigations into the industrialization of such an energy plant and the CRLAF facility are taking place to increase feasibility. Approximately 30–50 people are estimated to be employed, but expansion to the employment of over 300–500 people is expected due to the additive effect of the secondary and tertiary industrial sectors, known as the so-called sixth industry in which employment in the processing and service industries is combined. In particular, the aquaculture facilities can be built from timber from wood thinning and related enterprises, which intends to simultaneously boost local employment and save on construction. Operational management for the entity and visitor center is based in the central part of the complex, as the energy balance of the entity is composed of several end users and several power plants, requiring energy circulation by the smart community as well as the whole-community management. The total amount of investment is estimated to be around 30 billion yen, which includes around 10 billion yen for energy plants, another 10 billion for aquaculture facilities, and the rest for infrastructure such as roads. Detailed inspections will be required for this total investment, but by doing so it would become possible to deepen the understanding of both

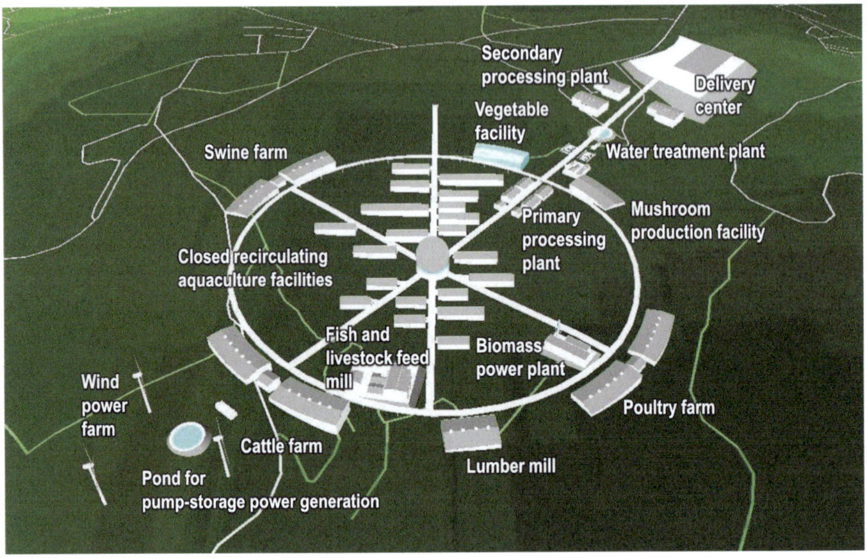

Fig. 14.15 Graphic of proposed eco-farm constructed in the northern Tohoku region in Japan. The eco-farm is small community of various food production systems based on closed recirculating aquaculture units with renewable energy power plants

experts in CRAS and those involved in fishery, livestock farming, urban development, NPOs, and administration, leading to developments that make it possible to spread this enterprise throughout not just the Tohoku region but the entire country and abroad.

Together with the development of technologies, definition of the implementing body is essential for industrialization. However, it is difficult to investigate the profitability of the current state of CRAS in the aquaculture industry in isolation, making it unlikely for an independent corporation to discuss investment in a new business favorably. Further reduction in expenses through cuts in facility running and construction costs and the application of energy-saving and energy-creating technologies is necessary for the future of the CRAS. Especially for the CRLAF performed in the present report, which uses exhaust heat from biomass power generation, it is necessary to set up a business model on the premise that it is a novel and integrative business established through the cooperation of multiple elements, such as fisheries, the livestock industry, forestry, and the waste disposal power generation industry. It is also crucial to deepen the understanding of potential business owners through the presentation of concrete examples, for which investigations of the synergetic effects of novel industrial cooperation must be presented in specific figures to industries and governments, as well as increase the accuracy of figures by carrying out demonstrations. In particular, the issue is not which side benefits, but rather to understand who gains the profits and to address this issue by clarifying the risks. A business proposal that would be welcomed after sorting the advantages and risks based on adequate data will be crucial. This is one such proposal, which should be implemented upon gaining the understanding of and achieving consensus among potential business owners and regional inhabitants.

In this business plan, it is possible for the implementing body to evaluate integrative business feasibility, as the business owners of aquaculture and energy providers follow the same premises. For the business feasibility of the CRAS, factors such as the unit price of a fish species, catch, labor costs, and energy costs must be individually evaluated. It is also necessary to adapt flexibly to changes in the state of society; however, it is not easy to respond to variations in energy supply and fuel costs. Concerning the issue of energy cost in a CRAS, the use of livestock excrement as energy will allow for the reduction of cost burden for the livestock farmers. Furthermore, profits can be expected from the provision of produced heat energy and electricity.

14.6 Summary

Evaluation of the feasibility of a CRLAF that utilizes the exhaust heat from biomass power generation was conducted in the Tohoku region. Considering Kuji City, Hirono Town, and Noda Village as subjects, the presence and availability of biomass were investigated, and the feasibility of implementation under the presumed scales of biomass plants was evaluated. From this, power generation from poultry manure was found to be highly feasible, with abundantly available materials. The business feasibility of this clean energy has a proven track record, as power plants in south Kyushu implements the same process. However, there is only one example of a business model for stand-alone power generation from poultry manure, and there are only two examples of businesses providing heat utilization support to poultry business owners in the vicinity.

The business model with CRLAF as a stable demander of thermal energy was built based on these examples. It became clear that a stable supply of poultry manure and the active entry of poultry business owners into CRLAF will lead to good prospects of a continuous profit structure, enabling a stable supply of thermal energy, which will in turn greatly contribute to the improvement of CRLAF's economic efficiency. This project is an evaluation of the system integration of the fusion of the energy sector, fishery trade, and livestock industry. It is therefore of importance to proceed by integrating the knowledge of experts in each of these respective fields.

References

Endo M (2008) Closed recirculating aquaculture system, evolving aquaculture technologies. Food Packag 49(8):446–452. (In Japanese)

Endo M (2014) Environment control technology taking advantage of the characteristics in aquaculture facility: growth enhancement, reproduction and teste control. Aqua Cult Bus 51 (7):48–51. (In Japanese)

Itoh T, Inoue F, Osakabe M (2010) Improvement of separation efficiency at low velocity region of liquid cyclone separator. Mar Eng 45:737–742. (In Japanese with English abstract)
Itoh T, Inoue F, Osakabe M (2011) Improvement of separation efficiency at low velocity region of liquid cyclone separator: effect of separator size. Mar Eng 46:763–768. (In Japanese with English abstract)
Iwate Prefecture (1998) Aiming for the harmony between environment and energy. In: Division resource and energy, promotion policy department, Iwate prefecture (ed) A new energy vision for iwate prefecture, Iwate, p 25. (In Japanese)
Kikuchi K (2006) Present status of research and production of tiger puffer Takifugu rubripes in Japan. In: Cruz- Suárez LE, Ricque- Marie D, Tapia-Salazar M, Nieto- López MG, Villarreal-Cavazos DA, Puello- Cruz A. C, Garcia-Ortega A (eds) Memorial book of VIII international symposium on nutrition for aquaculture. The Autonomous University of Nuevo León, Mexico, pp 20–28
Mishiro K (1997) Tiger puffer (best guide of formulated fish diets) (particulars 3-growing diet and optimum feeding plan by fish species). Aqua Cult Mag 34:138–142. (In Japanese)
Takii K (2005) Tiger puffer. In: Kumai H (ed) Aquaculture system 1: marine fish. Kouseisha Kouseikaku, Tokyo, pp 111–142. (In Japanese)
Takuma Corporation (2011) Chicken/cow manure incineration boilers. Takuma CSR report, pp 23–24

Index

T. Takeuchi (ed.), *Application of Recirculating Aquaculture Systems in Japan*, Fisheries Science Series, https://doi.org/10.1007/978-4-431-56585-7

P

R

S

T

MIX
Papier aus verantwortungsvollen Quellen
Paper from responsible sources
FSC® C105338

If you have any concerns about our products,
you can contact us on
ProductSafety@springernature.com

In case Publisher is established outside the EU,
the EU authorized representative is:
Springer Nature Customer Service Center GmbH
Europaplatz 3, 69115 Heidelberg, Germany

Printed by Libri Plureos GmbH
in Hamburg, Germany